Practical Industrial Data Networks: Design, Installation and Troubleshooting

Titles in the series

Practical Cleanrooms: Technologies and Facilities (David Conway)

Practical Data Acquisition for Instrumentation and Control Systems (John Park, Steve Mackay)

Practical Data Communications for Instrumentation and Control (Steve Mackay, Edwin Wright, John Park)

Practical Digital Signal Processing for Engineers and Technicians (Edmund Lai)

Practical Electrical Network Automation and Communication Systems (Cobus Strauss)

Practical Embedded Controllers (John Park)

Practical Fiber Optics (David Bailey, Edwin Wright)

Practical Industrial Data Networks: Design, Installation and Troubleshooting (Steve Mackay, Edwin Wright, John Park, Deon Reynders)

Practical Industrial Safety, Risk Assessment and Shutdown Systems for Instrumentation and Control (Dave Macdonald)

Practical Modern SCADA Protocols: DNP3, 60870.5 and Related Systems (Gordon Clarke, Deon Reynders)

Practical Radio Engineering and Telemetry for Industry (David Bailey)

Practical SCADA for Industry (David Bailey, Edwin Wright)

Practical TCP/IP and Ethernet Networking (Deon Reynders, Edwin Wright)

Practical Variable Speed Drives and Power Electronics (Malcolm Barnes)

Practical Industrial Data Networks: Design, Installation and Troubleshooting

Steve Mackay CPEng, BSc(ElecEng), BSc(Hons), MBA

Edwin Wright MIPENZ, BSc(Hons), BSc(Elec Eng)

Deon Reynders Pr.Eng, BSc(ElecEng)(Hons), MBA

John Park ASD

All with IDC Technologies, Perth, Australia

AMSTERDAM • BOSTON • HEIDELBERG • LONDON • NEW YORK • OXFORD
PARIS • SAN DIEGO • SAN FRANCISCO • SINGAPORE • SYDNEY • TOKYO

Newnes is an imprint of Elsevier

Newnes

Newnes is an imprint of Elsevier
Linacre House, Jordan Hill, Oxford OX2 8DP, UK
30 Corporate Drive, Suite 400, Burlington, MA 01803, USA

First edition 2004
Reprinted 2007

Notice
No responsibility is assumed by the publisher for any injury and/or damage to persons
or property as a matter of products liability, negligence or otherwise, or from any use
or operation of any methods, products, instructions or ideas contained in the material
herein. Because of rapid advances in the medical sciences, in particular, independent
verification of diagnoses and drug dosages should be made

British Library Cataloguing in Publication Data
A catalogue record for this book is available from the British Library

Library of Congress Cataloging-in-Publication Data
A catalog record for this book is available from the Library of Congress

ISBN: 978-0-7506-5807-2

For information on all Newnes publications
visit our website at www.newnespress.com

Transferred to Digital Printing in 2009

Working together to grow
libraries in developing countries

www.elsevier.com | www.bookaid.org | www.sabre.org

ELSEVIER   BOOK AID
            International   Sabre Foundation

Contents

13 ProfiBus PA/DP/FMS overview 181

14 Foundation Fieldbus overview 200

Preface

This is a comprehensive book covering the essentials of troubleshooting and problem solving of industrial data communications systems including areas such as RS-232, RS-485, industrial protocols such as Modbus, fiber optics, industrial Ethernet, TCP/IP, DeviceNet and Fieldbus protocols such as Profibus and Foundation Fieldbus. It can be used very beneficially in conjunction with the IDC Technologies two-day workshop on the topic.

We have taken all the key troubleshooting and problem-solving skills from experienced engineers and distilled these into one hard-hitting course to enable the user to solve real industrial communications problems.

The overall objective of the book is to help identify, prevent and fix common industrial communications problems. The focus is 'outside the box'. The emphasis is practical and on material that goes beyond typical communications issues. Also, on theory and focuses on providing with the necessary toolkit of skills in solving industrial communications problems whether it be RS-232/RS-485, Modbus, Fieldbus and DeviceNet or a local area network such as Ethernet. Industrial communications systems are being installed throughout the plant today from connecting simple instruments to programmable logic controllers to PCs throughout the business part of the enterprise. Communications problems range from simple wiring problems to intermittent transfer of protocol messages.

The communications system on the plant underpins the entire operation. It is critical that there should be the knowledge and tools to quickly identify and fix problems as they occur, to ensure that there is a secure and reliable system. No compromise is possible here. This book distills all the tips and tricks learnt with the benefit of many years of experience. It offers a common approach covering all of the sections listed below with each standard/protocol having the following structure:

- Quick overview of the standard
- Common problems and faults that can occur
- Description of tools used
- Each of the faults and ways of fixing them are then discussed in detail

The aim is to provide enough knowledge to troubleshoot and fix problems, as quickly as possible.

At the conclusion of this book, you will be able to:

- Identify, prevent and troubleshoot industrial communications problems
- Fix over 60 of the most common problems that occur in RS-232/RS-485, industrial protocols (incl. Modbus/ Data Highway Plus), industrial Ethernet, TCP/IP, DeviceNet and Fieldbus (such as Profibus and Foundation Fieldbus)
- Gain a practical toolkit of skills to troubleshoot industrial communications systems from RS-232, RS-485, fiber optics, Fieldbus to Ethernet
- Analyze most industrial communications problems and fix them
- Fault find your Ethernet and TCP/IP network problems

This book is intended for engineers and technicians who are:

- Instrumentation and control engineers/technicians
- Process control engineers
- Electrical engineers
- System integrators

- Designers
- Design engineers
- Systems engineers
- Network planners
- Test engineers
- Electronic technicians
- Consulting engineers
- Plant managers
- Shift electricians

A basic knowledge of data communications is useful but not essential.

The structure of the book is listed below. Each chapter is broken down into:

- Fundamentals of the standard or protocol
- Troubleshooting

Chapter 1: Introduction. An introduction to industrial data communications and the various standards and protocols in use as discussed in this book.

Chapter 2: Overall methodology. A review of the typical symptoms and problems that occur in industrial data communications. Typical approaches to follow in isolating and correcting problems that occur.

Chapter 3: EIA-232 overview. A review of the fundamentals of EIA-232 and ways of fixing problems here.

Chapter 4: EIA-485 overview. A discussion on RS-485 as far as cabling, common mode voltage, converters, isolation, idle state terminations and control.

Chapter 5: Current loop and EIA-485 converters overview. Problems with cabling and isolation and how to fix them.

Chapter 6: Fiber optics overview. A review of splicing, interface to cable, connectors, multimode, monomode, laser versus LED transmitters, driver incompatibility, bending radius and shock and installation issues.

Chapter 7: Modbus overview. A discussion on the problems with Modbus such as no response, exception reports, noise and radio interfaces and the lack of physical and application layers in the definition.

Chapter 8: Modbus Plus protocol overview. Review of typical problems such as cabling, grounding, shielding, terminators and token passing.

Chapter 9: Data Highway Plus/DH485 overview. Review of typical problems such as cabling, grounding, shielding, terminators and token passing.

Chapter 10: HART overview. Discussion on cabling, configuration and intrinsic safety.

Chapter 11: AS-interface (AS-i) overview. Essentials of cabling, connections and gateways.

Chapter 12: DeviceNet overview. Review of typical problems such as topology, power and earthing, signal voltage levels, common mode voltage, terminations, cabling, noise, node communications problems and creeping errors.

Chapter 13: ProfiBus PA/DP/FMS overview. Discussion on cabling, fiber, shielding, grounding, segmentation, color coding, addressing, token bus, unsolicited messages, fine tuning of impedance terminations, drop line lengths, GSD files and intrinsic safety.

Chapter 14: Foundation Fieldbus overview. Review of wiring, grounds, shielding, wiring polarity, power, terminations, intrinsic safety, voltage drop, power conditioning, surge protection and configuration.

Chapter 15: Industrial Ethernet overview. Discussion on problems such as noise, thin and thick coaxial cable, UTP cabling, wire types, components, incorrect media selection, jabber, too many nodes, excessive broadcasting, bad frames, 10/100 Mbps mismatch, faulty hubs and loading.

Chapter 16: TCP/IP overview. Review of Internet protocol, transmission control protocol, addressing, subnet mask, routers, tripe handshake and incorrect port usage.

Chapter 17: Radio and wireless communications overview. Discussions on reliability, noise, interference, power, distance, license, frequency and over and under modulation.

1

Introduction

Objectives

When you have completed study of this chapter you will be able to:

- Describe the modern instrumentation and control system
- List the main industrial communications systems
- Describe the essential components of industrial communications systems

1.1 Introduction

Data communications involves the transfer of information from one point to another. In this book, we are specifically concerned with digital data communication. In this context, 'data' refers to information that is represented by a sequence of zeros and ones, the same sort of data handled by computers. Many communications systems handle analog data; examples are telephone systems, radio and television. Modern instrumentation is almost wholly concerned with the transfer of digital data.

Any communications system requires a transmitter to send information, a receiver to accept it and a link between the two. Types of link include copper wire, optical fiber, radio and microwave.

Some short-distance links use parallel connections, meaning that several wires are required to carry a signal. This type of connection is confined to devices such as local printers. Virtually all modern data communications use serial links in which the data is transmitted in sequence over a single circuit.

Digital data is sometimes transferred using a system that is primarily designed for analog communication. A modem, for example, works by using a digital data stream to modulate an analog signal that is sent over a telephone line. Another modem demodulates the signal to reproduce the original digital data at the receiving end. The word 'modem' is derived from *modulator* and *demodulator*.

There must be mutual agreement on how data is to be encoded, that is, the receiver must be able to understand what the transmitter is sending. The structure in which devices communicate is known as a protocol.

In the past decade, many standards and protocols have been established, which allow data communications technology to be used more effectively in industry. Designers and users are beginning to realize the tremendous economic and productivity gains possible with the integration of systems that are already in operation.

Protocols are the structure used within a communications system so that, for example, a computer can talk to a printer. Traditionally, developers of software and hardware platforms have developed protocols that only their own products can use. In order to develop more integrated instrumentation and control systems, standardization of these communication protocols was required.

Standards may evolve from the wide use of one manufacturer's protocol (a *de facto* standard) or may be specifically developed by bodies that represent specific industries. Standards allow manufacturers to develop products that will communicate with equipment already in use. For the customer this simplifies the integration of products from different sources.

The industrial communications market is characterized by a lack of standardization. There are, however, a few dominant standards. Modbus has been a *de facto* standard for the last fifteen years and the tried and tested physical standards such as EIA-232 and EIA-485 have been widely used. The area that has caused a considerable amount of angst (and dare we say – irritation) amongst vendors and users is the choice of an acceptable Fieldbus, which would tie together instruments to programmable logic controllers and personal computers. This effort has resulted in a few dominant but competing standards such as Profibus, ASi, DeviceNet and Foundation Fieldbus being used in various areas of industry.

The standard that has created an enormous amount of interest in the past few years is Ethernet. After initially being rejected as being non-deterministic, which means there cannot be guarantee of a critical message being delivered within a defined time, this thorny problem has been solved with the latest standards in Ethernet and the use of switching technology. The other protocol, which fits onto Ethernet extremely well is TCP/IP. Being derived from the Internet, it is very popular and widely used.

1.2   Modern instrumentation and control systems

In an instrumentation and control system, data is acquired by measuring instruments and transmitted to a controller, typically a computer. The controller then transmits data (control signals) to control devices, which act upon a given process.

The integration of systems with each other enables data to be transferred quickly and effectively between different systems in a plant along a data communications link. This eliminates the need for expensive and unwieldy wiring looms and termination points.

Productivity and quality are the principal objectives in the good management of any production activity. Management can be substantially improved by the availability of accurate and timely data. From this, we can surmise that a good instrumentation and control system can facilitate both quality and productivity.

The main purpose of an instrumentation and control system, in an industrial environment, is to provide the following:

Control of the processes and alarms

Traditionally, control of processes such as temperature and flow was provided by analog controllers operating on standard 4–20 mA loops. The 4–20 mA standard is utilized by equipment from a wide variety of suppliers and it is common for equipment from various sources to be mixed in the same control system. Stand-alone controllers and instruments have largely been replaced by integrated systems such as distributed control systems (DCS), described below.

Control of sequencing, interlocking and alarms

Typically, this was provided by relays, timers and other components hardwired into control panels and motor control centers. The sequence control, interlocking and alarm requirements have largely been replaced by PLCs.

An operator interface for display and control

Traditionally, process and manufacturing plants were operated from local control panels by several operators, each responsible for a portion of the overall process. Modern control systems tend to use a central control room to monitor the entire plant. The control room is equipped with computer-based operator workstations, which gather data from the field instrumentation and use it for graphical display to control processes, to monitor alarms, to control sequencing and for interlocking.

Management information

Management information was traditionally provided by taking readings from meters, chart recorders, counters and transducers and from samples taken from the production process. This data is required to monitor the overall performance of a plant or process and to provide the data necessary to manage the process. Data acquisition is now integrated into the overall control system. This eliminates the gathering of information and reduces the time required to correlate and use the information to remove bottlenecks. Good management can achieve substantial productivity gains.

The ability of control equipment to fulfil these requirements has depended on the major advances that have taken place in the fields of integrated electronics, microprocessors and data communications. The four devices that have made the most significant impact on how plants are controlled are:

- Distributed control systems (DCSs)
- Programmable logic controllers (PLCs)
- SCADA (supervisory control and data acquisition) systems
- Smart instruments

Distributed control systems (DCSs)

A DCS is a hardware- and software-based (digital) process control and data acquisition system. The DCS is based on a data highway and has a modular, distributed, but integrated architecture. Each module performs a specific dedicated task such as the operator interface/analog or loop control/digital control. There is normally an interface unit situated on the data highway allowing easy connection to other devices such as PLCs and supervisory computer devices.

Programmable logic controllers (PLCs)

PLCs were developed in the late sixties to replace collections of electromagnetic relays, particularly in the automobile manufacturing industry. They were primarily used for sequence control and interlocking with racks of on/off inputs and outputs, called digital I/O. They are controlled by a central processor using easily written 'ladder logic' type programs. Modern PLCs now include analog and digital I/O modules as well as sophisticated programming capabilities similar to a DCS, e.g. PID loop programming.

High-speed inter-PLC links are also available, such as 10/100 Mbps Ethernet. A diagram of a typical PLC system is given in Figure 1.1.

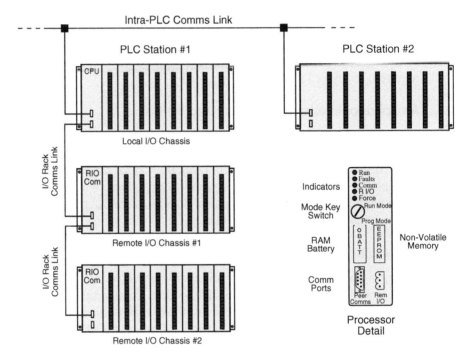

Figure 1.1
A typical PLC system

Supervisory control and data acquisition (SCADA) system

This refers to a system comprising a number of remote terminal units (RTUs) collecting field data and connected back to a master station via a communications system.

A diagram below gives an example of this.

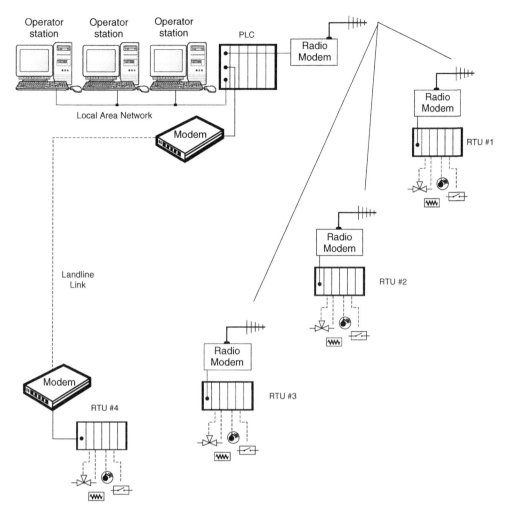

Figure 1.2
Diagram of a typical SCADA system

Smart instrumentation systems

In the 1960s, the 4–20 mA analog interface was established as the *de facto* standard for instrumentation technology. As a result, the manufacturers of instrumentation equipment had a standard communication interface on which to base their products. Users had a choice of instruments and sensors from a wide range of suppliers, which could be integrated into their control systems.

With the advent of microprocessors and the development of digital technology, the situation has changed. Most users appreciate the many advantages of digital instruments. These include more information being displayed on a single instrument, local and remote display, reliability, economy, self-tuning and diagnostic capability. There is a gradual shift from analog to digital technology.

There are a number of intelligent digital sensors with digital communications capability for most traditional applications. These include sensors for measuring temperature, pressure, levels, flow, mass (weight), density and power system parameters. These new intelligent digital sensors are known as 'smart' instrumentation.

The main features that define a 'smart' instrument are:
- Intelligent, digital sensors
- Digital data communications capability
- Ability to be multidropped with other devices

There is also an emerging range of intelligent, communicating, digital devices that could be called 'smart' actuators. Examples of these are devices such as variable speed drives, soft starters, protection relays and switchgear control with digital communication facilities.

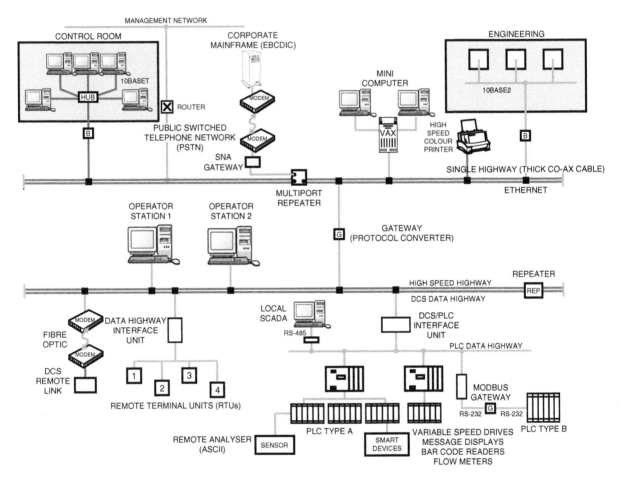

Figure 1.3
Graphical representation of data communication

1.3   Open systems interconnection (OSI) model

The OSI model, developed by the International Organization for Standardization, is rapidly gaining industry support. The OSI model reduces every design and communication problem into a number of layers as shown in Figure 1.4. A physical interface standard such as EIA-232 would fit into the layer 1, while the other layers relate to the protocol software.

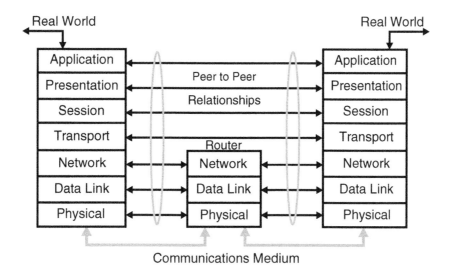

Figure 1.4
Representation of the OSI model

Messages or data are generally sent in packets, which are simply a sequence of bytes. The protocol defines the length of the packet. Each packet requires a source address and a destination address so that the system knows where to send it, and the receiver knows where it came from. A packet starts at the top of the protocol stack, the application layer, and passes down through the other software layers until it reaches the physical layer. It is then sent over the link. When travelling down the stack, the packet acquires additional header information at each layer. This tells the corresponding layers at the next stack what to do with the packet. At the receiving end, the packet travels up the stack with each piece of header information being stripped off on the way. The application layer at the receiver only receives the data sent by the application layer at the transmitter.

The arrows between layers indicate that each layer reads the packet as coming from, or going to the corresponding layer at the opposite end. This is known as peer-to-peer communication, although the actual packet is transported via physical link. The middle stack in Figure 1.4 (representing a router) has only the three lower layers, which is all that is required for the correct transmission of a packet between two devices in this particular case.

The OSI model is useful in providing a universal framework for all communication systems. However, it does not define the actual protocol to be used at each layer. It is anticipated that groups of manufacturers in different areas of industry will collaborate to define software and hardware standards appropriate to their particular industry. Those seeking an overall framework for their specific communications requirements have enthusiastically embraced this OSI model and used it as a basis for their industry specific standards.

1.4 Protocols

As previously mentioned, the OSI model provides a framework within which a specific protocol may be defined. A protocol, in turn, defines a frame format that might be made up as follows: The first byte(s) can be a string of ones and zeros to synchronize the receiver or flags to indicate the start of the frame (for use by the receiver). The second byte could contain the destination address detailing where the message is going. The third byte could contain the source address noting where the message originated. The bytes in the middle of the message could be the actual data that has to be sent from transmitter to receiver. The final byte(s) are end-of-frame indicators, which can be error detection codes and/or ending flags.

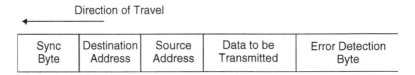

Figure 1.5
Basic structure of an information frame defined by a protocol

Protocols vary from the very simple (such as ASCII-based protocols) to the very sophisticated (such as TCP/IP), which operate at high speeds transferring megabits of data per second. There is no right or wrong protocol, the choice depends on a particular application.

1.5   Standards

A brief discussion is given below on the most important approaches that are covered in this book.
These are the following:

- RS-232 (EIA-232)
- RS-485 (EIA-485)
- Fiber optics
- Modbus
- Modbus Plus
- Data Highway Plus /DH485
- HART
- ASi
- DeviceNet
- Profibus
- Foundation Fieldbus
- Industrial Ethernet
- TCP/IP
- Radio and wireless communications

EIA-232 interface standard

The EIA-232C interface standard was issued in the USA in 1969 to define the electrical and mechanical details of the interface between data terminal equipment (DTE) and data communications equipment (DCE), which employ serial binary data interchange.
In serial data communications, the communications system might consist of:

- The DTE, a data sending terminal such as a computer, which is the source of the data (usually a series of characters coded into a suitable digital form)
- The DCE, which acts as a data converter (such as a modem) to convert the signal into a form suitable for the communications link e.g. analog signals for the telephone system
- The communications link itself, for example, a telephone system

- A suitable receiver, such as a modem, also a DCE, which converts the analog signal back to a form suitable for the receiving terminal
- A data receiving terminal, such as a printer, also a DTE, which receives the digital pulses for decoding back into a series of characters

Figure 1.6 illustrates the signal flows across a simple serial data communications link.

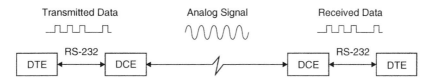

Figure 1.6
A typical serial data communications link

The EIA-232C interface standard describes the interface between a terminal (DTE) and a modem (DCE) specifically for the transfer of serial binary digits. It leaves a lot of flexibility to the designers of the hardware and software protocol. With the passage of time, this interface standard has been adapted for use with numerous other types of equipment such as personal computers (PCs), printers, programmable controllers, programmable logic controllers (PLCs), instruments and so on. To recognize these additional applications, the latest version of the standard, EIA-232E has expanded the meaning of the acronym DCE from 'data communications equipment' to the more general 'data circuit-terminating equipment'.

EIA-232 has a number of inherent weaknesses that make it unsuitable for data communications for instrumentation and control in an industrial environment. Consequently, other EIA interface standards have been developed to overcome some of these limitations. The most commonly used among them for instrumentation and control systems are EIA-423, EIA-422 and EIA-485.

EIA-485 interface standard

The EIA-485 is a balanced system with the same range as EIA-422 but with increased data rates and up to 32 'standard' transmitters and receivers per line.

The EIA-485 interface standard is very useful for instrumentation and control systems, where several instruments or controllers may be connected together on the same multi-point network.

A simple diagram of a typical RS-485 system is indicated in Figure 1.7.

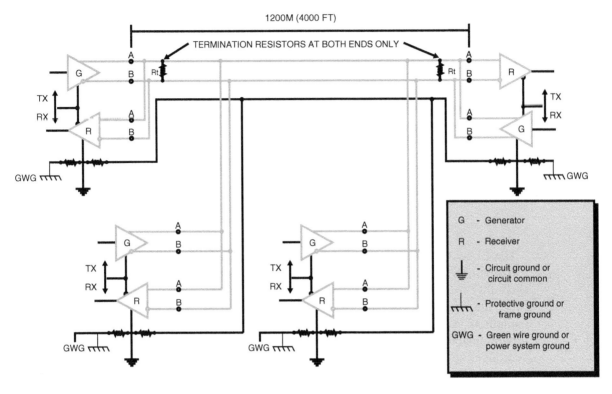

Figure 1.7
Typical two-wire multidrop network for EIA-485

Fiber optics

There are two main approaches possible with fiber optic cables:
- Single mode (monomode) cabling
- Multimode cabling

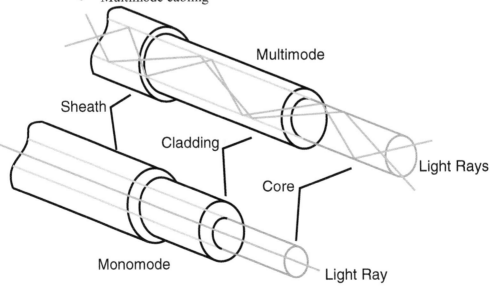

Figure 1.8
Monomode and multimode optic fibers

This is widely used throughout industrial communications systems for two main reasons:

- Immunity to electrical noise and
- Optical isolation from surges and transients

As a result, fiber is tending to dominate in all new installations that require reasonable levels of traffic.

Modbus

This protocol was developed by Modicon (now part of Schneider Electric) for process control systems. This standard only refers to the data link and application layers; so that any physical transport method can be used. It is a very popular standard with some estimates indicating that over 40% of industrial communications systems use Modbus. It operates as a master–slave protocol with up to 247 slaves.

Address Field	Function Field	DATA Data Field	Error Check Field
1 Byte	1 Byte	Variable	2 Bytes

Figure 1.9
Format of Modbus message frame

The address field refers to the number of the specific slave device being accessed. The function field indicates the operation that is being performed, for example, read or write of an analog or digital point in the slave device. The data field is the data that is being transferred from the slave device back to the master or from the master to the slave device (a write operation). Finally, the error check field is to ensure that the receiver can confirm the integrity of the protocol; it could almost be considered to be a unique fingerprint.

Modbus Plus

This builds on Modbus and incorporates the protocol within a token passing operation. This protocol was generally confined to Modicon programmable logic controllers and never really enjoyed support as an open protocol.

Data Highway Plus /DH485

This protocol formed the backbone of the Allen Bradley (now Rockwell Automation) range of programmable logic controllers. It is a protocol defining all three layers of the OSI model – physical layer, data link layer and application layer. A diagram of how it is structured is given below.

For a User Application Program

DST	CMD	STS	Command Data

For Common Application Routines

DST	SRC	CMD	STS	TNS	Data (From Application Data)

Data Link Layer Packet

DLE	STX	Data (From Application Data)	DLE	EXT	CRC

Figure 1.10
Data Highway Plus protocol structure

There are two addresses (DST and SRC) in this protocol message, indicating destination and source addresses. This is the result of using a token passing system where each station on the network has the ability to be the master for a short period of time.

HART

The highway addressable remote transducer (HART) protocol is a typical smart instrumentation Fieldbus that can operate in a hybrid 4–20 mA digital fashion. It has become popular as it is compatible with the 4–20 mA standard. However, it is not a true Fieldbus standard and has been superseded by Foundation Fieldbus. A typical diagram of how it operates is shown below.

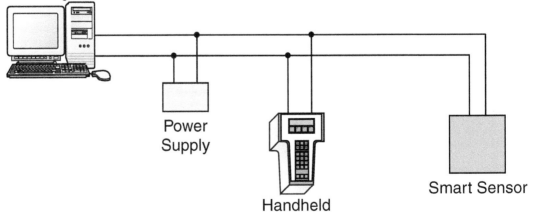

Power
Supply

Handheld

Smart Sensor

Figure 1.11
HART point-to-point communication

AS-i

This must be one of the most robust standards for simple digital control. It is almost idiot proof from a wiring point of view with a very simple wiring philosophy. It is a master–slave network, which can achieve transfer rates of up to 167 kbps where, for example, with 31 slaves and 124 I/O points connected, a 5 ms scan time can be achieved.

DeviceNet

DeviceNet, developed by Allen Bradley (now Rockwell Automation), is a low-level oriented network focusing on the transfer of digital points. It defines three layers as indicated below and can support up to 64 nodes with as many as 2048 total devices. The cabling is straightforward and simple and enables power to be carried for the instruments as well.

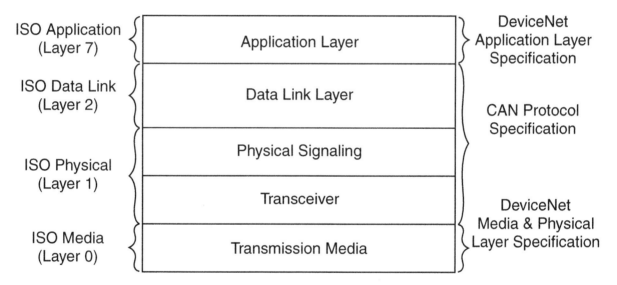

Figure 1.12
DeviceNet and the OSI model

Profibus

Although initially spawned by the German standards association, this standard based on the EIA-485 standard (Profibus DP) and the IEC 61158 standard (Profibus PA) for the physical layer, has become a very popular international standard. A typical configuration is shown in Figure 1.13.

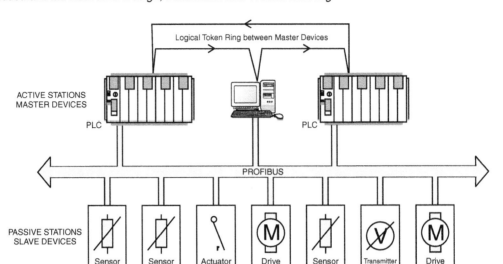

Figure 1.13
Typical architecture of a Profibus system

Profibus uses a combination strategy of token passing and master–slave to achieve its communications results. It defines three layers of the OSI model; namely being the physical layer, the data link layer and the application layer. It also added an 8th layer, the so-called 'user layer'.

Foundation Fieldbus

Foundation Fieldbus is probably the newest Fieldbus standard to connect instruments to the programmable logic controller or RTU of the distributed control system (DCS). A considerable amount of technical thought has gone into the standard that defines four layers with a low-speed version entitled H1 and a high-speed version called HSE. Three layers of the OSI model (the physical, data link and application) and an additional 8th layer, the user layer (where the function blocks are situated) are defined. It is eminently suitable for use with analog parameters where there is a minimum 100-msec update time required on the H1 standard (31.25 kbps). For the high-speed version (HSE), fast Ethernet (100 Mbps) is used. The HSE version, albeit different to the H1 version at OSI layers 1 and 2, is otherwise compatible with the H1 standard on the application layer and the user layer.

Industrial Ethernet

Industrial Ethernet is rapidly growing in importance after initially being dismissed as not being reliable enough. One of the main reasons for its success is its simplicity and low cost. Originally, Ethernet used only CSMA/CD (carrier sense multiple access with collision detection) as its media access control method. This is a non-deterministic method, not ideal for process control applications. Although all modern versions of Ethernet (100 Mbps and up) conform with CSMA/CD requirements for the sake of adherence to the IEEE 802.3 standard, they also allow full-duplex operation. Most modern industrial Ethernet systems are 100 Mbps full-duplex systems and allow switch ports to be prioritized, resulting in very deterministic behavior. This is far simpler than the token passing method of communications. A typical example of the 100BaseTX topology is given below.

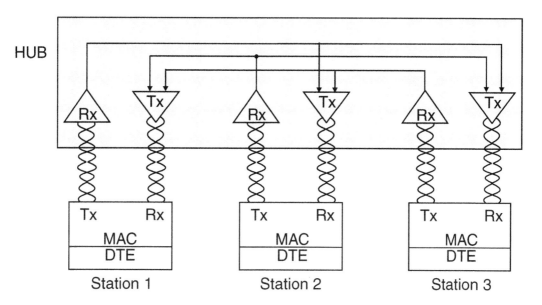

Figure 1.14
100BaseTX star topology

TCP/IP

A child of the Internet, the transmission control protocol (TCP)/Internet protocol (IP) is also becoming popular when used in conjunction with Ethernet. It really defines three layers.

- Process/application layer
  (equivalent to upper three layers in the OSI model)
- Services layer (host-to-host) layer
  (equivalent to the transport layer in the OSI model)
- Internetwork layer
  (equivalent to the network layer of the OSI model)

It is a very low cost protocol with wide support due its use on the Internet. Arguably it is an overkill for some industrial communications applications. However, its low cost and wide support make it very attractive.

Radio (or wireless) communications

The use of radio in the industrial context commenced with the use of radio modems as indicated in the diagram below where for example, Modbus could be used over the specific radio modem data link layer. The use of the latest wireless LAN standards such as IEEE 802.11b or IEEE 802.11a (and IEEE 802.15 'Bluetooth') are making this a reliable and low cost form of communications.

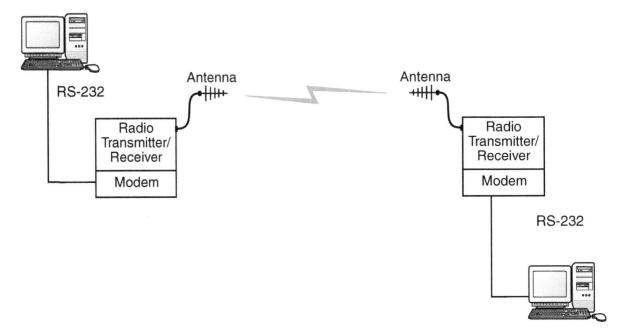

Figure 1.15
Radio modem configuration

2

Overall methodology

Objectives

When you have completed study of this chapter, you will be able to:

- List typical problems that can occur with industrial communications systems
- List the typical steps to follow in fixing industrial communications faults
- Describe the main issues with earthing/grounding/shielding and noise problems

2.1 Introduction

When troubleshooting a communications system, the engineer or the technician tries to use some standard format to arrive at a quicker solution. Industrial communications systems do not always follow the tried and tested rules, which worked with hardwired inputs and outputs. These tried and tested rules have some subtleties, which will be discussed in this chapter.

This chapter is broken down into the following sections:

- Common problems and solutions
- General comments on troubleshooting
- A specific methodology
- Grounding/shielding and noise

2.2 Common problems and solutions

A typical list of causes of industrial communications problems is:

- No power to the station on the network resulting in it stopping communications
- Cable failure due to damage with a resultant interruption in communications
- Earthing and grounding problems with intermittent failure of communications

- Electrostatic damage to the communications port
- Software crash on one of the stations resulting in communications failure
- High levels of electrostatic/electromagnetic interference on the communications link or port
- High traffic loads on the link resulting in intermittent communications
- Electrical surge or transient through the communications system resulting in significant damage

The impact on the communications system ranges from outright failure (with no communications possible) to intermittent communications depending on the level of interference or traffic level. It is generally considered that intermittent failure is the worst problem to have, as it is very difficult to diagnose and fix.

2.3 General comments on troubleshooting

Obviously, there is no cut and dried method of testing. It depends on the environment and the history of the system. However, a few rules are useful in troubleshooting a communications system effectively.

Documentation is critical

Gather documentation on the cabling and layout of the communications system, as much as possible. This means wiring diagrams/number of parameters (points) transferred across the link. Ensure you know exactly, which items are connected on this particular link or network and you know all the interface points to the next (separate) network.

Baseline report

Refer to the baseline report you have on your communications system. This obviously means that you have to build up a historical record of the operation of the network with information on packets sent/response times/utilization rates of the network, so that you can compare apples with apples. For example, you may have a few stations on the network with problems in communicating with each other. You then look at the network and discover that it has an average utilization of 25% and 100 packets in error every second. You are obviously not sure what to look for and compare these values to the baseline values of average utilization of 10% and 100 packets in error every second. You can then be sure that whilst there is a potential problem with packets in error, the more likely problem is that the network is a lot busier and some of the stations are probably timing out. So, you need to look at reducing the traffic or lengthening the time out for stations on the network.

Simplify the network

To simplify the network as much as possible, try to remove any devices that are not required. For example, if you are having two devices communicating with each other and you have a repeater connected on the network from some previously long disconnected communications link, remove this. In this particular case, it was found that this repeater was jabbering away on the network at random times creating major congestion on the network and was the cause of the problem. But even if it was not, this can eliminate another potential cause of a problem from the network.

Watch out for deviations

Watch out for deviations in the network. If the number of packets from a device suddenly goes up against the baseline, it is worth examining why.

2.4   A specific methodology

When troubleshooting your communications system, the following steps should be taken:

- Check that all stations and network/communications devices are powered up and operational (for example, by looking at the green status lights). Look for any devices, which may appear to be dead (for example, the heartbeat LED is not operating).
- Check all cabling for clean connections. Sometimes, a connection is inadvertently broken causing the inevitable communications problem. In addition, check all hardware taps off the main bus for any loose connections.
- Check grounding and earthing setups. Has the earth connection changed in any way? Has the ground resistance changed due to changes in the plant environment?
- Some new devices operating in the same power supply may be the cause of the problem. This ranges from power factor capacitor banks to new variable speed drives installed to a large radio source.
- Check whether there has been any changes or damage to screening of the cables. The higher the data rate, the more critical this becomes with massive errors, which are possible only with a few mm of screening removed.
- Check whether the voltages are correct. For example, if it is a hub, check that the voltage powering it is correct. If it is DeviceNet, check that the voltages have not dropped below the threshold for successful operation.
- Use the hardware diagnostic tools provided (ranging from the DeviceNet detective to a break out box for RS-232 to RS-485) and ascertain if there are any problems here.
- Look at the diagnostics packages provided as part of the system to look at the number of packets transmitted to packets dropped. Compare these figures to baseline figures if available.
- Commence removing devices, which are not critical to the problem. For example, if a PLC is connected to the system and is merely reading information off the network and is not critical to the operation, then remove it.
- Do simple diagnostic tests using simple utilities provided such as ping or netstat to identify what is happening on the network.
- Put on more sophisticated devices such as protocol analyzers to identify what is happening with the packets on network. This is one of the last stages to look and it requires some knowledge of the protocol.

While this book focuses on the communications problems, an often neglected area is the sources of electrical noise and methods of fixing them. This is discussed in the next section.

2.5 Grounding/shielding and noise

2.5.1 Sources of electrical noise

Typical sources of noise are devices, which produce quick changes (or spikes) in voltage or current, such as:

- Large electrical motors being switched on
- Fluorescent lighting tubes
- Lightning strikes
- High voltage surging due to electrical faults
- Welding equipment

From a general point of view, there must be three contributing factors for the existence of an electrical noise problem. They are:

- A source of electrical noise
- A mechanism coupling the source to the affected circuit
- A circuit conveying the sensitive communication signals

2.5.2 Electrical coupling of noise

There are four forms of coupling of electrical noise into the sensitive data communications circuits. They are:

- Impedance coupling (sometimes referred to as conductance coupling)
- Electrostatic coupling
- Magnetic or inductive coupling
- Radio frequency radiation (a combination of electrostatic and magnetic)

Each of these noise forms will be discussed in some detail in the following sections. Although the order of discussion is indicative of the frequency of problems, this will obviously depend on the specific application.

Impedance coupling (or common impedance coupling)

For situations where two or more electrical circuits share common conductors, there can be some coupling between the different circuits with deleterious effects on the connected circuits. Essentially, this means that the signal current from the one circuit proceeds back along the common conductor resulting in an error voltage along the return bus which affects all the other signals. The error voltage is due to the capacitance, inductance and resistance in the return wire. This situation is shown in the Figure 2.1.

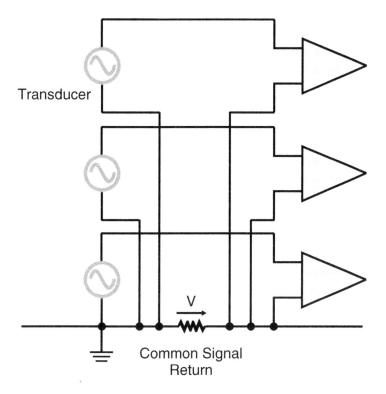

Figure 2.1
Impedance coupling

Obviously, the quickest way to reduce the effects of impedance coupling is to minimize the impedance of the return wire. The best solution is to use a balanced circuit with separate returns for each individual signal.

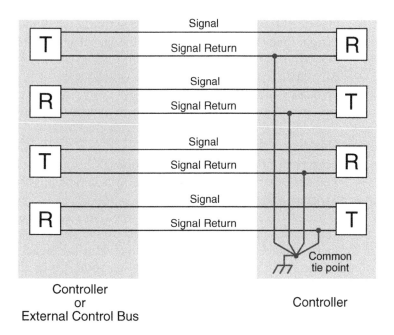

Figure 2.2
Impedance coupling eliminated with balanced circuits

Electrostatic or capacitive coupling

This form of coupling is proportional to the capacitance between the noise source and the signal wires. The magnitude of the interference depends on the rate of change of the noise voltage and the capacitance between the noise circuit and the signal circuit.

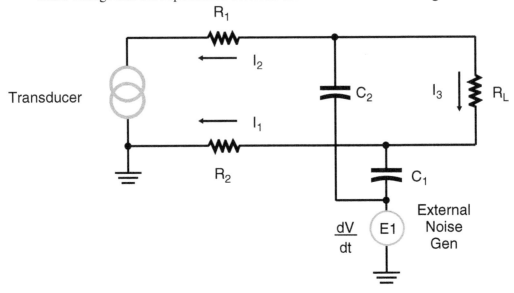

Figure 2.3
Electrostatic coupling

In the Figure 2.3, the noise voltage is coupled into the communication signal wires through two capacitors, C_1 and C_2, and a noise voltage is produced across the resistances in the circuit. The size of the noise (or error) voltage in the signal wires is proportional to the:

- Inverse of the distance of noise voltage from each of the signal wires
- Length (and hence impedance) of the signal wires into which the noise is induced
- Amplitude (or strength) of the noise voltage
- Frequency of the noise voltage

There are four methods for reducing the noise induced by electrostatic coupling. They are:

- Shielding of the signal wires
- Separating from the source of the noise
- Reducing the amplitude of the noise voltage (and possibly the frequency)
- Twisting of the signal wires

Figure 2.4 indicates the situation that occurs when an electrostatic shield is installed around the signal wires. The currents generated by the noise voltages prefer to flow down the lower impedance path of the shield rather than the signal wires. If one of the signal wires and the shield are tied to the earth at one point, which ensures that the shield and the signal wires are at an identical potential, then reduced signal current flows between the signal wires and the shield.

Note: The shield must be of a low resistance material such as aluminum or copper. For a loosely braided copper shield (85% braid coverage), the screening factor is about 100 times or 20 dB i.e. C_3 and C_4 are about 1/100 C_1 or C_2. For a low resistance multilayered screen, this screening factor can be 35 dB or 3000 times.

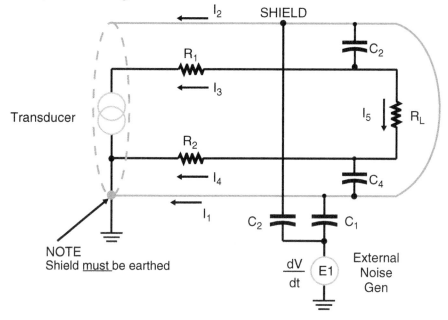

Figure 2.4
Shield to minimize electrostatic coupling

Twisting of the signal wires provides a slight improvement in reducing the induced noise voltage by ensuring that C_1 and C_2 are closer together in value; thus ensuring that any noise voltages induced in the signal wires tend to cancel one another out.

Note: Provision of a shield by a cable manufacturer ensures the capacitances between the shield and the wires are equal in value (thus eliminating any noise voltages by cancellation).

Magnetic or inductive coupling

This depends on the rate of change of the noise current and the mutual inductance between the noise system and the signal wires. Expressed slightly differently, the degree of noise induced by magnetic coupling will depend on the:

- Magnitude of the noise current
- Frequency of the noise current
- Area enclosed by the signal wires (through which, the noise current magnetic flux cuts)
- Inverse of the distance from the disturbing noise source to the signal wires.

The effect of magnetic coupling is shown in Figure 6.12.

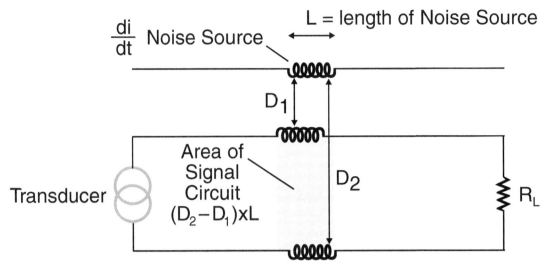

Figure 2.5
Magnetic coupling

The easiest way of reducing the noise voltage caused by magnetic coupling is to twist the signal conductors. This results in lower noise due to the smaller area for each loop. This means less magnetic flux to cut through the loop and consequently, a lower induced noise voltage. In addition, the noise voltage that is induced in each loop tends to cancel out the noise voltages from the next sequential loop. Hence an even number of loops will tend to have the noise voltages canceling each other out. It is assumed that the noise voltage is induced in equal magnitudes in each signal wire due to the twisting of the wires giving a similar separation distance from the noise voltage. See Figure 2.6.

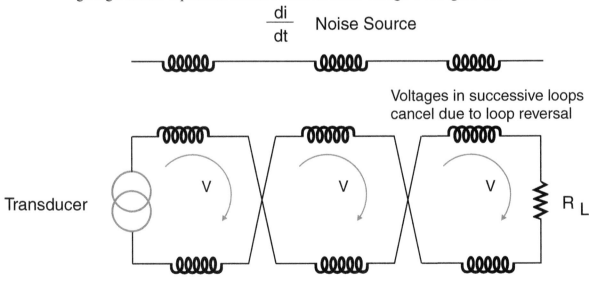

Figure 2.6
Twisting of wires to reduce magnetic coupling

The second approach is to use a magnetic shield around the signal wires. The magnetic flux generated from the noise currents induces small eddy currents in the magnetic shield. These eddy currents then create an opposing magnetic flux $\varnothing_1$ to the original flux $\varnothing_2$. This means a lesser flux $(\varnothing_2 - \varnothing_1)$ reaches our circuit!

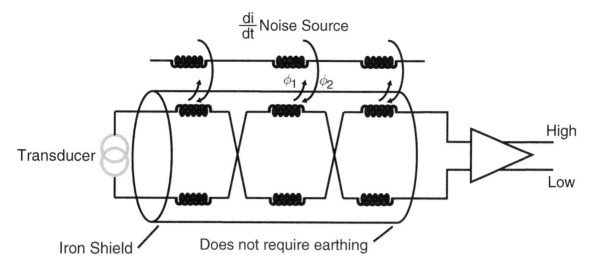

Figure 2.7
Use of magnetic shield to reduce magnetic coupling

Note: The magnetic shield does not require earthing. It works merely by being present. High permeability steel makes best magnetic shields for special applications. However, galvanized steel conduit makes quite an effective shield.

Radio frequency radiation

The noise voltages induced by electrostatic and inductive coupling (discussed above) are manifestations of the near field effect, which is electromagnetic radiation close to the source of the noise. This sort of interference is often difficult to eliminate and it requires close attention of grounding of the adjacent electrical circuit and the earth connection is only effective for circuits in close proximity to the electromagnetic radiation. The effects of electromagnetic radiation can be neglected unless the field strength exceeds 1 volt/meter. This can be calculated by the formula:

$$\text{Field Strength} = \frac{0.173\sqrt{\text{Power}}}{\text{Distance}}$$

where:

- Field strength in volt/meter
- Power in kilowatt
- Distance in km

The two most commonly used mechanisms to minimize electromagnetic radiation are:

- Proper shielding (iron)
- Capacitors to shunt the noise voltages to earth

Any incompletely shielded conductors will perform as a receiving aerial for the radio signal and hence care should be taken to ensure good shielding of any exposed wiring.

2.5.3   Shielding

It is important that electrostatic shielding is only earthed at one point. More than one earth point will cause circulating currents. The shield should be insulated to prevent inadvertent contact with multiple points, which behave as earth points resulting in circulating currents. The shield should never be left floating because that would tend to allow capacitive coupling, rendering the shield useless.

Two useful techniques for isolating one circuit from the other are by the use of opto-isolation as shown in the Figure 2.8, and transformer coupling as shown in Figure 2.9.

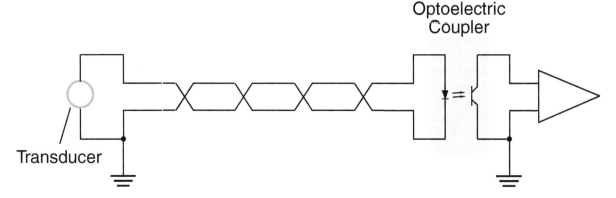

Figure 2.8
Opto-isolation of two circuits

Although opto-isolation does isolate one circuit from the other, it does not prevent noise or interference being transmitted from one circuit to another.

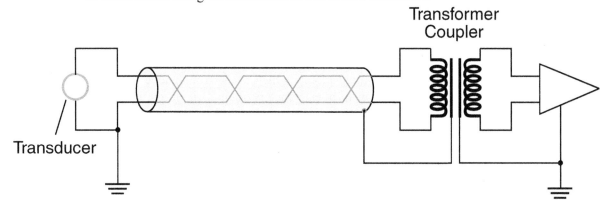

Figure 2.9
Transformer coupling

Transformer coupling can be preferable to optical isolation when there are very high speed transients in one circuit. There is some capacitive coupling between the LED and the base of the transistor, which is in the opto-coupler, can allow these types of transients to penetrate one circuit from another. This is not the case with transformer coupling.

Good shielding performance ratios

The use of some form of low resistance material covering the signal conductors is considered good shielding practice for reducing electrostatic coupling. When comparing shielding with no protection, this reduction can vary from copper braid (85% coverage), which returns a noise reduction ratio of 100:1 to aluminum mylar tape with drain wire, with a ratio of 6000:1.

Twisting the wires to reduce inductive coupling reduces the noise (in comparison to no twisting) by ratios varying from 14:1 (for four inch lay) to 141:1 (for one inch lay).   In comparison, putting parallel (untwisted) wires into steel conduit only gives a noise reduction of 22:1.

On very sensitive circuits with high levels of magnetic and electrostatic coupling the approach is to use coaxial cables. Double-shielded cable can give good results for very sensitive circuits.

Note: With double shielding, the outer shield could be earthed at multiple points to minimize radio frequency circulating loops. This distance should be set at intervals of less than ⅛ of the wavelength of the radio frequency noise.

2.5.4 Cable ducting or raceways

These are useful in providing a level of attenuation of electric and magnetic fields. These figures are 60 Hz for magnetic fields and 100 kHz for electric fields.

Typical screening factors are:

- **5 cm (2 inch) aluminum conduit with 0.154 inch thickness**
  - Magnetic fields 1.5:1
  - Electric fields 8000:1
- **Galvanized steel conduit 5 cm (2 inch), wall thickness 0.154 inch width**
  - Magnetic fields 40:1
  - Electric fields 2000:1

2.5.5 Cable spacing

In situations where there are a large number of cables varying in voltage and current levels, the IEEE 518-1982 standard has developed a useful set of tables indicating separation distances for various classes of cables. There are four classification levels of susceptibility for cables. Susceptibility, in this context, is understood to be an indication of how well the signal circuit can differentiate between the undesirable noise and required signal. It follows a data communication physical standard such as RS-232E that would have a high susceptibility and a 1000 volt, 200 amp AC cable that has a low susceptibility.

The four susceptibility levels defined by the IEEE 518-1982 standard are briefly:

- **Level 1 – High**

  This is defined as analog signals less than 50 volt and digital signals less than 15 volt. This would include digital logic buses and telephone circuits. Data communication cables fall into this category.

- **Level 2 – Medium**

  This category includes analog signals greater than 50 volt and switching circuits.

- **Level 3 – Low**

  This includes switching signals greater than 50 volt and analog signals greater than 50 volt. Currents less than 20 amp are also included in this category.

- **Level 4 – Power**

  This includes voltages in the range 0 – 1000 volt and currents in the range 20 – 800 amp. This applies to both AC and DC circuits.

The IEEE 518 also provides for three different situations when calculating the separation distance required between the various levels of susceptibility.

In considering the specific case where one cable is a high susceptibility cable and the other cable has a varying susceptibility, the required separation distance would vary as follows:

- **Both cables contained in a separate tray**
  - Level 1 to Level 2 – 30 mm
  - Level 1 to Level 3 – 160 mm
  - Level 1 to Level 4 – 670 mm
- **One cable contained in a tray and the other in conduit**
  - Level 1 to Level 2 – 30 mm
  - Level 1 to Level 3 – 110 mm
  - Level 1 to Level 4 – 460 mm
- **Both cables contained in separate conduit**
  - Level 1 to Level 2 – 30 mm
  - Level 1 to Level 3 – 80 mm
  - Level 1 to Level 4 – 310 mm

Figures are approximate as the original standard is quoted in inches.

A few words need to be said about the construction of the trays and conduits. It is expected that the trays are manufactured from metal and firmly earthed with complete continuity throughout the length of the tray. The trays should also be fully covered preventing the possibility of any area being without shielding.

2.5.6   Earthing and grounding requirements

This is a contentious issue and a detailed discussion laying out all the theory and practice is possibly the only way to minimize the areas of disagreement. The picture is further complicated by different national codes, which whilst not actively disagreeing with the basic precepts of other countries, tend to lay down different practical techniques in the implementation of a good earthing system.

A typical design should be based around two separate electrically insulated earth systems. The two earth systems are:

- The equipment earth
- The instrumentation (and data communications) earth

The aims of these two earthing systems are:

- To minimize the electrical noise in the system
- To reduce the effects of fault or earth loop currents on the instrumentation system
- To minimize the hazardous voltages on equipment due to electrical faults

Earth (or ground) is defined as a common reference point for all signals in equipment situated at zero potential. Below 10 MHz, the principle of a single point earthing system is the optimum solution. Two key concepts to be considered when setting up an effective earthing system are:

- To minimize the effects of impedance coupling between different circuits (i.e. when three different currents, for example, flow through a common impedance)

- To ensure that earth or ground loops are not created (for example, by mistakenly tying the screen of a cable at two points to earth)

There are three types of earthing system possible as shown in Figure 2.10. The series single point is perhaps the more common; while the parallel single point is the preferred approach with a separate earthing system for groups of signals:

- Safety or power earth
- Low level signal (or instrumentation) earth
- High level signal (motor controls) earth
- Building earth

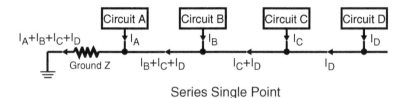

Series Single Point

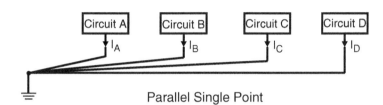

Parallel Single Point

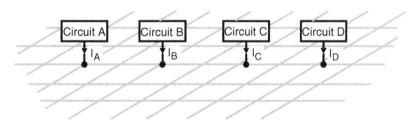

Three Dimensional Ground Plane
(eg Power Station)

Figure 2.10
Various earthing configurations

2.5.7   Suppression techniques

It is often appropriate to approach the problem of electrical noise proactively by limiting the noise at the source. This requires knowledge of the electrical apparatus that is causing the noise and then attempting to reduce the noise caused here. The two main approaches are shown in Figure 2.11.

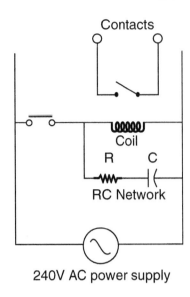

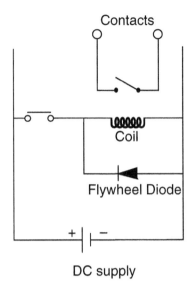

Suppression Network for AC   Suppression Network for DC

Figure 2.11
RC network in parallel across coil

In Figure 2.11, the inductance will generate a back emf across the contacts when the voltage source applied to it is switched off. This RC network then takes this back emf and thus reduces damage to the contacts.

The voltage can be limited by various combinations of devices (depending on whether the circuit is AC or DC).

The user of these techniques should be aware that the response time of the coil can be reduced by a significant time. For example, the drop out time of a coil can be increased by a factor of ten. Hence this should be approached with caution, where quick response is required from regular switched circuits (apart from the obvious deleterious impact on safety due to slowness of operation).

Two other areas to consider are:

Silicon controlled rectifiers (or SCRs) and triacs

Generate considerable electrical noise due to the switching of large currents. A possible solution is to place a correctly sized inductor placed in series with the switching device.

Lightning protection

Can be effected by the use of voltage limiters (suitably rated for the high level of current and voltage) connected across the power lines.

2.5.8   Filtering

Filtering should be done as close to the source of noise as possible. A table below summarizes some typical sources of noise and possible filtering means.

Typical sources of noise	Filtering remedy	Comments
AC voltage varies	Improved ferroresonant transformer	Conventional ferroresonant transformer fails
Notching of AC wave form	Improved ferroresonant transformer	Conventional ferroresonant transformer fails
Missing half cycle in AC wave form	Improved ferroresonant transformer	Conventional ferroresonant transformer fails
Notching in DC line	Storage capacitor	For extreme cases, active power line filters are required
Random excessively high voltage spikes or transients	Non-linear filters	Also called limiters
High frequency components	Filter capacitors across the line	Called low-pass filtering. Great care should be taken with high frequency vs performance of 'capacitors' at this frequency
Ringing of filters	Use T filters	From switching transients or high level of harmonics
60 Hz or 50 Hz interference	Twin-T RC notch filter networks	Sometimes low-pass filters can be suitable
Common mode voltages	Avoid filtering (isolation transformers or commonmode filters)	Opto isolation is preferred – eliminates ground loop
Excessive noise	Auto or cross correlation techniques	This extracts the signal spectrum from the closely overlapping noise spectrum

Table 2.1
Typical noise sources and some possible means of filtering

3

EIA-232 overview

Objectives

When you have completed study of this chapter, you will be able to:

- List the main features of the EIA-232 standard
- Fix the following problems:
    - Incorrect EIA-232 cabling
    - Male/female D-type connector confusion
    - Wrong DTE/DCE configuration
    - Handshaking
    - Incorrect signaling voltages
    - Excessive electrical noise
    - Isolation

3.1 EIA-232 interface standard (CCITT V.24 interface standard)

The EIA-232 interface standard was developed for the single purpose of interfacing data terminal equipment (DTE) and data circuit terminating equipment (DCE) employing serial binary data interchange. In particular, EIA-232 was developed for interfacing data terminals to modems.

The EIA-232 interface standard was issued in the USA in 1969 by the engineering department of the EIA. Almost immediately, minor revisions were made and EIA-232C was issued. EIA-232 was originally named RS-232, (recommended standard), which is still in popular usage. The prefix 'RS' was superseded by 'EIA/TIA' in 1988. The current revision is EIA/TIA-232E (1991), which brings it into line with the international standards ITU V.24, ITU V.28 and ISO 2110.

Poor interpretation of EIA-232 has been responsible for many problems in interfacing equipment from different manufacturers. This had led some users to dispute as to whether it is a 'standard.' It should be emphasized that EIA-232 and other related EIA standards define the electrical and mechanical details of the interface (layer 1 of the OSI model) and do not define a protocol.

The EIA-232 interface standard specifies the method of connection of two devices – the DTE and DCE. DTE refers to data terminal equipment, for example, a computer or a printer. A DTE device communicates with a DCE device. DCE, on the other hand, refers to data communications equipment such as a modem. DCE equipment is now also called data circuit-terminating equipment in EIA/TIA-232E. A DCE device receives data from the DTE and retransmits to another DCE device via a data communications link such as a telephone link.

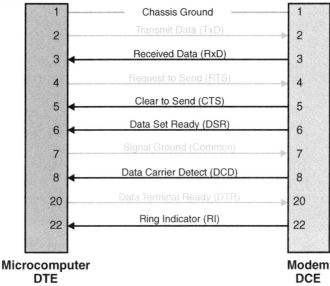

Figure 3.1
Connections between the DTE and the DCE using DB-25 connectors

3.1.1   The major elements of EIA-232

The EIA-232 standard consists of three major parts, which define:
- Electrical signal characteristics
- Mechanical characteristics of the interface
- Functional description of the interchange circuits

Electrical signal characteristics

EIA-232 defines electrical signal characteristics such as the voltage levels and grounding characteristics of the interchange signals and associated circuitry for an unbalanced system.

The EIA-232 transmitter is required to produce voltages in the range +/– 15 to +/– 25 V as follows:

- Logic 1: –5 V to –25 V
- Logic 0: +5 V to +25 V
- Undefined logic level: +5 V to –5 V

At the EIA-232 receiver, the following voltage levels are defined:

- Logic 1: –3 V to –25 V
- Logic 0: +3 V to +25 V
- Undefined logic level: –3 V to +3 V

Note: The EIA-232 transmitter requires a slightly higher voltage to overcome voltage drop along the line.

The voltage levels associated with a microprocessor are typically 0 V to +5 V for transistor–transistor logic (TTL). A line driver is required at the transmitting end to adjust the voltage to the correct level for the communications link. Similarly, a line receiver is required at the receiving end to translate the voltage on the communications link to the correct TTL voltages for interfacing to a microprocessor. Despite the bipolar input voltage, TTL compatible EIA-232 receivers operate on a single +5 V supply.

Modern PC power supplies usually have a standard +12 V output that could be used for the line driver.

The control or 'handshaking' lines have the same range of voltages as the transmission of logic 0 and logic 1, except that they are of opposite polarity. This means that:

- A control line asserted or made active by the transmitting device has a voltage range of +5 V to +25 V. The receiving device connected to this control line allows a voltage range of +3 V to +25 V.
- A control line inhibited or made inactive by the transmitting device has a voltage range of –5 V to –25 V. The receiving device of this control line allows a voltage range of –3 V to –25 V.

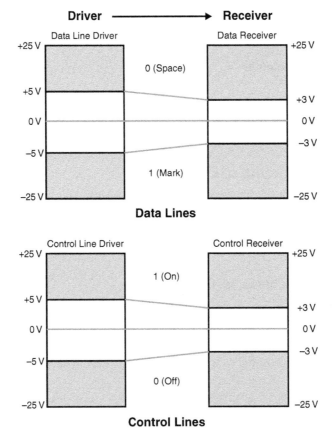

Figure 3.2
Voltage levels for EIA-232

At the receiving end, a line receiver is necessary in each data and control line to reduce the voltage level to the 0 V and +5 V logic levels required by the internal electronics.

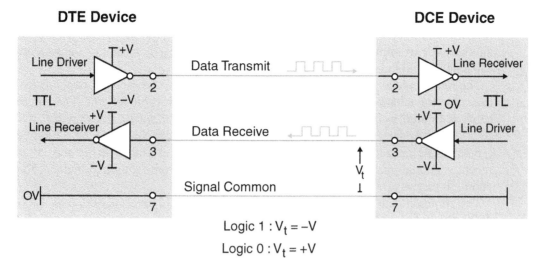

Figure 3.3
EIA-232 transmitters and receivers

The EIA-232 standard defines 25 electrical connections. The electrical connections are divided into four groups viz:
- Data lines
- Control lines
- Timing lines
- Special secondary functions

Data lines are used for the transfer of data. Data flow is designated from the perspective of the DTE interface. The transmit line, on which the DTE transmits and the DCE receives, is associated with pin 2 at the DTE end and pin 2 at the DCE end for a DB-25 connector. These allocations are reversed for DB-9 connectors. The receive line, on which the DTE receives, and the DCE transmits, is associated with pin 3 at the DTE end and pin 3 at the DCE end. Pin 7 is the common return line for the transmit and receive data lines. The allocations are illustrated in Table 3.2.

Control lines are used for interactive device control, which is commonly known as hardware handshaking. They regulate the way in which data flows across the interface. The four most commonly used control lines are:
- RTS: Request to send
- CTS: Clear to send
- DSR: Data set ready (or DCE ready in EIA-232D/E)
- DTR: Data terminal ready (or DTE ready in EIA-232D/E)

It is important to remember that with the handshaking lines, the enabled state means a positive voltage and the disabled state means a negative voltage.

Hardware handshaking is the cause of most interfacing problems. Manufacturers sometimes omit control lines from their EIA-232 equipment or assign unusual applications to them. Consequently, many applications do not use hardware handshaking but, instead, use only the three data lines (transmit, receive and signal common ground) with some form of software handshaking. The control of data flow is then part of the application program. Most of the systems encountered in data communications for instrumentation and control use some sort of software-based protocol in preference to hardware handshaking.

There is a relationship between the allowable speed of data transmission and the length of the cable connecting the two devices on the EIA-232 interface. As the speed of data transmission increases, the quality of the signal transition from one voltage level to another, for example, from –25 V to +25 V, becomes increasingly dependent on the capacitance and inductance of the cable.

The rate at which voltage can 'slew' from one logic level to another depends mainly on the cable capacitance and the capacitance increases with cable length. The length of the cable is limited by the number of data errors acceptable during transmission. The EIA-232 D&E standard specifies the limit of total cable capacitance to be 2500 pF. With typical cable capacitance having improved from around 160 pF/m to only 50 pF/m in recent years, the maximum cable length has extended from around 15 meters (50 feet) to about 50 meters (166 feet).

The common data transmission rates used with EIA-232 are 110, 300, 600, 1200, 2400, 4800, 9600 and 19200 bps. For short distances, however, transmission rates of 38 400, 57 600 and 115 200 can also be used. Based on field tests, Table 3.1 shows the practical relationship between selected baud rates and maximum allowable cable length, indicating that much longer cable lengths are possible at lower baud rates. Note that the achievable speed depends on the transmitter voltages, cable capacitance (as discussed above) as well as the noise environment.

In the context of the NRZ-type of coding used for asynchronous transmission on EIA-232 links, 1 baud = 1 bit per second.

Baud rate	Cable length (meters)
110	850
300	800
600	700
1200	500
2400	200
4800	100
9600	70
19200	50
115 K	20

Table 3.1
Demonstrated maximum cable lengths with EIA-232 interface

Mechanical characteristics of the interface

EIA-232 defines the mechanical characteristics of the interface between the DTE and the DCE. This dictates that the interface must consist of a plug and socket and that the socket will normally be on the DCE.

Although not specified by EIA-232C, the DB-25 connector (25-pin, D-type) is closely associated with EIA-232 and is the *de facto* standard with revision D. Revision E formally specifies a new connector in the 26-pin alternative connector (known as the ALT A connector). This connector supports all 25 signals associated with EIA-232. ALT A is physically smaller than the DB-25 and satisfies the demand for a smaller connector suitable for modern computers. Pin 26 is not currently used. On some EIA-232 compatible equipment, where little or no handshaking is required, the DB-9 connector

(9-pin, D-type) is common. This practice originated when IBM decided to make a combined serial/parallel adapter for the AT&T personal computer. A small connector format was needed to allow both interfaces to fit onto the back of a standard ISA interface card. Subsequently, the DB-9 connector has also became an industry standard to reduce the wastage of pins. The pin allocations commonly used with the DB-9 and DB-25 connectors for the EIA-232 interface are shown in Table 3.2. The pin allocation for the DB-9 connector is not the same as the DB-25 and often traps the unwary.

The data pins of DB-9 IBM connector are allocated as follows:

- Data transmit pin 3
- Data receive pin 2
- Signal common pin 5

Pin no. DTE	DB-9 connector IBM pin assignment	DB-25 connector EIA-232 pin assignment	DB-25 connector EIA-530 pin assignment
1	Received line signal	Shield	Shield
2	Received data	Transmitted data	Transmitted data (A)
3	Transmitted data	Received data	Received data (A)
4	DTE ready	Request to send	Request to send (A)
5	Signal/Common ground	Clear to send	Clear to send (A)
6	DCE ready	DCE ready	DCE ready (A)
7	Request to send	Signal/Common ground	Signal/Common ground
8	Clear to send	Received line signal	Received line signal (A)
9	Ring indicator	+Voltage (testing)	Receiver signal DCE element timing (B)
10		-Voltage (testing)	Received line (B)
11		Unassigned	Transmitter signal DTE element timing (B)
12		Sec received line signal detector/data signal	Transmitter signal DCE element timing
13		Sec clear to send	Clear to send (B)
14		Sec transmitted data	Transmitted data (B)
15		Transmitter signal DCE element timing	Transmitter signal DCE element timing (A)
16		Sec received data	Received data (B)
17		Receiver signal DCE element timing	Receiver signal DCE element timing (A)
18		Local loopback	Local loopback
19		Sec request to send	Request to send (B)
20		DTE ready	DTE ready (A)
21		Remote loopback/signal quality detector	Remote loopback
22		Ring indicator	DCE ready (B)
23		Data signal rate	DTE ready (B)
24		Transmit signal DTE element timing	Transmitter signal DTE element timing (A)
25		Test mode	Test mode

Table 3.2
Common DB-9 and DB-25 pin assignments for EIA-232 and EIA/TIA-530
(often used for EIA-422 and EIA-485)

Functional description of the interchange circuits

EIA-232 defines the function of the data, timing and control signals used at the interface of the DTE and DCE. However, very few of the definitions are relevant to applications for data communications for instrumentation and control.

The circuit functions are defined with reference to the DTE as follows:

- **Protective ground (shield)**

  The protective ground ensures that the DTE and DCE chassis are at equal potentials (remember that this protective ground could cause problems with circulating earth currents).

- **Transmitted data (TxD)**

  This line carries serial data from the DTE to the corresponding pin on the DCE. The line is held at a negative voltage during periods of line idle.

- **Received data (RxD)**

  This line carries serial data from the DCE to the corresponding pin on the DTE.

- **Request to send (RTS)**

  (RTS) is the request to send hardware control line. This line is placed active (+V) when the DTE requests permission to send data. The DCE then activates (+V) the CTS (clear to send) for hardware data flow control.

- **Clear to send (CTS)**

  When a half-duplex modem is receiving, the DTE keeps RTS inhibited. When it is the DTE's turn to transmit, it advises the modem by asserting the RTS pin. When the modem asserts the CTS, it informs the DTE that it is now safe to send data.

- **DCE ready**

  Formerly called data set ready (DSR). The DTE ready line is an indication from the DCE to the DTE that the modem is ready.

- **Signal ground (common)**

  This is the common return line for all the data transmit and receive signals and all other circuits in the interface. The connection between the two ends is always made.

- **Data carrier detect (DCD)**

  This is also called the received line signal detector. It is asserted by the modem when it receives a remote carrier and remains asserted for the duration of the link.

- **DTE ready (data terminal ready)**

  Formerly called data terminal ready (DTR). DTE ready enables but does not cause, the modem to switch onto the line. In originate mode, DTE ready must be asserted in order to auto dial. In answer mode, DTE ready must be asserted to auto answer.

- **Ring indicator**

  This pin is asserted during a ring voltage on the line.

- **Data signal rate selector (DSRS)**

  When two data rates are possible, the higher is selected by asserting DSRS; however, this line is not used much these days.

Pin no.	CCITT no.	Circuit	Description	Circuit direction
1	-	-	Shield	To DCE
2	103	BA	Transmitted data	To DCE
3	104	BB	Received data	From DCE
4	105/133	CA/CJ	Request to send/ready for receiving	To DCE
5	106	CB	Clear to send	From DCE
6	107	CC	DCE ready	From DCE
7	102	AB	Signal common	-
8	109	CF	Received line signal detector	From DCE
9	-	-	Reserved for testing	-
10	-	-	Reserved for testing	-
11	126	See Note	Unassigned	-
12	122/112	SCF/CI	Secondary received line signal detector/data signal rate selector	From DCE
13	121	SCB	Secondary clear to send	From DCE
14	118	SBA	Secondary transmitted data	To DCE
15	114	DB	Transmitter signal element timing (DTE source)	From DCE
16	119	SBB	Secondary received data	From DCE
18	141	LL	Local loopback	To DCE
19	120	SCA	Secondary request to send	To DCE
20	108/112	CD	DTE ready	To DCE
21	140/110	RL/CG	Remote loopback/signal quality detector	T/F DCE
22	125	CE	Ring indicator	From DCE
23	111/112	CH/CI	Data signal rate selector (DTE/DCE source)	T/F DCE
24	113	DA	Transmit signal element timing (DTE source)	To DCE
25	142	TM	Test mode	From DCE
26		None	(Alt A connector) No connection at this time	

Table 3.3
ITU-T V24 pin assignment (ISO 2110)

3.2   Half-duplex operation of the EIA-232 interface

The following description of one particular operation of the EIA-232 interface is based on half-duplex data interchange. The description encompasses the more generally used full-duplex operation.

Figure 3.4 shows the operation with the initiating user terminal, DTE, and its associated modem DCE on the left of the diagram and the remote computer and its modem on the right.

The following sequence of steps occurs when a user sends information over a telephone link to a remote modem and computer:

- The initiating user manually dials the number of the remote computer.
- The receiving modem asserts the ring indicator line (RI) in a pulsed ON/OFF fashion reflecting the ringing tone. The remote computer already has its data terminal ready (DTR) line asserted to indicate that it is ready to receive calls. Alternatively, the remote computer may assert the DTR line after a few rings. The remote computer then sets its request to send (RTS) line to ON.
- The receiving modem answers the phone and transmits a carrier signal to the initiating end. It asserts the DCE ready line after a few seconds.
- The initiating modem asserts the data carrier detect (DCD) line. The initiating terminal asserts its DTR, if it is not already high. The modem responds by asserting its DTE ready line.
- The receiving modem asserts its clear to send (CTS) line, which permits the transfer of data from the remote computer to the initiating side.
- Data is transferred from the receiving DTE (transmitted data) to the receiving modem. The receiving remote computer then transmits a short message to indicate to the originating terminal that it can proceed with the data transfer. The originating modem transmits the data to the originating terminal.
- The receiving terminal sets its request to send (RTS) line to OFF. The receiving modem then sets its clear to send (CTS) line to OFF.
- The receiving modem switches its carrier signal OFF.
- The originating terminal detects that the data carrier detect (DCD) signal has been switched OFF on the originating modem and switches its RTS line to the ON state. The originating modem indicates that transmission can proceed by setting its CTS line to ON.
- Transmission of data proceeds from the originating terminal to the remote computer.
- When the interchange is complete, both carriers are switched OFF and in many cases, the DTR is set to OFF. This means that the CTS, RTS and DCE ready lines are set to OFF.

Full-duplex operation requires that transmission and reception occur simultaneously. In this case, there is no RTS/CTS interaction at either end. The RTS line and CTS line are left ON with a carrier to the remote computer.

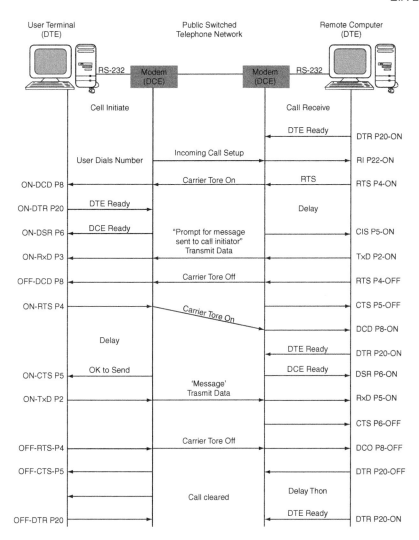

Figure 3.4
Half-duplex operational sequence of EIA-232

3.3   Summary of EIA/TIA-232 revisions

A summary of the main differences between EIA-232 revisions, C, D and E are discussed below.

Revision D – EIA-232D

The 25-pin D-type connector was formally specified. In revision C, reference was made to the D-type connector in the appendices and a disclaimer was included revealing that it was not intended to be part of the standard; however, it was treated as the *de facto* standard.

The voltage ranges for the control and data signals were extended to a maximum limit of 25 V from the previously specified 15 V in revision C.

The 15 meter (50 foot) distance constraint, implicitly imposed to comply with circuit capacitance, was replaced by 'circuit capacitance shall not exceed 2500 pF' (standard EIA-232 cable has a capacitance of 50 pF/ft).

Revision E – EIA-232E

Revision E formally specifies the new 26-pin alternative connector, the ALT A connector. This connector supports all 25 signals associated with EIA-232, unlike the 9-pin connector, which has become associated with EIA-232 in recent years. Pin 26 is currently not used. The technical changes implemented by EIA-232F do not present compatibility problems with equipment confirming to previous versions of EIA-232.

This revision brings the EIA-232 standard into line with international standards CCITT V.24, V.28 and ISO 2110.

3.4   Limitations

In spite of its popularity and extensive use, it should be remembered that the EIA-232 interface standard was originally developed for interfacing data terminals to modems. In the context of modern requirements, EIA-232 has several weaknesses. Most have arisen as a result of the increased requirements for interfacing other devices such as PCs, digital instrumentation, digital variable speed drives, power system monitors and other peripheral devices in industrial plants.

The main limitations of EIA-232 when used for the communications of instrumentation and control equipment in an industrial environment are:

- The point-to-point restriction, a severe limitation when several 'smart' instruments are used
- The distance limitation of 15 meters (50 feet) end-to-end, too short for most control systems
- The 20 kbps rate, too slow for many applications
- The –3 to –25 V and +3 to +25 V signal levels, not directly compatible with modern standard power supplies.

Consequently, a number of other interface standards have been developed by the EIA which overcome some of these limitations. The EIA-485 interface standards are increasingly being used for instrumentation and control systems.

3.5   Troubleshooting

3.5.1   Introduction

Since EIA-232 is a point-to-point system, installation is fairly straightforward and simple and all EIA-232 devices use either DB-9 or DB-25 connectors. These connectors are used because they are cheap and allow multiple insertions. None of the 232 standards define which device uses a male or female connector, but traditionally the male (pin) connector is used on the DTE and the female type connector (socket) is used on DCE equipment. This is only traditional and may vary on different equipment. It is often asked why a 25-pin connector is used when only 9 pins are needed. This was done because EIA-232 was used before the advent of computers. It was therefore used for hardware control (RTS/CTS). It was originally thought that, in the future, more hardware control lines would be needed hence the need for more pins.

When doing an initial installation of an EIA-232 connection it is important to note the following.

- Is one device a DTE and the other a DCE?
- What is the sex and size of connectors at each end?
- What is the speed of the communication?

- What is the distance between the equipment?
- Is it a noisy environment?
- Is the software setup correctly?

3.5.2   Typical approach

When troubleshooting a serial data communications interface, one needs to adopt a logical approach in order to avoid frustration and wasting many hours. A procedure similar to that outlined below is recommended:

1. Check the basic parameters. Are the baud rate, stop/start bits and parity set identically for both devices? These are sometimes set on DIP switches in the device. However, the trend is towards using software, configured from a terminal, to set these basic parameters.
2. Identify which is DTE or DCE. Examine the documentation to establish what actually happens at pins 2 and 3 of each device. On the 25-pin DTE device, pin 2 is used for transmission of data and should have a negative voltage (mark) in idle state, whilst pin 3 is used for the receipt of data, (passive) and should be approximately at 0 volts. Conversely, at the DCE device, pin 3 should have a negative voltage, whilst pin 2 should be around 0 volts. If no voltage can be detected on either pin 2 or 3, then the device is probably not EIA-232 compatible and could be connected according to another interface standard, such as EIA-422, EIA-485, etc.

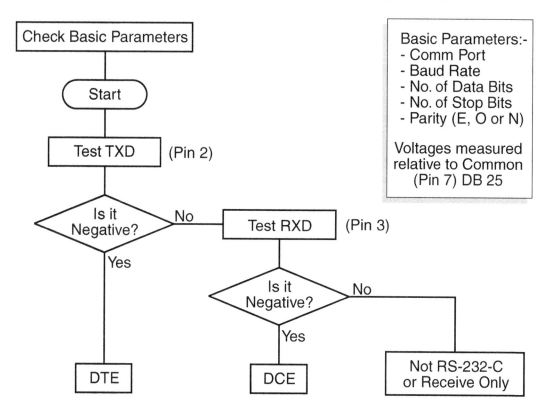

Figure 3.5
Flowchart to identify an EIA-232 device as either a DTE or DCE

3. Clarify the needs of the hardware handshaking when used. Hardware handshaking can cause the greatest difficulties and the documentation should be carefully studied to yield some clues about the handshaking sequence. Ensure all the required wires are correctly terminated in the cables.

4. Check the actual protocol used. This is seldom a problem but, when the above three points do not yield an answer, it is possible that there are irregularities in the protocol structure between the DCE and DTE devices.

5. Alternatively, if software handshaking is utilized, make sure both have compatible application software. In particular, check that the same ASCII character is used for XON and XOFF.

3.5.3   Test equipment

From a testing point of view, section 2.1.2 in the EIA-232-E interface standard states that:

'The generator on the interchange circuit shall be designed to withstand an open circuit, a short circuit between the conductor carrying that interchange circuit in the interconnecting cable and any other conductor in that cable including signal ground, without sustaining damage to itself or its associated equipment.'

In other words, any pin may be connected to any other pin, or even earth, without damage and, theoretically, one cannot blow up anything! This does not mean that the EIA-232 interface cannot be damaged. The incorrect connection of incompatible external voltages can damage the interface, as can static charges.

If a data communication link is inoperable, the following devices may be useful when analyzing the problem:

- A digital multimeter

  Any cable breakage can be detected by measuring the continuity of the cable for each line. The voltages at the pins in active and inactive states can also be ascertained by the multimeter to verify its compatibility to the respective standards.

- A LED

  The use of LED is to determine which are the asserted lines or whether the interface conforms to a particular standard. This is laborious and accurate pin descriptions should be available.

- A breakout box
- PC-based protocol analyzer (including software)
- Dedicated hardware protocol analyzer (e.g. Hewlett Packard)

The breakout box

The breakout box is an inexpensive tool that provides most of the information necessary to identify and fix problems on data communications circuits, such as the serial EIA-232, EIA-422, EIA-423 and EIA-485 interfaces and also on parallel interfaces.

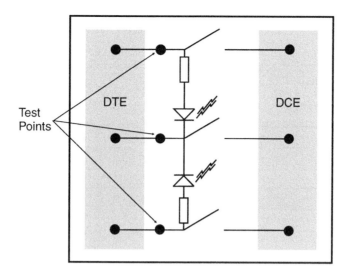

Figure 3.6
Breakout box showing test points

A breakout box is connected to the data cable, to bring out all conductors in the cable to accessible test points. Many versions of this equipment are available on the market, from the 'homemade' using a back-to-back pair of male and female DB-25 sockets to fairly sophisticated test units with built-in LEDs, switches and test points.

Breakout boxes usually have a male and a female socket and by using two standard serial cables, the box can be connected in series with the communication link. The 25 test points can be monitored by LEDs, a simple digital multimeter, an oscilloscope or a protocol analyzer. In addition, a switch in each line can be opened or closed while trying to identify the problem.

The major weakness of the breakout box is that while one can interrupt any of the data lines, it does not help much with the interpretation of the flow of bits on the data communication lines. A protocol analyzer is required for this purpose.

Null modem

Null modems look like DB-25 'through' connectors and are used when interfacing two devices of the same gender (e.g. DTE to DTE, DCE to DCE) or devices from different manufacturers with different handshaking requirements. A null modem has appropriate internal connections between handshaking pins that 'trick' the terminal into believing conditions are correct for passing data. A similar result can be achieved by soldering extra loops inside the DB-25 plug. Null modems generally cause more problems than they cure and should be used with extreme caution and preferably avoided.

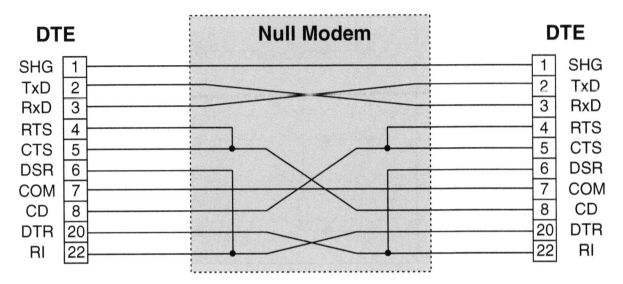

Figure 3.7
Null modem connections

Note that the null modem may inadvertently connect pins 1 together, as in Figure 3.7. This is an undesirable practice and should be avoided.

Loop back plug

This is a hardware plug which loops back the transmit data pin to receive data pin and similarly for the hardware handshaking lines. This is another quick way of verifying the operation of the serial interface without connecting to another system.

Protocol analyzer

A protocol analyzer is used to display the actual bits on the data line, as well as the special control codes, such as STX, DLE, LF, CR, etc. The protocol analyzer can be used to monitor the data bits, as they are sent down the line and compared with what should be on the line. This helps to confirm that the transmitting terminal is sending the correct data and that the receiving device is receiving it. The protocol analyzer is useful in identifying incorrect baud rate, incorrect parity generation method, incorrect number of stop bits, noise, or incorrect wiring and connection. It also makes it possible to analyze the format of the message and look for protocol errors.

When the problem has been shown not to be due to the connections, baud rate, bits or parity, then the content of the message will have to be analyzed for errors or inconsistencies. Protocol analyzers can quickly identify these problems.

Purpose built protocol analyzers are expensive devices and it is often difficult to justify the cost when it is unlikely that the unit will be used very often. Fortunately, software has been developed that enables a normal PC to be used as a protocol analyzer. The use of a PC as a test device for many applications is a growing field, and one way of connecting a PC as a protocol analyzer is shown in Figure 3.8.

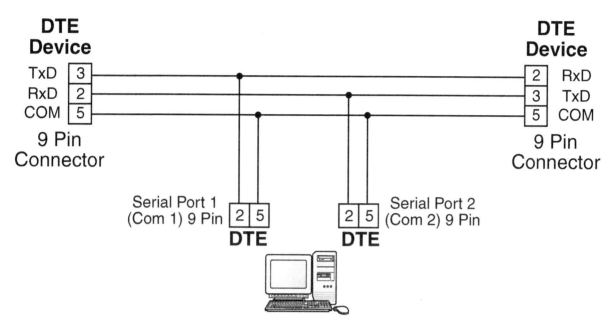

Figure 3.8
Protocol analyzer connection

The above figure has been simplified for clarity and does not show the connections on the control lines (for example, RTS and CTS).

3.5.4 Typical EIA-232 problems

Table 3.4 lists typical EIA-232 problems, which can arise because of inadequate interfacing. These problems could equally apply to two PCs connected to each other or to a PC connected to a printer.

To determine whether the devices are DTE or DCE, connect a breakout box at one end and note the condition of the TX light (pin 2 or 3) on the box. If pin 2 is ON, then the device is probably a DTE. If pin 3 is ON, it is probably a DCE. Another clue could be the sex of the connector, male are typically DTEs and females are typically DCEs, but not always.

Figure 3.9
A 9-pin EIA-232 connector on a DTE

When troubleshooting an EIA-232 system, it is important to understand that there are two different approaches. One approach is followed if the system is new and never been run

Problem	Probable cause of problem
Garbled or lost data	Baud rates of connecting ports may be different
	Connecting cables could be defective
	Data formats may be inconsistent (Stop bit/parity/ number of data bits)
	Flow control may be inadequate
	High error rate due to electrical interference
	Buffer size of receiver inadequate
First characters garbled	The receiving port may not be able to respond quickly enough. Precede the first few characters with the ASCII (DEL) code to ensure frame synchronization
No data communications	Power for both devices may not be on
	Transmit and receive lines of cabling may be incorrect
	Handshaking lines of cabling may be incorrectly connected
	Baud rates mismatch
	Data format may be inconsistent
	Earth loop may have formed for EIA-232 line
	Extremely high error rate due to electrical interference for transmitter and receiver
	Protocols may be inconsistent/intermittent communications
	Intermittent interference on cable
ASCII data has incorrect spacing	Mismatch between 'LF' and 'CR' characters generated by transmitting device and expected by receiving device

Table 3.4
A list of typical EIA-232 problems

before and the other if the system has been operating and for some reason does not communicate at present. New systems that have never worked have more potential problems than a system that has been working before and now has stopped. If a system is new it can have three main problems viz. mechanical, setup or noise. A previously working system usually has only one problem, viz. mechanical. This assumes that no one has changed the setup and noise has not been introduced into the system. In all systems, whether having previously worked or not, it is best to check the mechanical parts first. This is done by:

- Verifying that there is power to the equipment
- Verifying that the connectors are not loose
- Verifying that the wires are correctly connected
- Checking that a part, board or module has not visibly failed

3.5.4.1 Mechanical problems

Often, mechanical problems develop in EIA-232 systems because of incorrect installation of wires in the D-type connector or because strain reliefs were not installed correctly. The following recommendations should be noted when building or installing EIA-232 cables:

- Keep the wires short (20 meters maximum)
- Stranded wire should be used instead of solid wire (solid wire will not flex)
- Only one wire should be soldered in each pin of the connector
- Bare wire should not be showing out of the pin of the connector
- The back shell should reliably and properly secure the wire

The speed and distance of the equipment will determine if it is possible to make the connection at all. Most engineers try to stay less than 50 feet or about 16 meters at 115 200 bits per second. This is a very subjective measurement and will depend on the cable, voltage of the transmitter and the amount and noise in the environment. The transmitter voltage can be measured at each end when the cable has been installed. A voltage of at least +/– 5 V should be measured at each end on both the TX and RX lines.

An EIA-232 breakout box is placed between the DTE and DCE to monitor the voltages placed on the wires by looking at pin 2 on the breakout box. Be careful here because it is possible that the data is being transmitted so fast that the light on the breakout box doesn't have time to change. If possible, lower the speed of the communication at both ends to something like 2 bps.

Figure 3.10
Measuring the voltage on EIA-232

Once it has been determined that the wires are connected as DTE to DCE and that the distance and speed are not going to be a problem, the cable can be connected at each end. The breakout box can still be left connected with the cable and both pin 2 and 3 lights on the breakout box should now be on.

The color of the light depends on the breakout box. Some breakout boxes use red for a one and others use green for a one. If only one light is on then that may mean that a wire is broken or there is a DTE to DTE connection. A clue to a possible DTE to DTE connection would be that the light on pin 3 would be off and the one on pin 2 would be on. To correct this problem, first check the wires for continuity then turn switches 2 and 3 off on the breakout box and use jumper wires to swap them. If the TX and RX lights come on, a null modem cable or box will need to be built and inserted in-line with the cable.

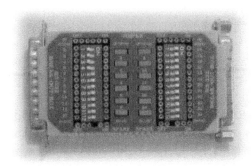

Figure 3.11
An EIA-232 breakout box

If the pin 2 and pin 3 lights are on, one end is transmitting and the control is correct, then the only thing left is the protocol or noise. Either a hardware or software protocol analyzer will be needed to troubleshoot the communications between the devices. On new installations, one common problem is mismatched baud rates. The protocol analyzer will tell exactly what the baud rates are for each device. Another thing to look for with the analyzer is the timing. Often, the transmitter waits some time before expecting a proper response from the receiver. If the receiver takes too long to respond or the response is incorrect, the transmitter will 'time out'. This is usually denoted as a 'communications error or failure'.

3.5.4.2 Setup problems

Once it is determined that the cable is connected correctly and the proper voltage is being received at each end, it is time to check the setup. The following circumstances need to be checked before trying to communicate:

- Is the software communications setup at both ends for either 8N1, 7E1 or 7O1?
- Is the baud rate the same at both devices? (1200, 4800, 9600, 19 200 etc)
- Is the software setup at both ends for binary, hex or ASCII data transfer?
- Is the software setup for the proper type of control?

Although the 8 data bits, no parity and 1 stop bit is the most common setup for asynchronous communication, often 7 data bits even parity with 1 stop bit is used in industrial equipment. The most common baud rate used in asynchronous communications is 9600. Hex and ASCII are commonly used as communication codes.

If one device is transmitting but the other receiver is not responding, then the next thing to look for is what type of control the devices are using. The equipment manual may define whether hardware or software control is being used. Both ends should be setup either for hardware control, software control or none.

3.5.4.3 Noise problems

EIA-232, being a single ended (unbalanced) type of circuit, lends itself to receiving noise. There are three ways that noise can be induced into an EIA-232 circuit.

- Induced noise on the common ground
- Induced noise on the TX or RX lines
- Induced noise on the indicator or control lines

Ground induced noise

Different ground voltage levels on the ground line (pin 7) can cause ground loop noise. Also, varying voltage levels induced on the ground at either end by high power equipment can cause intermittent noise. This kind of noise can be very difficult to reduce. Sometimes, changing the location of the ground on either the EIA-232 equipment or the high power equipment can help, but this is often not possible. If it is determined that the noise problem is caused by the ground it may be best to replace the EIA-232 link with a fiber optic or RS-422 system. Fiber optic or EIA-422 to EIA-232 adapters are relatively cheap, readily available and easy to install. When the cost of troubleshooting the system is included, replacing the system often is the cheapest option.

Induced noise on the TX or RX lines

Noise from the outside can cause the communication on an EIA-232 system to fail, although this voltage must be quite large. Because EIA-232 voltages in practice are usually between +/–7 and +/–12, the noise voltage value must be quite high in order to induce errors. This type of noise induction is noticeable because the voltage on the TX or RX will be outside of the specifications of EIA-232. Noise on the TX line can also be induced on the RX line (or *vice versa*) due to the common ground in the circuit. This type of noise can be detected by comparing the data being transmitted with the received communication at the other end of the wire (assuming no broken wire). The protocol analyzer is plugged into the transmitter at one end and the data monitored. If the data is correct, the protocol analyzer is then plugged into the other end and the received data monitored. If the data is corrupt at the receiving end, then noise on that wire may be the problem. If it is determined that the noise problem is caused by induced noise on the TX or RX lines, it may be best to move the EIA-232 line and the offending noise source away from each other. If this doesn't help, it may be necessary to replace the EIA-232 link with a fiber optic or RS-485 system.

Induced noise on the indicator or control lines

This type of noise is very similar to the previous TX/RX noise. The difference is that noise on these wires may be harder to find. This is because the data is being received at both ends, but there still is a communication problem. The use of a voltmeter or oscilloscope would help to measure the voltage on the control or indicator lines and therefore locate the possible cause of the problem, although this is not always very accurate. This is because the effect of noise on a system is governed by the ratio of the power levels of the signal and the noise, rather than a ratio of their respective voltage levels. If it is determined that the noise is being induced on one of the indicator or control lines, it may be best to move the EIA-232 line and the offending noise source away from each other. If this doesn't help, it may be necessary to replace the EIA-232 link with a fiber optic or RS-485 system.

3.5.5 Summary of troubleshooting

Installation

- Is one device a DTE and the other a DCE?
- What is the sex and size of the connectors at each end?
- What is the speed of the communications?

- What is the distance between the equipment?
- Is it a noisy environment?
- Is the software setup correctly?

Troubleshooting new and old systems

- Verify that there is power to the equipment
- Verify that the connectors are not loose
- Verify that the wires are correctly connected
- Check that a part, board or module has not visibly failed

Mechanical problems on new systems

- Keep the wires short (20 meters maximum)
- Stranded wire should be used instead of solid wire (stranded wire will flex)
- Only one wire should be soldered in each pin of the connector
- Bare wire should not be showing out of the connector pins
- The back shell should reliably and properly secure the wire

Setup problems on new systems

- Is the software communications setup at both ends for either 8N1, 7E1 or 7O1?
- Is the baud rate the same for both devices? (1200, 4800, 9600, 19 200 etc)
- Is the software setup at both ends for binary, hex or ASCII data transfer?
- Is the software setup for the proper type of control?

Noise problems on new systems

- Noise from the common ground
- Induced noise on the TX or RX lines
- Induced noise on the indicator or control lines

4

EIA-485 overview

Objectives

When you have completed study of this chapter, you will be able to:

- Describe the EIA-485 standard
- Remedy the following problems:
  - Incorrect EIA-485 wiring
  - Excessive common mode voltage
  - Faulty converters
  - Isolation
  - Idle state problems
  - Incorrect or missing terminations
  - RTS control via hardware or software

4.1   The EIA-485 interface standard

The EIA-485A standard is one of the most versatile of the EIA interface standards. It is an extension of EIA-422 and allows the same distance and data speed but increases the number of transmitters and receivers permitted on the line. EIA-485 permits a 'multidrop' network connection on 2 wires and allows reliable serial data communication for:

- Distances of up to 1200 m (4000 feet, same as EIA-422)
- Data rates of up to 10 Mbps (same as EIA-422)
- Up to 32 line drivers on the same line
- Up to 32 line receivers on the same line.

The maximum bit rate and maximum length can, however, not be achieved at the same time. For 24 AWG twisted pair cable the maximum data rate at 4000 ft (1200 m) is approximately 90 kbps. The maximum cable length at 10 Mbps is less than 20 ft (6 m). Better performance will require a higher-grade cable and possibly the use of active (solid state) terminators in the place of the 120-ohm resistors.

According to the EIA-485 standard, there can be 32 'standard' transceivers on the network. Some manufacturers supply devices that are equivalent to ½ or ¼ standard device, in which case this number can be increased to 64 or 128. If more transceivers are required, repeaters have to be used to extend the network.

The two conductors making up the bus are referred to as A in B in the specification. The A conductor is alternatively known as A–, TxA and Tx+. The B conductor, in similar fashion, is called B+, TxB and Tx–. Although this is rather confusing, identifying the A and B wires is not difficult. In the MARK or OFF state (i.e. when the EIA-232 TxD pin is LOW (e.g. minus 8 V), the voltage on the A wire is more negative than that on the B wire.

The differential voltages on the A and B outputs of the driver (transmitter) are similar (although not identical) to those for EIA-422, namely:

- –1.5 V to –6 V on the A terminal with respect to the B terminal for a binary 1 (MARK or OFF) state, and
- +1.5 V to +6 V on the A terminal with respect to the B terminal for a binary 0 (SPACE or ON state).

As with EIA-422, the line driver for the EIA-485 interface produces a ±5 V differential voltage on two wires.

The major enhancement of EIA-485 is that a line driver can operate in three states called tri-state operation:

- Logic 1
- Logic 0
- High-impedance

In the high-impedance state, the line driver draws virtually no current and appears not to be present on the line. This is known as the 'disabled' state and can be initiated by a signal on a control pin on the line driver integrated circuit. Tri-state operation allows a multidrop network connection and up to 32 transmitters can be connected on the same line, although only one can be active at any one time. Each terminal in a multidrop system must be allocated a unique address to avoid conflicting with other devices on the system. EIA-485 includes current limiting in cases where contention occurs.

The EIA-485 interface standard is very useful for systems where several instruments or controllers may be connected on the same line. Special care must be taken with the software to coordinate which devices on the network can become active. In most cases, a master terminal, such as a PC or computer, controls which transmitter/receiver will be active at a given time.

The two-wire data transmission line does not require special termination if the signal transmission time from one end of the line to the other end (at approximately 200 meters per microsecond) is significantly smaller than one quarter of the signal's rise time. This is typical with short lines or low bit rates. At high bit rates or in the case of long lines, proper termination becomes critical. The value of the terminating resistors (one at each end) should be equal to the characteristic impedance of the cable. This is typically 120 ohms for twisted pair wire.

Figure 4.1 shows a typical two-wire multidrop network. Note that the transmission line is terminated on both ends of the line but not at drop points in the middle of the line.

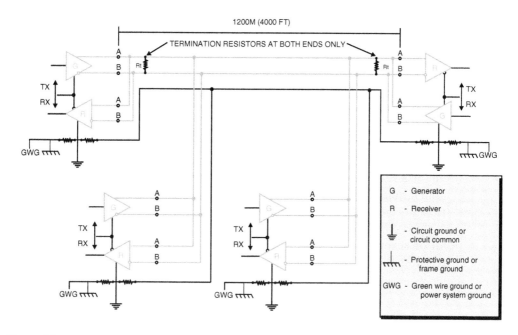

Figure 4.1
Typical two-wire multidrop network

An EIA-485 network can also be connected in a four-wire configuration as shown in Figure 4.2. In this type of connection it is necessary that one node is a master node and all others slaves. The master node communicates to all slaves, but a slave node can communicate only to the master. Since the slave nodes never listen to another slave's response to the master, a slave node can not reply incorrectly to another slave node. This is an advantage in a mixed protocol environment.

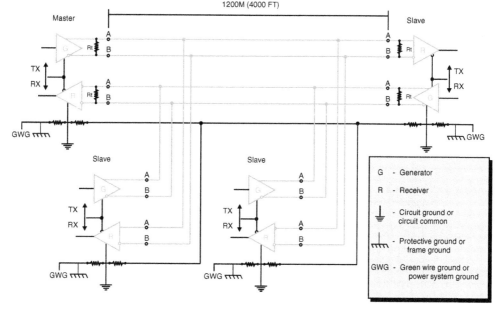

Figure 4.2
Four-wire network configuration

During normal operation there are periods when all EIA-485 drivers are off, and the communications lines are in the idle, high-impedance state. In this condition the lines are susceptible to noise pick up, which can be interpreted as random characters on the communications line. If a specific EIA-485 system has this problem, it should incorporate bias resistors, as indicated in Figure 4.3. The purpose of the bias resistors is not only to reduce the amount of noise picked up, but to keep the receiver biased in the IDLE state when no input signal is received. For this purpose the voltage drop across the 120 ohm termination resistor must exceed 200 mV AND the A terminal must be more negative than the B terminal. Keeping in mind that the two 120 ohm resistors appear in parallel, the bias resistor values can be calculated using Ohm's law. For a +5 V supply and 120 ohm terminators, a bias resistor value of 560 ohm is sufficient. This assumes that the bias resistors are only installed on ONE node.

Some commercial systems use higher values for the bias resistors, but then assume that all or several nodes have bias resistors attached. In this case the value of all the bias resistors in parallel must be small enough to ensure 200 mV across the A and B wires.

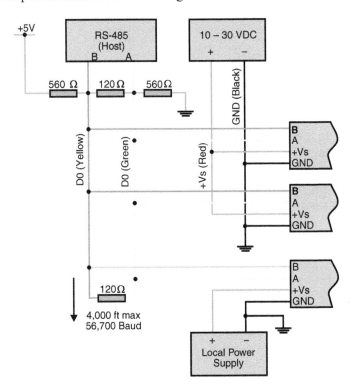

Figure 4.3
Suggested installation of resistors to minimize noise

EIA-485 line drivers are designed to handle 32 nodes. This limitation can be overcome by employing an EIA-485 repeater connected to the network. When data occurs on either side of the repeater, it is transmitted to the other side. The EIA-485 repeater transmits at full voltage levels, consequently another 31 nodes can be connected to the network. A diagram for the use of EIA-485 with a bi-directional repeater is given in Figure 4.4.

The 'gnd' pin of the EIA-485 transceiver should be connected to the logic reference (also known as circuit ground or circuit common), either directly or through a 100-ohm ½ watt resistor. The purpose of the resistor is to limit the current flow if there is a significant potential difference between the earth points. This is not shown in Figure 4.2.

In addition, the logic reference is to be connected to the chassis reference (protective ground or frame ground) through a 100 ohm ½ watt resistor. The chassis reference, in turn, is connected directly to the safety reference (green wire ground or power system ground).

If the grounds of the nodes are properly interconnected, then a third wire running in parallel with the A and B wires is technically speaking not necessary. However, this is often not the case and thus a third wire is added as in Figure 4.2. If the third wire is added, a 100 ohm ½ watt resistor is to be added at each end as shown in Figure 4.2.

The 'drops' or 'spurs' that interconnect the intermediate nodes to the bus need to be as short as possible since a long spur creates an impedance mismatch, which leads to unwanted reflections. The amount of reflection that can be tolerated depends on the bit rate. At 50 kbps a spur of, say, 30 meters could be in order, whilst at 10 Mbps the spur might be limited to 30 cm. Generally speaking, spurs on a transmission line are 'bad news' because of the impedance mismatch (and hence the reflections) they create, and should be kept as short as possible.

Some systems employ EIA-485 in a so-called 'star' configuration. This is not really a star, since a star topology requires a hub device at its center. The 'star' is in fact a very short bus with extremely long spurs, and is prone to reflections. It can therefore only be used at low bit rates.

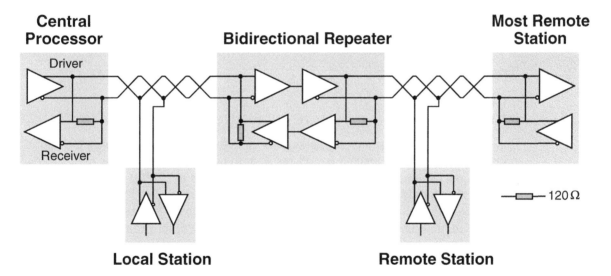

Figure 4.4
EIA-485 used with repeaters

The 'decision threshold' of the EIA-485 receiver is identical to that of both EIA-422 and EIA-423 receivers (not discussed as they have been superseded by EIA-423) at 400 mV (0.4 V), as indicated in Figure 4.5.

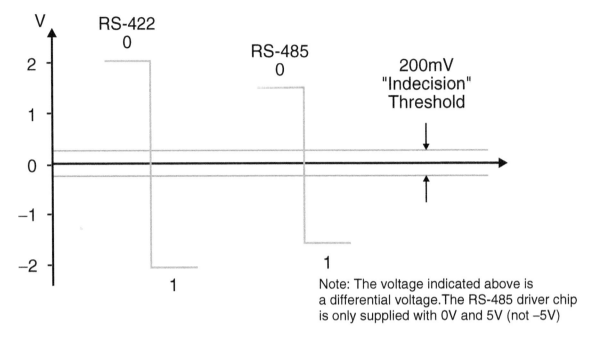

Figure 4.5
EIA-485/422 and 423 receiver sensitivities

4.2   Troubleshooting

4.2.1   Introduction

EIA-485 is the most common asynchronous voltage standard in use today for multidrop communication systems since it is very resistant to noise, can send data at high speeds (up to 10 Mbps), can be run for long distances (5 km at 1200 bps, 1200 m at 90 kbps), and is easy and cheap to use.

The EIA-485 line drivers/receivers are differential chips. This means that the TX and RX wires are referenced to each other. A one is transmitted, for example, when one of the lines is +5 volts and the other is 0 volts. A zero is then transmitted when the lines reverse and the line that was + 5 volts is now 0 volts and the line that was 0 volts is now +5 volts. In working systems the voltages are usually somewhere around +/– 2 volts with reference to each other. The indeterminate voltage levels are +/– 200 mV. Up to 32 devices can be connected on one system without a repeater. Some systems allow the connection of five legs with four repeaters and get 160 devices on one system.

Figure 4.6
EIA-485 chip

Resistors are sometimes used on EIA-485 systems to reduce noise, common mode voltages and reflections.

Bias resistors of values from 560 ohms to 4 k ohms can sometimes be used to reduce noise. These resistors connect the B+ line to + 5 volts and the A– line to ground. Higher voltages should not be used because anything over +12 volts will cause the system to fail. Unfortunately, sometimes these resistors can increase the noise on the system by allowing a better path for noise from the ground. It is best not to use bias resistors unless required by the manufacturer.

Common mode voltage resistors usually have a value between 100 k and 200 k ohms. The values will depend on the induced voltages on the lines. They should be equal and as high as possible and placed on both lines and connected to ground. The common mode voltages should be kept less then +7 volts, measured from each line to ground. Again, sometimes these resistors can increase the noise on the system by allowing a better path for noise from the ground. It is best not to use common mode resistors unless required by the manufacturer or as needed.

The termination resistor value depends on the cable used and is typically 120 ohms. Values less than 110 ohms should not be used since the driver chips are designed to drive a load resistance not less than 54 ohms, being the value of the two termination resistors in parallel plus any other stray resistance in parallel. These resistors are placed between the lines (at the two furthest ends, not on the stubs) and reduce reflections. If the lines are less than 100 meters long and speeds are 9600 baud or less, the termination resistor usually becomes redundant, but having said that, you should always follow the manufacturers' recommendations.

4.2.2   EIA-485 vs EIA-422

In practice, EIA-485 and EIA-422 are very similar to each other and manufacturers often use the same chips for both. The main working difference is that EIA-485 is used for 2-wire multidrop half-duplex systems and EIA-422 is for 4-wire point-to-point full-duplex systems. Manufacturers often use a chip like the 75154 with two EIA-485 drivers on board as an EIA-422 driver. One driver is used as a transmitter and the other is dedicated as a receiver. Because the EIA-485 chips have three states, TX, RX and high-impedance, the driver that is used as a transmitter can be set to high-impedance mode when the driver is not transmitting data. This is often done using the RTS line from the EIA-232 port. When the RTS goes high (+ voltage) the transmitter is effectively turned off by being put the transmitter in the high-impedance mode. The receiver is left on all the time, so data can be received when it comes in. This method can reduce noise on the line by having a minimum of devices on the line at a time.

4.2.3   EIA-485 installation

Installation rules for EIA-485 vary per manufacturer and since there are no standard connectors for EIA-485 systems, it is difficult to define a standard installation procedure. Even so, most manufacture procedures are similar. The most common type of connector used on most EIA-485 systems is either a one-part or two-part screw connector. The preferred connector is the two-part screw connector with the sliding box under the screw (phoenix type). Other connectors use a screw on top of a folding tab. Manufacturers sometimes use the DB-9 connector instead of a screw connector to save money. Unfortunately, the DB-9 connector has problems when used for multidrop connections. The problem is that the DB-9 connectors are designed so that only one wire can be inserted per pin. EIA-485 multidrop systems require the connection of two wires so that the wire can continue down the line to the next device. This is a simple matter with screw connectors, but it is not so easy with a DB-9 connector. With a screw connector, the two wires are twisted together and inserted in the connector under the screw. The screw is then tightened down and the connection is made. With the DB-9 connector, the two wires must be soldered together with a third wire. The third wire is then soldered to the single pin on the connector.

Note: When using screw connectors, the wires should NOT be soldered together. Either the wires should be just twisted together or a special crimp ferrule should be used to connect the wires before they are inserted in the screw connector.

Figure 4.7
A bad EIA-485 connection

Serious problems with EIA-485 systems are rare (that is one reason it is used) but having said that, there are some possible problems that can arise in the installation process:

1. The wires get reversed (e.g. black to white and white to black)
2. Loose or bad connections due to improper installation
3. Excessive electrical or electronic noise in the environment
4. Common mode voltage problems
5. Reflection of the signal due to missing or incorrect terminators
6. Shield not grounded, grounded incorrectly or not connected at each drop
7. Starring or tee-ing of devices (i.e. long stubs)

To make sure the wires are not reversed, check that the same color is connected to the same pin on all connectors. Check the manufacturer's manual for proper wire color-codes.

Verifying that the installers are informed of the proper installation procedures can reduce loose connections. If the installers are provided with adjustable torque screwdrivers, then the chances of loose or over-tightened screw connections can be minimized.

4.2.4   Noise problems

EIA-485, being a differential type of circuit, is resistant to receiving common mode noise. There are five ways that noise can be induced into an EIA-485 circuit.

- Induced noise on the A/B lines
- Common mode voltage problems
- Reflections
- Unbalancing the line
- Incorrect shielding

4.2.4.1   Induced noise

Noise from the outside can cause communication on an EIA-485 system to fail. Although the voltages on an EIA-485 system are small (+/–5 volts), because the output of the receiver is the difference of the two lines, the voltage induced on the two lines must be different. This makes EIA-485 very tolerant to noise. The communications will also fail if the voltage level of the noise on either or both lines is outside of the minimum or maximum EIA-485 specification. Noise can be detected by comparing the data communication being transmitted out of one end with the received communication at the other (assuming no broken wire.) The protocol analyzer is plugged into the transmitter at one end and the data monitored. If the data is correct, the protocol analyzer is then plugged into the other end and the received data monitored. If the data is corrupt at the received end, then the noise on that wire may be the problem. If it is determined that the noise problem is caused by induced noise on the A or B lines it may be best to move the EIA-485 line or the offending noise source away from each other.

Excessive noise is often due to the close proximity of power cables. Another possible noise problem could be caused by an incorrectly installed grounding system for the cable shield. Installation standards should be followed when the EIA-485 pairs are installed close to other wires and cables. Some manufacturers suggest biasing resisters to limit noise on the line while others dissuade the use of bias resistors completely. Again, the procedure is to follow the manufacturer's recommendations. Having said that, it is usually found that biasing resisters are of minimal value, and that there are much better methods of reducing noise in an EIA-485 system.

4.2.4.2   Common mode noise

Common mode noise problems are usually caused by a changing ground level. The ground level can change when a high current device is turned on or off. This large current draw causes the ground level as referenced to the A and B lines to rise or decrease. If the voltages on the A or B line are raised or lowered outside of the minimum or maximum as defined by the manufacturer's specifications, it can prohibit the line receiver from operating correctly. This can cause a device to float in and out of service. Often, if the common mode voltage gets high enough, it can cause the module or device to be damaged. This voltage can be measured using a differential measurement device like a handheld digital voltmeter. The voltage between A and ground and then B to ground is measured. If the voltage is outside of specifications then resistors of values between 100 k ohm and 200 k ohm are placed between A and ground and B and ground. It is best to start with the larger value resistor and then verify the common mode voltage. If it is still too high, try a lower resistor value and recheck the voltage. At idle the voltage on the A line should be close to 0 and the B line should be between 2 and 6 volts. It is not uncommon for an EIA-485 manufacturer to specify maximum common voltage values of +12 and –7 volts, but it is best to have a system that is not near these levels. It is important to follow the manufacturer's recommendations for the common mode voltage resistor value or whether they are needed at all.

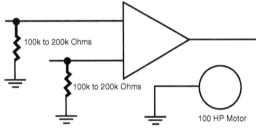

Figure 4.8
Common mode resistors

Note: When using bias resistors, neither the A nor the B line on the EIA-485 system should ever be raised higher than +12 volts or lower than –7 volts. Most EIA-485 driver chips will fail if this happens. It is important to follow the manufacturer's recommendations for bias resistor values or whether they are needed at all.

4.2.4.3   Reflections or ringing

Reflections are caused by the signal reflecting off the end of the wire and corrupting the signal. It usually affects the devices near the end of the line. It can be detected by placing a balanced ungrounded oscilloscope across the A and B lines. The signal will show ringing superimposed on the square wave. A termination resistor of typically 120 ohms is placed at each end of the line to reduce reflections. This is more important at higher speeds and longer distances.

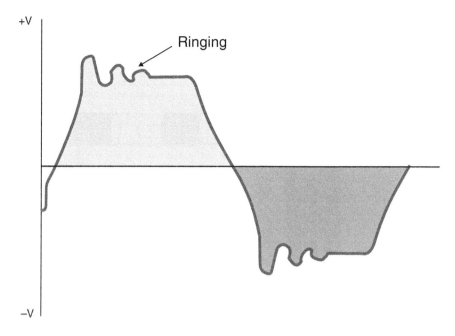

Figure 4.9
Ringing on an EIA-485 signal

4.2.4.4   Unbalancing the line

Unbalancing the line does not actually induce noise, but it does make the lines more susceptible to noise. A line that is balanced will have a ballpark balance between the capacitance and inductance on it. If this balance is disrupted, the lines then become affected by noise more easily. There are a few ways most EIA-485 lines become unbalanced:

- Using a star topology
- Using a 'tee' topology
- Using unbalanced cable
- Damaged transmitter or receiver

There should, ideally, be no stars or tees in the EIA-485-bus system. If another device is to be added in the middle, a two-pair cable should be run out and back from the device. The typical EIA-485 system would have a topology that would look something like the following:

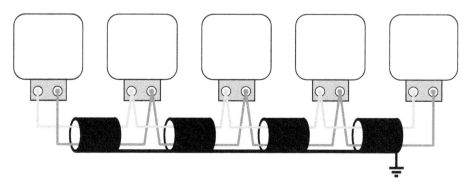

Figure 4.10
A typical EIA-485

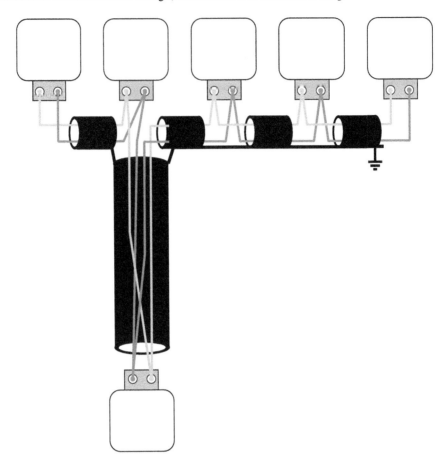

Figure 4.11
Adding a new device to a EIA-485 bus

The distance between the end of the shield and the connection in the device should be no more than 10 mm or ½ inch. The end of the wires should be stripped only far enough to fit all the way into the connector, with no exposed wire outside the connector. The wire should be twisted tightly before insertion into the screw connector. Often, installers will strip the shield from the wire and connect the shields together at the bottom of the cabinet. This is incorrect, as there would be from one to two meters of exposed cable from the terminal block at the bottom of the cabinet to the device at the top. This exposed cable will invariably receive noise from other devices in the cabinet. The pair of wires should be brought right up to the device and stripped as mentioned above.

4.2.4.5   Shielding

The choices of shielding for an EIA-485 installation are:

- Braided
- Foil (with drain wire)
- Armored

From a practical point of view, the noise reduction differences between the first two are minimal. Both the braided and the foil will provide the same level of protection against capacitive noise. The third choice, armored cable, has the distinction of protecting against magnetic induced noise. Armored cable is much more expensive than the first two and therefore braided and the foil types of cable are more popular. For most installers, it is a matter of personal choice when deciding to use either braided or foil shielded wire.

With the braided shield, it is possible to pick the A and B wires between the braids of the shield without breaking the shield. If this method is not used, then the shields of the two wires should be soldered or crimped together. A separate wire should be run from the shield at the device down to the ground strip in the bottom of the cabinet, but only one per bus, not per cabinet. It is incorrect in most cases to connect the shield to ground in each cabinet, especially if there are long distances between cabinets.

4.2.5   Test equipment

When testing or troubleshooting an EIA-485 system, it is important to use the right test equipment. Unfortunately, there is very little in generic test equipment specifically designed for EIA-485 testing. The most commonly used are the multimeter, oscilloscope and the protocol analyzer. It is important to remember that both of these types of test equipment must have floating differential inputs. The standard oscilloscope or multimeter each have their specific uses in troubleshooting an EIA-485 system.

4.2.5.1   Multimeter

The multimeter has three basic functions in troubleshooting or testing an EIA-485 system:

1.   Continuity verification
2.   Idle voltage measurement
3.   Common mode voltage measurement

Continuity verification

The multimeter can be used before startup to check that the lines are not shorted or open. This is done as follows:

4.   Verify that the power is off
5.   Verify that the cable is disconnected from the equipment
6.   Verify that the cable is connected for the complete distance
7.   Place the multimeter in the continuity check mode
8.   Measure the continuity between the A and B lines
9.   Verify that it is open
10.   Short the A and B at the end of the line
11.   Verify that the lines are now shorted
12.   Un-short the lines when satisfied that the lines are correct

If the lines are internally shorted before they are manually shorted as above, then check to see if an A line is connected to a B line. In most installations the A line is kept as one color wire and the B is kept as another. This procedure keeps the wires away from accidentally being crossed.

The multimeter is also used to measure the idle and common mode voltages between the lines.

Idle voltage measurement

At idle the master usually puts out a logical '1' and this can be read at any station in the system. It is read between A and B lines and is usually somewhere between -1.5 volts and -5 volts (A with respect to B). If a positive voltage is measured, it is possible that the leads on the multimeter need to be reversed. The procedure for measuring the idle voltage is as follows:

13. Verify that the power is on
14. Verify that all stations are connected
15. Verify that the master is not polling
16. Measure the voltage difference between the A– and B+ lines starting at the master
17. Verify and record the idle voltage at each station

If the voltage is zero, then disconnect the master from the system and check the output of the master alone. If there is idle voltage at the master, then plug in each station one at a time until the voltage drops to or near zero. The last station probably has a problem.

Common mode voltage measurement

Common mode voltage is measured at each station, including the master. It is measured from each of the A and B lines to ground. The purpose of the measurement is to check if the common mode voltage is getting close to maximum tolerance. It is important therefore to know what the maximum common mode voltage is for the system. In most cases, it is +12 and –7 volts. A procedure for measuring the common mode voltage is:

18. Verify that the system is powered up
19. Measure and record the voltage between the A and ground and the B and ground at each station
20. Verify that voltages are within the specified limits as set by the manufacturer

If the voltages are near or out of tolerance, then either contact the manufacturer or install resistors between each line to ground at the station that has the problem. It is usually best to start with a high value such as 200 k ohms ¼ watt and then go lower as needed. Both resistors should be of the same value.

4.2.5.2   Oscilloscope

Oscilloscopes are used for:

21. Noise identification
22. Ringing
23. Data transfer

Noise identification

Although the oscilloscope is not the best device for noise measurement, it is good for detection of some types of noise. The reason the oscilloscope is not that good at noise detection is that it is a two-dimensional voltmeter; whereas the effect of the noise is seen in the ratio of the power of a signal vs the power of the noise. But having said that, the oscilloscope is useful for determining noise that is constant in frequency. This can be a signal such as 50/60 Hz hum, motor induced noise or relays clicking on and off. The oscilloscope will not show intermittent noise, high frequency radio waves or the power ratio of the noise vs the signal.

Ringing

Ringing is caused by the reflection of signals at the end of the wires. It happens more often on higher baud rate signals and longer lines. The oscilloscope will show this ringing as a distorted square wave. As mentioned before, the 'fix' for ringing is a termination resistor at each end of the line. Testing the line for ringing can be done using a two-channel oscilloscope in differential (A–B) mode as follows:

The troubleshooting sessions are listed below:

24. Connect the probes of the oscilloscope to the A and B lines. Do NOT use a single channel oscilloscope, connecting the ground clip to one of the wires will short that wire to ground and prevent the system from operating
25. Setup the oscilloscope for a vertical level of around 2 volts per division
26. Setup the oscilloscope for horizontal level that will show one square wave of the signal per division
27. Use an EIA-485 driver chip with a TTL signal generator at the appropriate baud rate. Data can be generated by allowing the master to poll, but because of the intermittent nature of the signal, the oscilloscope will not be able to trigger. In this case a storage oscilloscope will be useful
28. Check to see if the waveform is distorted

Data transfer

Another use for the oscilloscope is to verify that data is being transferred. This is done using the same method as described for observing ringing, and by getting the master to send data to a slave device. The only difference is the adjustment of the horizontal level. It is adjusted so that the screen shows complete packets. Although this is interesting, it is of limited value unless noise is noted or some other aberration is displayed.

4.2.5.3   Protocol analyzer

The protocol analyzer is a very useful tool for checking the actual packet information. Protocol analyzers come in two varieties, hardware and software. Hardware protocol analyzers are very versatile and can monitor, log and interpret many types of protocols. When the analyzer is hooked-up to the RS-485 system, many problems can be displayed such as:

29. Wrong baud rates
30. Bad data
31. The effects of noise
32. Incorrect timing
33. Protocol problems

The main problem with the hardware protocol analyzer is the cost and the relatively rare use of it. The devices can cost from US$ 5000 to US$ 10 000 and are often used only once or twice a year.

The software protocol analyzer, on the other hand, is cheap and has most of the features of the hardware type. It is a program that sits on the normal PC and logs data being transmitted down the serial link. Because it uses existing hardware, (the PC) it can be a much cheaper but useful tool. The software protocol analyzer can see and log most of the same problems a hardware type can. The following procedure can be used to analyze the data stream:

34. Verify that the system is on and the master is polling
35. Set up the protocol analyzer for the correct baud rate and other system parameters
36. Connect the protocol analyzer in parallel with the communication bus
37. Log the data and analyze the problem

4.2.6   Summary

Installation

- Are the connections correctly made?
- What is the speed of the communications?
- What is the distance between the equipment?
- Is it a noisy environment?
- Is the software setup correctly?
- Are there any tees or stars in the bus?

Troubleshooting new and old systems

- Verify that there is power to the equipment
- Verify that the connectors are not loose
- Verify that the wires are correctly connected
- Check that a part, board or module has not visibly failed

Mechanical problems on new systems

- Keep the wires short, if possible
- Stranded wire should be used instead of solid wire (stranded wire will flex)
- Only one wire should be soldered in each pin of the connector
- Bare wire should not be showing out of the pin of the connector
- The back shell should reliably and properly secure the wire

Setup problems on new systems

- Is the software communications setup at both ends for either 8N1, 7E1 or 7O1?
- Is the baud rate the same at both devices? (1200, 4800, 9600, 19 200 etc)
- Is the software setup at both ends for binary, hex or ASCII data transfer?
- Is the software setup for the proper type of control?

Noise problems on new systems

- Induced noise on the A or B lines?
- Common mode voltage noise?
- Reflection or ringing?

5

Current loop and EIA-485 converters overview

Objectives

When you have completed study of this chapter, you will be able to:

- Describe current loop hardware
- Describe EIA-232/EIA-485 interface converters
- Fix problems with cabling and isolation for interface converters

5.1   The 20 mA current loop

Another commonly used interface technique is the current loop. This uses a current signal rather than a voltage signal, employing a separate pair of wires for the transmitter current loop and receiver current loop.

A current level of 20 mA, or up to 60 mA, is used to indicate logic 1 and 0 mA or 4 mA to indicate logic 0. The use of a constant current signal enables a far greater separation distance to be achieved than with a standard EIA-232 voltage connection. This is due to the higher noise immunity of the 20 mA current loop, which can drive long lines of up to 1 km, but at reasonably slow bit rates. Current loops are mainly used between printers and terminals in the industrial environment. Figure 5.1 illustrates the current loop interface.

Device #1   **Device #2**

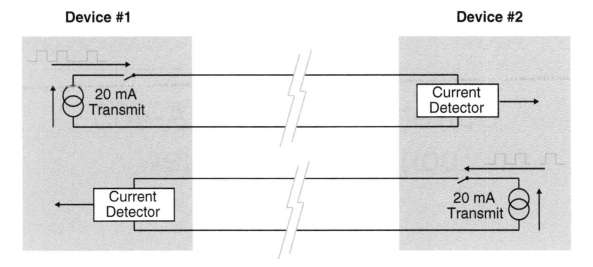

Figure 5.1
The 20 mA current loop interface

5.2   Serial interface converters

Serial interface converters are becoming increasingly important with the move away from EIA-232C to industrial standards such as EIA-422 and EIA-485. Since many industrial devices still use EIA-232 ports, it is necessary to use converters to interface a device to other physical interface standards. Interface converters can also be used to increase the effective distance between two EIA-232 devices.

The most common converters are:

- EIA-232/422
- EIA-232/485
- EIA-232/current loop

Figure 5.2 is a block diagram of an EIA-232/EIA-485 converter.

EIA-232   **EIA-232 to EIA-485**   **EIA-485**
                             Converter

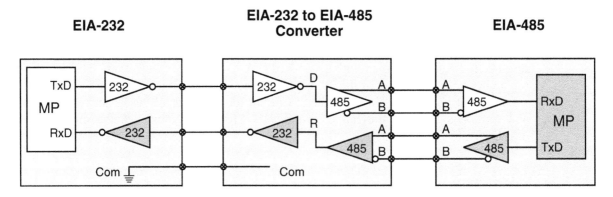

Figure 5.2
Block structure of EIA-232/EIA-485 converter

Figure 5.3 shows a circuit-wiring diagram.

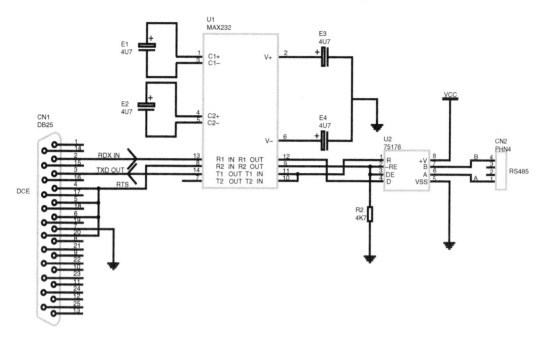

Figure 5.3
EIA-232/485 converter

The EIA-232/422 and EIA-232/485 interface converters are very similar and provide bi–directional full-duplex conversion for synchronous or asynchronous transfer between EIA-232 and EIA-485 ports. These converters may be powered from an external AC source, such as 240 V, or smaller units can be powered at 12 V DC from pins 9 and 10 of the EIA-232 port. For industrial applications, externally powered units are recommended. The EIA-232 standard was designed for communications, not as a power supply unit!

LEDs are provided to show the status of incoming signals from both EIA-232 and EIA-485.

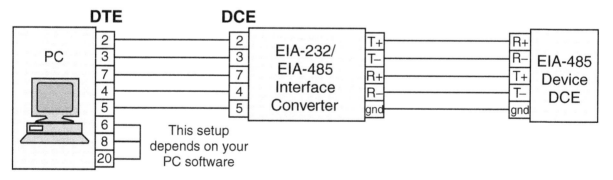

Figure 5.4
Wiring diagram for EIA-232/485 converter

When operating over long distances, a useful feature of interface converters is optical isolation. This is especially valuable in areas prone to lightning. Even if the equipment is not directly hit by lightning, the earth potential rise (EPR) in the surrounding area may be sufficient to damage the communications equipment at both ends of the link. Some specifications quote over 10 kV isolation, but these figures are seldom backed up with engineering data and should be treated with some caution.

Typical specifications for the EIA-232/422 or EIA-232/485 converters are:

- Data transfer rate of up to 1 Mbps
- DCE/DTE switch selectable
- Converts all data and control signals
- LEDs for status of data and control signals
- Powered from AC source
- Optically isolated (optional)
- DB-25 connector (male or female)
- DB-37 connector (male or female)

Typical specifications for the EIA-232/current loop converters are:

- 20 ma or 60 ma operation
- DCE/DTE, full/half-duplex selectable
- Active or passive loops supported
- Optically isolated (optional)
- Powered from AC source
- Data rates of up to 19 200 kbps over 3 km (10 000 feet)
- DB-25 connector (male or female)
- Current loop connector – 5 screw

5.3   Troubleshooting

5.3.1   Troubleshooting converters

The troubleshooting procedure is very similar to that discussed in the EIA-232 and EIA-485 sections with a few additional considerations. It should be emphasized that care should be taken in using non-isolated converters where there are any possibilities of electrical surges and transients. In addition, loop powered EIA-232/485 converters should also be avoided, as they can be unreliable. Using one of the handshaking pins of the EIA-232 port may sound clever and avoid the necessity of providing a separate power supply but EIA-232 was never intended to be a power supply.

A few suggestions for troubleshooting and the diagnostic LEDs used are outlined in the following sections:

Check that the converter is powered up

If there is no power to the converter or the power supply on the converter is damaged the LED on the converter will not light up. Typically it should be illuminated solid green.

In addition, on some converters there is provision for an isolated power LED, which should also be illuminated solid green.

With all the connections made, the converter should have the power light on, DCD, CTS, RTS LEDs on and the receive and transmit lights off. If you loop back the transmit and receive pins of the converter, and use a protocol analysis program such as PAT, you should see the transmit and receive lights flickering to indicate data flow and characters transmitted should be received. This will be the quickest test to indicate that all is well with the converter.

Network transmit

This LED will be lit up to indicate that the EIA-485 converter is active and driving the network. In a multidrop network, only one unit should be driving the line at one time otherwise, there will be contention with no resultant transmission at all.

Configuration jumpers

Although the configuration jumpers are often set internally, sometimes the user can also set them.

Choose the RTS transmit polarity correctly. Normally assertion of RTS is used to place the EIA-485 transmitter in transmit mode.

Failsafe termination

This jumper determines what happens if the EIA-232 host is disconnected from the converter. Conventionally, the EIA-485 transmitter should be disabled when the EIA-232 host is disconnected. To test this jumper, power up the system and disconnect the EIA-232 connection. The network transmit LED should be off.

DTE/DCE settings

These should be checked that they are correctly setup depending on the application. Note that no damage will occur if a mistake is made in setting them up. The system simply will not work.

Digital ground/shield

Sometimes, noise rejection performance can be improved by grounding the digital logic. But this is a very rare setting and other issues should be examined first before trying this configuration.

6

Fiber optics overview

Objectives

When you have completed study of this chapter, you will be able to:

- List the main features of fiber optics cabling
- Fix problems with:
  - Splicing
  - Laser and LED transmitters
  - Driver incompatibility
  - Incorrect bending radius in installation
  - Shock and other installation issues
  - Interface to cable connectors

6.1   Introduction

Fiber optic communication uses light signals guided through a fiber core. Fiber optic cables act as waveguides for light, with all the energy guided through the central core of the cable. The light is guided due to the presence of a lower refractive index cladding surrounding the central core. None of the energy in the signal is able to escape into the cladding and no energy is able to enter the core from any external sources. Therefore the transmissions are not subject to electromagnetic interference.

The core and the cladding will trap the light ray in the core, provided the light ray enters the core at an angle greater than the critical angle. The light ray will then travel through the core of the fiber, with minimal loss in power, by a series of total internal reflections. Figure 6.1 illustrates this process.

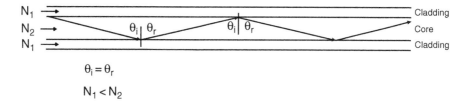

Figure 6.1
Light ray traveling through an optical fiber

Little of the light signal is absorbed in the glass core. Therefore fiber optic cables can be used for longer distances before the signal must be amplified or repeated. Some fiberoptic segments can be many kilometers long before a repeater is needed. Data transmission using a fiber optic cable is many times faster than with electrical methods and speeds of over 10 Gbps are possible. Fiber optic cables deliver more reliable transmissions over greater distances, although at a somewhat greater cost. Cables of this type differ in their physical dimensions and composition and in the wavelength(s) of light with which the cable transmits.

6.1.1 Applications for fiber optic cables

Fiber optic cables offer the following advantages over other types of transmission media:
- Light signals are impervious to interference from EMI or electrical crosstalk.
- Light signals do not interfere with other signals. As a result, fiber optic connections can be used in extremely adverse environments, such as in lift shafts or assembly plants, where powerful motors produce lots of electrical noise.
- Optical fibers have a much wider, flat bandwidth than coaxial cables and equalization of the signals is not required.
- The fiber has a much lower attenuation, so signals can be transmitted much further than with coaxial or twisted pair cable before amplification is necessary.
- Optical fiber cables do not conduct electricity and so eliminate problems of ground loops, lightning damage and electrical shock when cabling in high-voltage areas.
- Fiber optic cables are generally much thinner and lighter than copper cable.
- Fiber optic cables have greater data security than copper cables.
- Licensing is not required, although a right-of way for laying the cable is needed.

6.2 Fiber optic cable components

The major components of a fiber optic cable are the core, cladding, buffer, strength members and jacket, as shown below. Some types of fiber optic cable even include a conductive copper wire that can be used to provide power to a repeater.

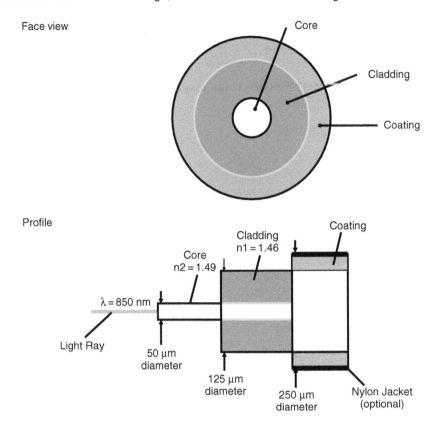

Figure 6.2
Fiber optic cable components

Fiber core

The core of fiber optic telecommunications cable consists of glass fibers through which the light signal travels. The most common core sizes are 50 and 62.5 micrometers (microns), which are used in multimode cables. 8.5 Micron fibers are used in single-mode systems.

Cladding

The core and cladding are actually manufactured as a single unit. The cladding is a protective layer with a lower index of refraction than the core. The lower index means any light that hits the core walls will be redirected back to continue on its path. The cladding diameter is typically 125 microns.

Fiber optic buffer

The buffer of a fiber optic cable is made of one or more layers of plastic surrounding the cladding. The buffer helps strengthen the cable, thereby decreasing the likelihood of micro cracks, which can eventually break the fiber. The buffer also protects the core and cladding from potential invasion by water or other materials in the operating environment. The buffer typically doubles the diameter of the fiber.

A buffer can be tight or loose, as shown below. A tight buffer fits snugly around the fiber. A tight buffer can protect the fibers from stress due to pressure and impact but not from changes in temperature. A loose buffer is a rigid tube of plastic with one or more fibers (consisting of core and cladding) running through it. The fibers are longer than the tube so that the tube takes all the stresses applied to the cable, isolating the fiber from these stresses.

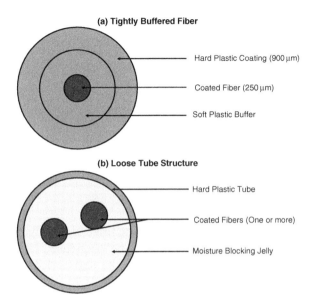

Figure 6.3
Fiber housing types

Strength members

Fiber optic cable also has strength members, which are strands of very tough material (such as steel, fiberglass, or Kevlar) that provide tensile strength for the cable. Each of these materials has advantages as well as drawbacks. For example, steel conducts electricity, making the cable vulnerable to lightning but it will not disrupt an optical signal. However, it may seriously damage the cable or equipment.

Cable sheath

The sheath of a fiber optic cable is an outer casing that provides primary mechanical protection, as with electrical cable.

6.3   Fiber optic cable parameters

Attenuation

The attenuation of a multimode fiber depends on the wavelength and the fiber construction and range from around 3–8 dB/km at 850 nm and 1–3 dB/km at 1300 nm. The attenuation of single-mode fiber ranges from around 0.4–0.6 dB/km at 1300 nm and 0.25–0.35 dB/km at 1550 nm.

Diameter

The fiber diameter is either 50 or 62.5 microns for multimode fiber or 8.5 microns for single-mode.

- **Multimode fibers (50 or 62.5 micron)**

  In multimode fibers, a beam of light has room to follow multiple paths through the core. Multiple modes in a transmission produce signal distortion at the receiving end due to the difference in arrival time between the fastest and slowest of the alternate light paths.

- **Single-mode fibers (8.5 micron)**

  In a single-mode fiber, the core is so narrow that the light can take only a single path through it. Single-mode fiber has the least signal attenuation, usually less than 0.5 dB/km. This type of cable is the most difficult to install because it requires precise alignment of the system components and the light sources; the detectors are also very expensive. However, transmission speeds of 50 Gbps and higher are possible.

Wavelength

Fiber optic systems today operate in one of the three wavelength bands; 850 nm, 1300 nm or 1550 nm. The shorter wavelengths have a greater attenuation than the longer wavelengths. Short haul systems tend to use the 850 or 1300 nm wavelengths with the multimode cable and light emitting diode (LED) light sources. The 1550 nm fibers are used almost exclusively with the long distance systems using single-mode fiber and laser light sources.

Bandwidth

The bandwidth of a fiber is given as the range of frequencies across which the output power is maintained within 3 dB of the nominal output. It is quoted as the product of the frequencies of bandwidth multiplied by distance; for example, 500 MHz–km. This means 500 MHz of bandwidth is available over a distance of one kilometer or 100 MHz of bandwidth over five kilometers.

Dispersion

Modal dispersion is measured as nanoseconds of pulse spread per kilometer (ns/km). The value also imposes an upper limit on the bandwidth, since the duration of a signal must be larger than the nanoseconds of a tail value. With step-index fiber, expect between 15 and 30 ns/km. Note that a modal dispersion of 20 ns/km yields a bandwidth of less than 50 Mbps. There is no modal dispersion in single-mode fibers, because only one mode is involved.

Chromatic dispersion occurs in single-mode cables and is measured as the spread of the pulses in picoseconds for each nanometer of spectral spread of the pulse and for each kilometer traveled. This is the only dispersion effect in single-mode cables and typical values are in the order of 3.5 ps/nm–km at 1300 nm and 20 ps/nm–km at 1550 nm.

6.4   Types of optical fiber

One reason for optical fiber making such a good transmission medium is because the different indexes of refraction for the cladding and core help to contain the light signal within the core, producing a waveguide for the light. Optical fiber can be constructed by changing abruptly from the core refractive index to that of the cladding, or this change can be made gradually. The two main types of multimode fiber differ in this respect:

- **Step-index cable**

    Cable with an abrupt change in refraction index is called step-index cable. In step-index cable, the change is made in a single step. Single-step multimode cable uses this method, and it is the simplest, least expensive type of fiber optic cable. It is also the easiest to install. The core is usually 50 or 62.5 microns in diameter; the cladding is normally 125 microns. The core width gives light quite a bit of room to bounce around in, and the attenuation is high (at least for fiber-optic cable): between 10 and 50 dB/km. Transmission speeds up to 10 Mbps over 1 km are possible.

- **Graded-index cable**

    Cable with a gradual change in refraction index is called graded-index cable, or graded-index multimode. This fiber optic cable type has a relatively wide core, like single-step multimode cable. The change occurs gradually and involves several layers, each with a slightly lower index of refraction. A gradation of refraction indexes controls the light signal better than the step-index method. As a result, the attenuation is lower, usually less than 15 dB/km. Similarly, the modal dispersion can be 1 ns/km and lower, which allows more than ten times the bandwidth of step-index cable.

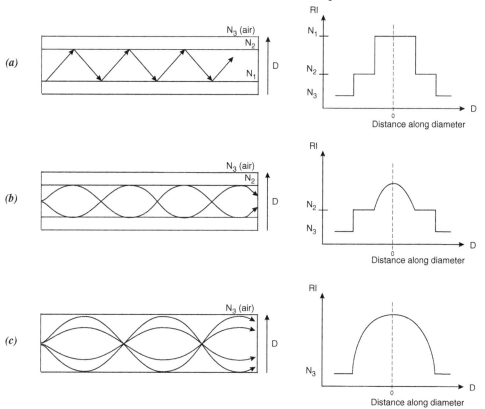

Figure 6.4
Fiber refractive index profiles

• **Fiber designations**

Optical fibers are specified in terms of their core, cladding and coating diameters. For example, a 62.5/125/250 fiber has a core diameter of 62.5 microns, a cladding of 125 microns and a coating of 250 microns.

6.5   Basic cable types

There are four broad application areas into which fiber optic cables can be classified: aerial cable, underground cable, sub-aqueous cable and indoor cable.   The special properties required for each of these applications will now be considered. Note that this list is not all encompassing and that some specialized cables need to combine the features of several of these classes.

6.5.1   Aerial cable

Aerial cables are literally exposed to the elements, more than any other application and as such, are exposed to many external forces and hazards. Aerial cables are installed between poles with the weight of the cable continuously supported by usually a steel messenger wire to which the cable can be directly lashed or by the strength members integral to the cable construction. Greatly increased tensile forces can be produced by the effects of combined wind and ice loadings. Other considerations are the wide variations in temperature to which the cable may be subjected, affecting the physical properties of the fibers and the attenuation of the fibers. The longitudinal cable profile is important for reducing the wind and ice loadings of such cables. Moisture barriers are essential, with jelly filled, loose buffered fiber cable configurations being predominant. Any water freezing within the fiber housings would expand and could produce excessive bending of fibers.

The cable sheath material is required to withstand the extremes of temperature and the intense ultraviolet light exposure from continuous exposure to sunlight. UV stabilized polyethylene is frequently used for this purpose.

The installed span length and sag requirements are important design parameters affecting the maximum cable tension and which dictates the type of cable construction to be used. Short span cables have less stringent tension requirements, which can be met by the use of integral Kevlar layers, whereas long span cables may need to utilize multiple stranded FRP rods to meet the required maximum tensions.

Advantages of aerial cable

• Useful in areas where it may be very difficult or too expensive to bury the cable or install it in ducts
• Also useful where temporary installations are required

Disadvantages

• System availability is not as high as compared to underground cables. Storms can disrupt these communication bearers, with cables damaged by falling trees, storm damage and blown debris. Roadside poles can be hit by vehicles and frustrated shooters seem unable to miss aerial cables!

6.5.2   Underground cable

Underground cable experiences less environmental extremes than aerial cables. Cables are usually pulled into ducts or buried directly in the ground, with the cables being placed in deep narrow trenches, which are backfilled with dirt or else ploughed directly into the ground.

Cable type

Loose buffering, using loose tube or slotted core construction, is generally used to isolate fibers from external forces including temperature variations.

Advantages

- Usually, the most cost effective method of installing cables outdoors
- Greater environmental protection than aerial cable
- Usually, more secure than aerial cable

Disadvantages

- Can be disrupted by earthquakes, digging, farming, flooding etc.
- Rodents biting cables can be a problem in some areas. This is overcome with the use of steel tape armor or steel braid, or the use of a plastic duct of more than 38mm OD for all dielectric cable installations. Also, the use of Teflon coatings on the sheath make the cable too slippery for the rodent to grip with its teeth.

6.5.3   Sub-aqueous cables

Sub-aqueous cables are basically outdoor cables designed for continuous immersion in water. While international telecommunications carriers use the most sophisticated cables for deep ocean communications, there are practical applications for sub-aqueous cables for smaller users. These include cabling along or across rivers, lakes, water races or channels, where alternatives are not cost effective. Sub-aqueous cable is a preferred option for direct buried cabling in areas subjected to flooding or with a high water table, where for example, if the cable were buried at say 1 meter depth, it would be permanently immersed in water. These cables are essentially outdoor cable constructions incorporating a hermetically sealed unit, using a welded metallic layer and encasing the fiber core.

Advantages

- Cheaper installation in some circumstances

Disadvantages

- Unit cost of cable is higher.

6.5.4   Indoor cables

Indoor cables are used inside buildings and have properties dictated by the fire codes. Such cables need to minimize the spread of fires and must comply with your relevant local fire codes, such as outlined in the National Electrical Codes (NEC) in USA. Outdoor cables generally contain oil-based moisture blocking compounds like petroleum jelly. These support combustion and so, their use inside buildings is strictly controlled. Outdoor cables are frequently spliced to appropriate indoor cables close to the building entry points to avoid the expense of encasing long runs of outdoor cable inside metallic conduit.

The fiber in indoor cables and the indoor cable itself is usually tightly buffered as was discussed in Section 6.2. The tight buffer provides adequate water resistance for indoor applications but such cables should not be used for long outdoor cable runs. The buffered fibers can be given sufficient strength to enable them to be directly connected to equipment from the fiber structure without splicing to patch cords.

6.6   Connecting fibers

This section will identify the main issues involved in connecting fibers together and to optical devices, such as sources and detectors. This can be done using splices or connectors. A splice is a permanent connection used to join two fibers and a connector is used where the connection needs to be connected and disconnected repeatedly, such as at patch panels. A device used to connect three or more fibers or devices is called a coupler.

6.6.1   Connection losses

The main parameter of concern when connecting two optical devices together is the attenuation – that fraction of the optical power lost in the connection process. This attenuation is the sum of losses caused by a number of factors, the main ones being:

- Lateral misalignment of the fiber cores
- Differences in core diameters
- Misalignment of the fiber axes
- Numerical aperture differences of the fibers
- Reflection from the ends of fibers
- Spacing of the ends of the fibers
- End finish and cleanliness of fibers

The most important of these loss mechanisms involved in connecting multimode fibers is the axial misalignment of the fibers.

With connectors, the minimum loss across the glass/air interface between them will always be about 0.35 dB unless index-matching gel is used.

6.6.2   Splicing fibers

Two basic techniques are used for splicing of fibers: fusion splicing or mechanical splicing. With the fusion splicing technique, the fibers are welded together, requiring expensive equipment but will produce consistently lower loss splices with low consumable costs. With mechanical splicing, the fibers are held together in an alignment structure, using an adhesive or mechanical pressure. Mechanical splicers require lower capital cost equipment but have a high consumable cost per splice.

- **Fusion splicing**

  Fusion splices are made by melting the end faces of the prepared fibers and fusing the fibers together. Practical field fusion splicing machines use an electric arc to heat the fibers. Factory splicing machines often use a small hydrogen flame. The splicing process needs to precisely pre-align the fibers, then heat their ends to the required temperature and move the softened fiber ends together sufficiently to form the fusion joint, whilst maintaining their precise alignment. Fusion splices have consistently very low losses and are the preferred method for joining fibers, particularly for single-mode systems. Modern single-mode fusion splicers utilize core alignment systems to ensure the cores of the two fibers are precisely aligned before splicing.

- **Mechanical splicing**

  Mechanical splicing involves many different approaches for bringing the two ends of the fibers into alignment and then clamping them within a jointing structure or gluing them together. Mechanical splices generally rely on aligning the outer diameters of the fiber cladding and assumes that the cores are concentric with the outside of the cladding. This is not the case always, particularly with monomode fibers. Various mechanical structures are used to align the fibers, including V-grooves, sleeves, 3-rods and various proprietary clamping structures.

6.6.3 Connectors

Connectors are used to make flexible interconnections between optical devices. Connectors have significantly greater losses than splices since it is much more difficult to repeatedly align the fibers with the required degree of precision. Active alignment, as was used to minimize some splice losses, is not possible. Axial misalignment of the fibers contributes most of the losses at any connection, consequently connector loss can be expected to be in the range from 0.2 to over 3 dB.

Most of the connector designs produce a butt joint with the fiber ends as close together as possible. The fiber is mounted in a ferrule with a central hole sized to closely match the diameter of the fiber cladding. The ferrule is typically made of metal or ceramic and its purpose is to center and align the fiber as well as to provide mechanical protection to the end of the fiber. The fiber is normally glued into the ferrule then the end is cut and polished to be flush with the face of the ferrule. The two most common connectors are the SC and ST, as detailed below. Although, many new proprietary connectors are now available for different types of equipment.

- **SC connector**

  This is built with a cylindrical ceramic ferrule, which mates with a coupling receptacle. The connector has a square cross-section for high packing density on equipment and has a push–pull latching mechanism. The ISO and TIA have adopted a polarized duplex version as standard and this is now being used as a low-cost FDDI connector. The SC connector has a specified loss of less than 0.6 dB (typically 0.3 dB) for both monomode and multimode fibers and a typical return loss of 45 dB.

- **ST connector**

  The ST connector is shown in Figure 6.5. This is an older standard used for data communications. This is also built with a cylindrical ceramic ferrule which mates with a coupling receptacle. The connector has a round cross-section and is secured by twisting to engage it in the spring-loaded bayonet coupling. Since it relies on the internal spring to hold the ferrules together, optical contact can be lost if a force greater than that of about one kilogram is applied to the connector.

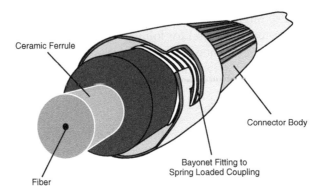

Ceramic Ferrule

Connector Body

Bayonet Fitting to
Spring Loaded Coupling

Fiber

Figure 6.5
ST connector

6.6.4   Connector handling

Most fiber optic connectors are designed for indoor use. Connectors for outdoor use require to be hermetically sealed. It is very important to protect the optical faces of the connectors from contamination. The optical performance can be badly degraded by the presence of dirt or dust on the fiber ends. A single dust particle could be 10 microns in diameter but it would either scatter or absorb the light and could totally disrupt a single-mode system. Connectors and patch panels are normally supplied with protective caps. These should always be fitted whenever the connectors are not mated. These not only protect from dust and dirt but also provide protection to the vulnerable, polished end of the fiber. Compressed air sprays are available for cleaning connectors and adapters without needing to physically touch the mating surfaces.

Take care not to touch the end of the connector ferrules, as the oil from your fingers can cause dirt to stick to the fiber end. Clean connectors with lint-free wipes and isopropyl alcohol.

Durability of connectors is important throughout their lifetime. Typical fiber connectors for indoor use are specified for 500 to 1000 mating cycles, and the attenuation is typically specified not to change by more than 0.2 dB throughout their lifetime. Repeated connection and disconnection of connectors can wear the mechanical components and introduce contamination to the optical path.

6.6.5   Optical couplers

Optical couplers or splitters and combiners are used to connect three or more fibers or other optical devices. These are devices that split the input power to a number of outputs. While the splitting of the light is done passively, active couplers include optical amplifiers, which boost the signal before or after the splitting process. Coupler configuration depends on the number of ports and whether each of these are unidirectional, so called directional couplers, or bi-directional. Most couplers are found within the equipment for monitoring purposes.

6.7   Splicing trays/organizers and termination cabinets

This section looks at different types of storage units that are used for housing optical fiber splices and end of cable terminations.

6.7.1   Splicing trays

Splices are generally located in units referred to as 'splicing centers', 'splicing trays' or 'splicing organizers'. The splicing tray is designed to provide a convenient location to store and to protect the cable and the splices. They also provide cable strain relief to the splices themselves.

Splicing trays can be located at intermediate points along a route where cables are required to be joined or at the termination and patch panel points at the end of the cable runs.

The incoming cable is brought into the splicing center where the sheath of the cable is stripped away. The fibers are then looped completely around the tray and into a splice holder. Different holders are available for different types of splices. The fibers are then spliced onto the outgoing cable if it is an intermediate point or on to pigtails if it is a termination point. These are also looped completely around the tray and then fed out of the tray. A typical splicing tray is illustrated below:

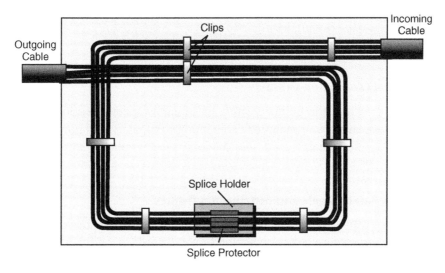

Figure 6.6
A typical splicing tray

The cables are physically attached to the splice tray to provide strain relief. The cables normally enter the tray on one side only to facilitate moving the tray/joint enclosure to a more accessible jointing location. The fibers are looped completely around the tray to provide slack, which may be required to accommodate any changes in the future, and also to provide tension relief on the splices.

Each splice joint is encased in a splice protector (plastic tube) or in heat shrink before it is clipped into the holder.

Splicing trays are available that have patching facilities. This allows different fibers to be cross-connected and to be looped back for testing purposes.

6.7.2   Splicing enclosures

The splicing trays are not designed to be left in the open environment and must be placed in some type of enclosure. The enclosure that is used will depend on the application. The following are examples of some enclosures used for splicing trays.

• **Direct buried cylinders**

At an intermediate point where two cables are joined to continue a cable run, the splices can be directly buried by placing the splice trays in a tightly sealed cylindrical enclosure that is generally made from heavy duty plastic or aluminum. The container is completely sealed from moisture ingress and contains desiccant packs to remove any moisture that may get in. A typical direct buried cylinder is illustrated below. Note that the cables normally enter the enclosure at one end only to allow the enclosure to be lifted from the ground for easier splicing access.

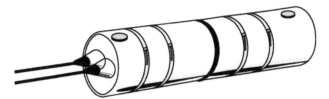

Figure 6.7
Direct buried splicing enclosure

• **Termination cabinets**

At junction points where a lot of cables meet, the splicing trays are stored in a larger wall mounted cabinet (approximately 500 mm × 500 mm × 100 mm) with a hinged door. For outdoor use, the cabinets must be sealed against bad weather conditions. Figure 6.8 illustrates a splicing tray in a termination cabinet.

• **Patch panels and distribution frames**

Splice trays can be installed in the back of patch panels and distribution frames used for connection of patch cords to the main incoming cable.

6.7.3 Termination in patch panels and distribution frames

There are three main methods of connecting an incoming cable into a patch panel or distribution frame. Firstly, if the incoming cable contains fibers that have a large minimum bending radius, then it is recommended to splice each fiber to a fiber patch cord that has a smaller bending radius. This reduces undue stress on the incoming fibers but it does introduce small losses into the link. This also replaces the more fragile glass of the incoming cable with the more flexible and stronger glass of the patch cords. This is illustrated in Figure 6.8.

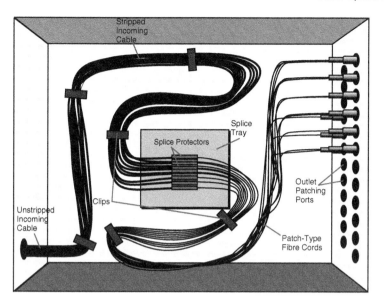

Figure 6.8
Termination cabinet for splicing trays

The second method is to place the fibers from the incoming cable into a breakout unit. The breakout unit separates the fibers and allows a plastic tube to be fitted over the incoming fibers to provide protection and strength as they are fed to the front of the patch panel. Note there are no splices, which therefore keep losses to a minimum. This is illustrated in Figure 6.9.

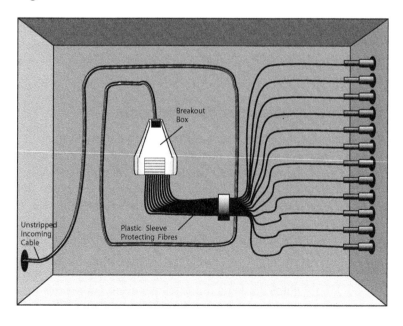

Figure 6.9
Patch panel with breakout box

If the incoming cable contains tight buffered fibers that are flexible and strong, then they can be taken directly to the front of the patch panel. This is referred to as direct termination, and is illustrated below:

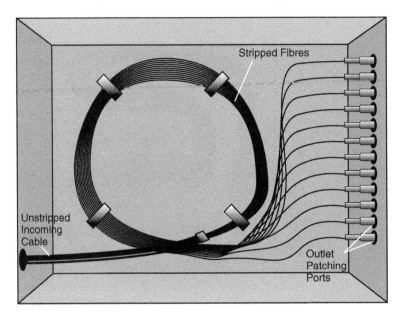

Figure 6.10
Direct termination of cables in a patch panel

6.8   Troubleshooting

6.8.1   Introduction

This section deals with problems on fiber optic cables. Problems can be caused by poor installation practices, where fibers are subjected to excessive tension or bending forces. This section also deals with the basic methods of testing fibers and how to locate faults on fiber optic systems.

6.8.2   Standard troubleshooting approach

The standard approach to troubleshooting fiber optic systems is as follows:
1. Observe the system status indicators and determine whether signals are being transmitted and received at both terminals.
2. Determine whether the appropriate fibers are functional by either a simple continuity test or a qualitative insertion loss measurement between the patch panels.
3. Once the faulty fiber is identified, clean the optical connectors and repeat the test.
4. If the fault remains, swap the system onto a spare fiber by rearranging the patch cords on the patch panels at both ends.
5. Update the records to indicate the faulty fiber.
6. When a link does not have sufficient spare fibers available to maintain system integrity, then attempt fault localization using an OTDR, if appropriate.
7. On short lengths of fiber, total replacement may be more cost effective than expensive location and subsequent repair. This is particularly appropriate where a spare duct is available for the cable replacement.

6.8.3   Tools required

Continuity tester

This device has a fiber optic transmitter with a suitable fiber optic connector that transmits a visible red light (650 nm). This can transmit visible light over several kilometers and is used for such applications as continuity testing, finding fractures in fibers or bad splices by observing light that may be leaking out and for identifying fibers at the end of a cable that has many fibers in it.

Optical source

This device has a calibrated fiber optic transmitter with a suitable fiber optic connector and is used with an optical power meter for insertion loss testing of fibers.

Optical power meter

This device has a fiber optic receiver with a suitable fiber optic connector and displays the received optical power levels. It is used with an optical source for insertion loss testing of fibers.

Optical time domain reflectometer (OTDR)

The OTDR sends a short pulse of light down the fiber and measures and records the light energy that is reflected back up the fiber. A reflection may be caused by the presence of a connector, splice, crack, impurity or break in the fiber. By measuring the time it takes for the reflected light to return to the source and knowing the refractive index of the fiber, it is possible to calculate the distance to the reflection point.

6.8.4   Fiber installation rules

The following section provides general installation rules that should be followed when installing fiber optic cabling systems to avoid long-term reliability problems. Fibers can break at any surface defects if subjected to excessive bending forces while under tension.

6.8.4.1   Cable bending radius

- The most important consideration when installing fiber optic cables is to ensure that at all times during an installation, the cable radius is not less than the manufacturer's recommended minimum installation bending radius.
- Avoidance of sharp bends along the installation route is absolutely essential. Sharp bends in cable trays or in conduits can cause macrobends or microbends in the fibers. This can lead to fiber breakage or adversely affect signal attenuation.
- Ensure that the conduit or the cable tray is constructed with no sharp edges. Use curved construction components (ducts or cable trays) and not right angled components.
- Ensure the cables are laid on to a flat surface, and that no heavy objects will be laid on to the cables in the future.
- Avoid putting kinks or twists into the cable. This is best achieved by pulling the cable directly off the reel and having a member of the

installation team carefully watching any cable slack for possible formation of kinks.
- Cable manufacturers will specify a minimum bending radius that applies to the long-term final installed cable. The long-term radius is significantly larger than the installation radius. If this long-term radius is exceeded, the macrobending in the cable will produce additional attenuation. Such unnecessary losses will not only damage the fiber, but also will be detrimental to cable performance. Once the cable has been installed and the tension has been released, ensure that the cable radius is not less than the long term installed radius at any point along the cable.

6.8.4.2   Cable tension

When a longitudinal tensile force is applied to an optical fiber, it can cause minute surface defects. These potential weak points can develop into microcracks, which may cause breakage of the fiber later on, when it is subjected to an equal or greater tension.

Optical fibers have some elasticity, stretching under light loads before returning to their original length when the load is removed. Under heavier loads, the fiber may theoretically stretch as much as 9% before breaking; however, it is considered advisable to limit this, in practice, to permanent strains of less than 0.2% to avoid promoting premature failures.

- Although modern fiber optic cables are generally stronger than copper cables, failure due to excess cable tension during installation is more catastrophic (i.e. fiber snapping rather than copper stretching).
- The maximum permissible installation cable tension is specified by the manufacturer; the cable tension should not exceed that limit at any time. A general rule of thumb sometimes used is that the maximum permissible cable tension during installation is approximately the weight of 1 km of the cable itself.
- When pulling the cable during installation, avoid sudden, short sharp jerking. These sudden forces could easily exceed the maximum cable tension. The cable should be pulled in an easy smooth process.
- When pulling cable off a large drum, ensure that the cable drum is smoothly rotated by one team member to feed off the cable. If the cable is allowed to jerk the drum around, the high moment of inertia of the drum can cause excessive tension in the cable.
- It is very important to minimize cable stress after the installation is complete. A slack final resting condition will help to ensure a long operating life of the fiber optic cable. It is recommended that the slack be left in the junction boxes at the completion of the installation to reduce overall stress in the cable.
- If there are a lot of bends in the cable route, it is recommended to use as many intermediate junction boxes as possible to reduce cable tension. The cable can be pulled through at these points, laid out in a large figure '8' pattern on the ground and then pulled into the next section. On long cable runs, pulling assistance can also be done at these intermediate access points. Laying the cable in a figure '8' pattern naturally avoids kinking and twisting of the cable. Curved guides or block systems may be used in the junction boxes where the cable changes direction.

6.8.5   Clean optical connectors

With the core of a single mode fiber being only 9 microns in diameter, a lot of dust particles can easily cover much of the core area on the end face of the connector. In addition, connector damage can occur if foreign particles are caught in the end face area of mated connectors. It is vital to make sure that cleaning of connectors is completed by all staff handling them, and that 'good housekeeping' practices are employed such as always installing the protective dust caps and never allowing any object other than the approved cleaners to contact the end face.

- The preferred method of connector cleaning is by wiping the end face and ferrule of the connector with some isopropyl alcohol on a lint free tissue. Cleaning tissues can be obtained in sealed packages already impregnated with alcohol.
- Another method of connector cleaning is the proprietary cassette type cleaners. These use a dry tape which is advanced every time the cassette is opened ensuring availability of a clean section of tape each time.
- Where proprietary connectors have bare fibers exposed such as Volition or MTRJ, the only acceptable method of cleaning is to use a can of compressed gas.

Through adapters are primarily kept clean by always cleaning the connectors prior to insertion and fitting of dust caps to all unused patch panel ports. A method of cleaning through adapters is the careful use of a can of compressed gas. An alternative is a pipe cleaner moistened with isopropyl alcohol. Remember that most pipe cleaners have a steel wire center and the end will damage any connector if it is inserted into an adapter that still has a connector inserted in the other side.

6.8.6   Locating broken fibers

6.8.6.1   Continuity testing

The most fundamental of all fiber optic cable tests that can be performed is to carry out a continuity test. The continuity test simply checks that the fiber is continuous from one end to the other. A light beam is inserted from a light source in one end of the fiber and is observed coming out of the other end of the fiber. This test provides little information about the condition of the fiber other than there are no complete breaks along the fiber length. This can be performed by shining a powerful torch beam (or laser pointer) into one end of the fiber and observing the light coming out of the other end. This is a cost-effective method for multimode fibers with large core diameters over distances up to approximately five hundred meters. It is not reliable over longer distances or with single-mode fibers because the small diameter core makes penetration into the fiber difficult for the light.

There is specific test equipment available that can be used as a continuity tester. The device has a fiber optic transmitter with a suitable fiber optic connector that transmits a 650 nm visible red light. This will transmit visible light over several kilometers and can be used for such applications as continuity testing, finding fractures in fibers or bad splices by observing light that may be leaking out and for identifying fibers at the end of a cable that has many fibers in it. It can also be used for identifying fibers along the route of a cable (where it is required to break into a cable for a system extension) by bending the fiber and watching the light leaking out of the bend. This type of fiber optic test has limited application, as it is of no use in finding faults in buried cables or aerial cables.

As a word of warning, the user should not look into fiber groups at the end of cables if any fibers on the system at any location are connected to lasers. The infra-red light from lasers cannot be seen by the human eye but can cause permanent eye damage.

6.8.6.2   Insertion loss testing

The most common qualitative test that is carried out on a fiber optic system is to measure the attenuation of the length of fiber. This figure will allow most elements of the system design to be verified.

Most insertion loss testing is carried out with a power source and a power meter. Firstly, the power meter is calibrated to the power source by connecting the two instruments together with a short piece of optic fiber approximately 2 m in length. Generally, the power source is set to transmit a level of –10 dBm and the power meter then adjusted accordingly to read –10 dBm. Ensure that the level used to calibrate the power meter is within the dynamic range of the power meter.

There are four important points to check before commencing insertion loss testing. Firstly, ensure that the optic fiber type used for calibration purposes is the same optic fiber type that is to be tested for insertion loss. Secondly, the power meter and the power source must operate at the same wavelength as the installed system equipment. Thirdly, the power meter and source must also use the same source and detector types (LED or laser) that the transmitter and receiver in the installed system are to use. Fourthly, to avoid possible incorrect calibration, ensure that the same connectors are used for calibration as are used in the installation.

Once the power meter has been calibrated, then the power meter and source are taken into the field and connected to the installed cable. The level that is read at the meter can be used to calculate the insertion loss through the cable section under test. This will include the losses caused by the optic fiber, splices and the connectors. The test procedure is illustrated in Figure 6.11.

If the power source and the power meter are calibrated in milliwatts, then the formula for converting the loss figure to decibels is:

$$\text{Attenuation (dB)} = -10 \text{ Log } (Po/Pi)$$

where:
Po is power out of the fiber
Pi is power into the fiber

To calculate the insertion loss, subtract the dBm reading at the power meter from the input power source value. For the example, shown in Figure 6.11 is the insertion loss in 9.3 dB.

It is recommended that the insertion loss measurement be performed in both directions of an installed cable. The losses measured in each direction tend to vary because connectors and splices sometimes connect unevenly and because the core diameter of fibers tends to vary slightly. For example, if the core diameters of two fibers spliced together are 49.5 mm and 50.5 mm, light waves traveling from the thinner fiber into the thicker fiber will all enter the thicker fiber. For light traveling from the larger diameter fiber into the smaller diameter fiber, a small amount will be lost around the edges of the interface between the two cores. A mismatch of this type could account for a difference in insertion loss in the two directions of 0.2 dB. If the fiber losses are different in each direction, then that fiber can be used in whichever direction gives it the best system performance.

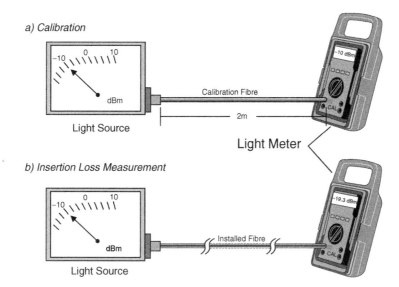

Figure 6.11
Insertion loss measurement

6.8.6.3   Optical time domain reflectometer

The only method of analyzing the losses along an individual fiber is to test it with an optical time domain reflectometer (OTDR). In particular, this allows the location of a broken fiber to be established. The OTDR sends a short pulse of light down the fiber and measures and records the light energy that is reflected back up the fiber. A reflection may be caused by the presence of a connector, splice, crack, impurity or break in the fiber. By measuring the time, it takes for the reflected light to return to the source and knowing the refractive index of the fiber, it is possible to calculate the distance to the reflection point.

Impurities in the glass will cause continuous low level reflection as the light travels through the glass fiber. This is due to Rayleigh scattering and is commonly referred to as backscatter. The strength of the backscattered signal received at the source gradually drops as the pulse moves away from the source. This is seen on an OTDR display as a near linear drop in the received reflected signal and the slope of this line is the attenuation of the fiber (dB per km). Figure 6.12 illustrates a typical reflection curve for an OTDR and notes the backscatter.

An OTDR will generally not provide accurate readings of irregularities and losses in the fiber for the first 15 m of the cable. This is because the pulse length and its rise time from the OTDR are comparatively large when compared to the time it takes for the pulse to travel the short distance to the point of reflection within this 15 m and back.

In general, for shorter local cable running less than 200 m, there is not a lot to be gained from carrying out an OTDR test unless there are connectors and splices along the cable route.

With reference to the OTDR plot in Figure 6.12, the Y-axis of the plot shows the relative amplitude of the light signal that has reflected back to the source and the X-axis represents time. The time base is directly translated and displayed as distance by the OTDR.

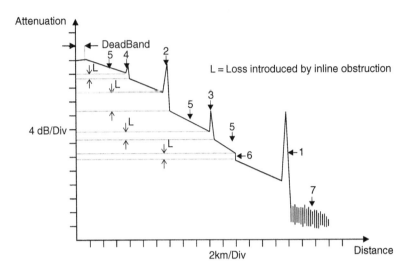

Figure 6.12
Trace from an OTDR

The sudden peaks that appear along the slope are the points where reflections have occurred and light that has reflected back to the source is stronger than the backscatter. There are four main reflection points illustrated in Figure 6.12.

In their order of decreasing magnitude they are:

1. Reflection from the un-terminated end of the fiber
2. Reflection from a connector
3. Reflection from a splice
4. Reflection from a hairline crack in the fiber
5. Backscatter

After each of the reflections, the slope of the attenuation curve drops suddenly. This drop represents the loss introduced by the connector, splice or imperfection in the fiber.

Point (6) noted in Figure 6.12 illustrates a splice where the cores of the fibers are well matched for light traveling in the direction away from the source. This splice has no reflection just a loss introduced by the splice. The type of drop at point (6) in the attenuation curve could also be caused by a sharp bend in the fiber where light escapes out of the fiber at the bend and is not reflected back. Some types of faults in the fiber will also cause a similar result.

Point (7) noted in Figure 6.12 shows the noise floor of the instrument. This is the lowest sensitivity of received signal that the device can accept. Measurements made close to this level are not very accurate.

OTDR testing can provide very accurate fault analysis over long lengths of fiber. On the better quality instruments, a resolution of 1 m for fault location and .01 dB for in line losses is available. Some instruments will operate with a range up to 200 km.

In general, OTDRs are relatively easy to use and special analysis software packages are available for downloading the test results and carrying out detailed analysis if required. The unfortunate downside with OTDR technology is that it is generally very expensive. Even small reduced feature units can be very expensive.

Care should be taken when interpreting the results from an OTDR. Where different fibers are joined together, this connection may represent a change in refractive index, core size, modal properties and/or material properties of the fiber. For example, after a splice or connector the OTDR may display what appears to be a signal gain on the screen. What has probably occurred is that the light has entered a fiber with greater impurities and there has been an increase in backscatter.

The OTDR test should be carried out on every optical fiber in a cable while it is still on the reel prior to installation to ensure that faulty fibers are not installed. The results of these tests should be stored in memory or as a print out. These pre-installation tests are generally carried out when the custody of the cable is passed from one party to the next; for example, when the cable is handed down from the purchaser to the installation contractor.

Once the cable has been installed, the OTDR tests should be carried out again on every optical fiber. The results of the 'as installed' tests can then be compared to the pre-installation test results to determine if the fibers have been damaged or poorly installed.

The results of the pre-installation and the post-installation tests should be kept as part of the commissioning documentation. If there is a fault with the system at a later date, then the commissioning test results can be used to help determine where the faults are located. For high integrity longer distance systems, it is worth carrying out an audit of the system after a number of years of operation and performing the OTDR tests again and comparing them to the commissioning test results to measure any deterioration in the cabling system since installation.

The OTDR can be used for attenuation measurements but it is generally not recommended as OTDR measurements are a relative measurement rather than an absolute measurement. Also, the OTDR does not account for poor quality OTDR connector to fiber end face connections. Therefore the insertion loss measurement procedure that was described above is a more appropriate and accurate technique.

Since the OTDR is not used for accurate attenuation measurements but only for relative measurements, the wavelength at which it operates is not important. Distance readings, splice losses, and connector losses are not affected by the small changes in wavelength associated with lasers and LEDs.

It was noted in the previous section that the loss of a connector or splice might be different when measured from each direction into the optic fiber. For cable sections of greater than 2 km, it is recommended that the OTDR tests be carried out from each end of the optical fiber.

Some fiber optic cables are constructed so that the fibers are laid in a helical fashion around the center of the cable. In this case, the length of the cable is not going to be the length of the fibers. This difference will make it inherently difficult to determine the distance to faults. To overcome this problem, the manufacturer will generally provide a ratio of fiber length to cable length. The ratio is then used to calculate the exact cable distance to the fault from the OTDR distance reading. If a ratio is not available, an OTDR measurement is performed on a known length of cable (generally 1 km) and the ratio is calculated:

$$\text{Fiber/Cable Ratio} = \frac{\text{Length of fiber in 1 km of cable}}{1\,\text{km}}$$

$$\text{Distance to fault} = \frac{\text{OTDR Distance Reading}}{\text{Fiber/Cable Ratio}}$$

7

Modbus overview

Objectives

When you have completed study of this chapter, you will be able to:

- List the main Modbus structures and frames used
- Identify and correct problems with:
  - No response to protocol messages
  - Exception reports
  - Noise

7.1   General overview

Modbus® is a transmission protocol (note – a *protocol* only), developed by Gould Modicon (now Schneider Electric) for process control systems. It is, however, regarded as a 'public' protocol and has become the *de facto* standard in multi-vendor integration. In contrast to other buses and protocols, no physical (OSI layer 1) interface has been defined.

Modbus is a simple, flexible, publicly published protocol, which allows devices to exchange discrete and analog data. End users are aware that specifying MODBUS as the required interface between subsystems is a way to achieve multi-vendor integration with the most purchasing options and at the lowest cost. Small equipment makers are also aware that they must offer MODBUS with EIA-232 and/or EIA-485 to sell their equipment to system integrators for use in larger projects.

System integrators know that MODBUS is a safe interface to commit to, as they can be sure of finding enough equipment on the market to both realize the required designs and handle the inevitable 'change orders,' which come along. However, Modbus suffers from the limitations imposed by EIA-232/485 serial links, including the following:

- Serial lines are relatively slow – 9600 to 115 000 baud means only 0.010 Mbps to 0.115 Mbps. Compare that to today's common 'control network' speeds of 5 to 16 Mbps – or even the new Ethernet speeds of 100 Mbps, and 1 Gbps and 10 Gbps!

- While it is easy to link 2 devices by EIA-232 and 20–30 devices by EIA-485, the only solution to link 500 devices with EIA-485 is a complex hierarchy of masters and slaves in a deeply nested tree structure. Such solutions are never simple or easy to maintain.
- Serial links with Modbus are inherently single-master designs. That means, only one device can talk to a group of slave devices – so only that one device (the master) is aware of all the current real-time data.

Designers share this data with multiple operator workstations, control systems, database systems, customized process optimizing workstations and all the other potential users of the data by ending up with complex, fragile hierarchies of master/slave groups shuffling data, up the ladder. Apart from the complexity involved, lower levels of the hierarchy (even expensive DCS systems) waste valuable time shuffling data solely for the benefit of higher levels of the hierarchy.

Even with all these limitations, Modbus has the advantage of wide acceptance among instrument manufacturers and users with many systems in operation. It can therefore be regarded as a *de facto* industrial standard with proven capabilities.

Certain characteristics of the Modbus protocol are fixed, such as frame format, frame sequences, handling of communications errors and exception conditions and the functions performed. Other characteristics are selectable. These are transmission medium, transmission characteristics and transmission mode, viz. RTU or ASCII. The user characteristics are set at each device and cannot be changed when the system is running.

The two transmission modes in which data is exchanged are:

- ASCII – readable; used, for example, for testing. (ASCII format)
- RTU – compact and faster; used for normal operation. (Hexadecimal format)

The RTU mode (sometimes also referred to as Modbus-B for Modbus Binary) is the preferred Modbus mode. The ASCII transmission mode (sometimes referred to as Modbus-A) has a typical message that is about twice the length of the equivalent RTU message.

Modbus also provides an error check for transmission and communication errors. Communication errors are detected by character framing, a parity check, a redundancy check or a sixteen bit cyclic redundancy check (CRC-16). The latter varies depending on whether the RTU or ASCII transmission mode is being used.

Modbus packets can also be sent over local area and wide area networks by encapsulating the Modbus data in a TCP/IP packet.

7.2 Modbus protocol structure

The following illustrates a typical Modbus message frame format.

Address field	Function field	Data field	Error check field
1 byte	1 byte	Variable	2 bytes

Table 7.1
Format of Modbus message frame

The first field in each message frame is the address field, which consists of a single byte of information. In request frames, this byte identifies the controller to which the request is being directed. The resulting response frame begins with the address of the responding device. Each slave can have an address field between 1 and 247; although practical limitations will limit the maximum number of slaves. A typical Modbus installation will have one master and two or three slaves.

The second field in each message is the function field, which also consists of a single byte of information. In a host request, this byte identifies the function that the target PLC is to perform.

If the target PLC is able to perform the requested function, the function field of its response will echo that of the original request. Otherwise, the function field of the request will be echoed with its most-significant bit set to one, thus signaling an exception response. Table 7.1 summarizes the typical functions used.

The third field in a message frame is the data field, which varies in length according to which function is specified in the function field. In a host request, this field contains information the PLC may need to complete the requested function. In a PLC response, this field contains any data requested by that host.

The last two bytes in a message frame comprise the error-check field. The numeric value of this field is calculated by performing a cyclic redundancy check (CRC-16) on the message frame. This error checking assures that devices do not react to messages that may have been damaged during transmission.

Table 7.2 lists the address range and offsets for these four data types, as well as the function codes that apply to each. The table also gives an easy reference to the Modbus data types.

Data type	Absolute addresses	Relative addresses	Function codes	Description
Coils	00001 to 09999	0 to 9998	01	Read coil status
Coils	00001 to 09999	0 to 9998	05	Force single coil
Coils	00001 to 09999	0 to 9998	15	Force multiple coils
Discrete inputs	10001 to 19999	0 to 9998	02	Read input status
Input registers	30001 to 39999	0 to 9998	04	Read input registers
Holding registers	40001 to 49999	0 to 9998	03	Read holding register
Holding registers	40001 to 49999	0 to 9998	06	Preset single register
Holding registers	40001 to 49999	0 to 9998	16	Preset multiple registers
–	–	–	07	Read exception status
–	–	–	08	Loopback diagnostic test

Table 7.2
Modicon addresses and function codes

7.3 Function codes

Each request frame contains a function code that defines the action expected for the target controller. The meaning of the request data fields is dependent on the function code specified.

The following paragraphs define and illustrate most of the popular function codes supported. In these examples, the contents of the message-frame fields are shown as hexadecimal bytes.

7.3.1 Read coil or digital output status (function code 01)

This function allows the host to obtain the ON/OFF status of one or more logic coils in the target device.

The data field of the request consists of the relative address of the first coil followed by the number of coils to be read. The data field of the response frame consists of a count of the coil bytes followed by that many bytes of coil data.

The coil data bytes are packed with one bit for the status of each consecutive coil (1=ON, 0=OFF). The least significant bit of the first coil data byte conveys the status of the first coil read. If the number of coils read is not an even multiple of eight, the last data byte will be padded with zeros on the high end. Note that if multiple data bytes are requested, the low order bit of the first data byte in the response of the slave contains the first addressed coil.

In the following example, the host requests the status of coils 000A (decimal 00011) and 000B (decimal 00012). The target device's response indicates both coils are ON.

Request Message

Address	Function Code	Initial Coil Offset		Number of Points		CRC
		Hi	Lo	Hi	Lo	
01	01	00	0A	00	02	9D C9

Response Frame

Address	Function Code	Byte Count	Coil Data	CRC
01	01	01	03	11 89

Figure 7.1
Example of read coil status

7.3.2 Read digital input status (function code 02)

This function enables the host to read one or more discrete inputs in the target device.

The data field of the request frame consists of the relative address of the first discrete input followed by the number of discrete inputs to be read. The data field of the response frame consists of a count of the discrete input data bytes followed by that many bytes of discrete input data.

The discrete-input data bytes are packed with one bit for the status of each consecutive discrete input (1=ON, 0=OFF). The least significant bit of the first discrete input data byte conveys the status of the first input read. If the number of discrete inputs read is not an even multiple of eight, the last data byte will be padded with zeros on the high end. The low order bit of the first byte of the response from the slave contains the first addressed digital input.

In the following example, the host requests the status of discrete inputs with offsets 0000 and 0001 hex i.e. decimal 10001 and 10002. The target device's response indicates that discrete input 10001 is OFF and 10002 is ON.

Request Message

Address	Function Code	Initial Coil Offset		Number of Points		CRC	
		Hi	Lo	Hi	Lo		
01	02	00	00	00	02	F9	CB

Response Frame

Address	Function Code	Byte Count	Input Data	CRC	
01	02	01	02	20	49

Figure 7.2
Example of read input status

7.3.3   Read holding registers (function code 03)

This function allows the host to obtain the contents of one or more holding registers in the target device.

The data field of the request frame consists of the relative address of the first holding register followed by the number of registers to be read. The data field of the response time consists of a count of the register data bytes followed by that many bytes of holding register data.

The contents of each requested register (16 bits) are returned in two consecutive data bytes (most significant byte first).

In the following example, the host requests the contents of holding register hexadecimal offset 0002 or decimal 40003. The controller's response indicates that the numerical value of the register's contents is hexadecimal 07FF or decimal 2047. The first byte of the response register data is the high order byte of the first addressed register.

Request Message

Address	Function Code	Starting Register		Register Count		CRC	
		Hi	Lo	Hi	Lo		
01	03	00	02	00	01	25	CA

Response Frame

Address	Function Code	Byte Count	Register Data		CRC	
			Hi	Lo		
01	03	02	07	FF	FA	34

Figure 7.3
Example of reading holding register

7.3.4 Reading input registers (function code 04)

This function allows the host to obtain the contents of one or more input registers in the target device.

The data field of the request frame consists of the relative address of the first input register followed by the number of registers to be read. The data field of the response frame consists of a count of the register data bytes followed by that many bytes of input register data.

The contents of each requested register are returned in two consecutive register data bytes (most significant byte first). The range for register variables is 0 to 4095.

In the following example, the host requests the contents of input register hexadecimal offset 000 or decimal 30001. The PLC's response indicates that the numerical value of that register's contents is 03FFH, which would correspond to a data value of 25 per cent (if the scaling of 0 to 100 per cent is adopted) and a 12 bit D to A converter with a maximum reading of 0FFFH is used.

Request Message

Address	Function Code	Starting Register		Register Count		CRC	
		Hi	Lo	Hi	Lo		
01	04	00	00	00	01	31	CA

Response Frame

Address	Function Code	Byte Count	Register Data		CRC	
			Hi	Lo		
01	04	02	03	FF	F9	80

Figure 7.4
Example of reading input register

7.3.5   Force single coil (function code 05)

This function allows the host to alter the ON/OFF status of a single logic coil in the target device.

The data field of the request frame consists of the relative address of the coil followed by the desired status for that coil. A hexadecimal status value of FF00 will activate the coil, while a status value of 0000H will deactivate it. Any other status value is illegal.

If the controller is able to force the specified coil to the requested state, the response frame will be identical to the request. Otherwise, an exception response will be returned.

If the address 00 is used to indicate broadcast mode, all attached slaves will modify the specified coil address to the state required.

The following example illustrates a successful attempt to force coil 11 (decimal) OFF.

Request Message

Address	Function Code	Coil Offset		New Coil Status		CRC	
		Hi	Lo	Hi	Lo		
01	05	00	0A	00	00	ED	C8

Response Frame

Address	Function Code	Coil Offset		New Coil Status		CRC	
		Hi	Lo	Hi	Lo		
01	05	00	0A	00	00	ED	C8

Figure 7.5
Example of forcing a single coil

7.3.6   Preset single register (function code 06)

This function enables the host to alter the contents of a single holding register in the target device.

The data field of the request frame consists of the relative address of the holding register followed by the new value to be written to that register (most significant byte first).

If the controller is able to write the requested new value to the specified register, the response frame will be identical to the request. Otherwise, an exception response will be returned.

The following example illustrates a successful attempt to change the contents of holding register 40003 to 3072 (0C00 hex).

When slave address is set to 00 (broadcast mode), all slaves will load the specified register with the value specified.

Request Message

Address	Function Code	Register Offset		Register Value		CRC	
		Hi	Lo	Hi	Lo		
01	06	00	02	0C	00	2D	0A

Response Frame

Address	Function Code	Register Offset		Register Value		CRC	
		Hi	Lo	Hi	Lo		
01	06	00	02	0C	00	2D	0A

Figure 7.6
Example of presetting a single register

7.3.7   Read exception status (function code 07)

This is a short message requesting the status of eight digital points within the slave device.

This will provide the status of eight predefined digital points in the slave. For example, this could be items such as the status of the battery, whether memory protect has been enabled or the status of the remote input/output racks connected to the system.

Request Message

Address	Function Code	CRC
11	07	.. ..

Response Frame

Address	Function Code	Coil Station	CRC
11	07	02	.. ..

Figure 7.7
Read exception status query message

7.3.8   Loopback test (function code 08)

The objective of this function code is to test the operation of the communications system without affecting the memory tables of the slave device. It is also possible to implement additional diagnostic features in a slave device (should this be considered necessary) such as number of CRC errors, number of exception reports etc.

The most common implementation will only be considered in this section; namely, a simple return of the query messages.

Request Frame

Address	Function Code	Data Diagnostic Code		Data		CRC
		Hi	Lo	Hi	Lo	
11	08	00	00	A5	37	.. ..

Response Frame

Address	Function Code	Data Diagnostic Code		Data		CRC
		Hi	Lo	Hi	Lo	
11	08	00	00	A5	37	.. ..

Figure 7.8
Loopback test message

7.3.9 Force multiple coils or digital outputs (function code 0F)

This forces a contiguous (or adjacent) group of coils to an ON or OFF state. The following example sets 10 coils starting at address 01 hex (at slave address 01) to the ON state. If slave address 00 is used in the request frame, broadcast mode will be implemented resulting in all slaves changing their coils at the defined addresses.

Request Frame

Address	Function Code	Address		Byte Count	Data Coil Status		CRC
		Hi	Lo		Hi	Lo	
01	0F	00	01	0F	FF	03	.. ..

Response Frame

Address	Function Code	Address		Number of Coils		CRC
		Hi	Lo	Hi	Lo	
01	0F	00	01	00	0A	.. ..

Figure 7.9
Example of forcing multiple coils

7.3.10   Force multiple registers (function code 10)

This is similar to the preset of a single register and the forcing of multiple coils. In the example below, a slave address 01 has 2 registers changed commencing at address 10.

Request Frame

Address	Function Code	Address		Quantity		Byte Count	First Register		Second Register		CRC
		Hi	Lo	Hi	Lo		Hi	Lo	Hi	Lo	
01	10	00	0A	00	02	04	00	0A	01	02	.. ..

Response Frame

Address	Function Code	Address		Quality		CRC
		Hi	Lo	Hi	Lo	
01	10	00	0A	00	02	.. ..

Figure 7.10
Example of presetting multiple registers

Table 7.3 lists the most important exception codes that may be returned.

Code	Name	Description
01	Illegal function	Requested function is not supported
02	Illegal data address	Requested data address is not supported
03	Illegal data value	Specified data value is not supported
04	Failure in associated device	Slave PLC has failed to respond to a message
05	Acknowledge	Slave PLC is processing the command
06	Busy, rejected message	Slave PLC is busy

Table 7.3
Abbreviated list of exception codes returned

An example of an illegal request and the corresponding exception response is shown below. The request in this example is to READ COIL STATUS of points 514 to 521 (eight coils beginning with an offset 0201H). These points are not supported in this PLC, so an exception report is generated indicating code 02 (illegal address).

Request Message

Address	Function Code	Starting Point	Number of Points	CRC
01	01	02 01	00 08	6D B4

Exception Response Message

Address	Function Code	Exception Code	CRC
01	81	02	C1 91

Figure 7.11
Example of an illegal request

7.4   Troubleshooting

7.4.1   Common problems and faults

No matter what extremes of care that you may have taken, there is hardly ever an installation that boasts of trouble-free setup and configuration. Some common problems related to Modbus installations are listed below. They can be categorized as either hardware or software problems.

Hardware problems include mis-wired communication cabling and faulty communication interfaces.

Software (protocol) related issues arise when the master application tries to access non-existent nodes or use invalid function codes, address non-existent memory locations in the slaves, or specify illegal data format types, which obviously the slave devices do not understand. These issues will be dealt with in detail later under the Modbus Plus protocol, as these issues are common to both the traditional Modbus protocol and the Modbus Plus protocol. They are summarized under software related problems.

Note that these issues are also applicable to the latest Modbus/TCP protocol, which is discussed in appendix D.

7.4.2   Description of tools used

In order to troubleshoot the problems listed above, you would require the use of a few tools, both hardware and software, to locate the errors. The most important tool of all would always be the instruction manuals of various components involved.

The hardware tools that may be required include EIA-232 breakout boxes, EIA-232 to EIA-485 converters, continuity testers, voltmeters, screwdrivers, pliers, crimping tools, spare cabling tools and other similar tools and tackle. These would generally be used to ensure that the cabling and terminations are proper and have been installed as per the recommended procedures detailed in the instruction manuals.

On the software tools front, you would need some form of protocol analyzer that is in a position to eavesdrop on the communications link between the master and slave modules. These could be either a dedicated hardware protocol analyzer, that is very expensive, or software-based protocol analyzer that could reside on a computer and act as the eavesdropping node on the network.

Obviously, this second option is more economical and also requires the relevant hardware component support in order to connect to the network.

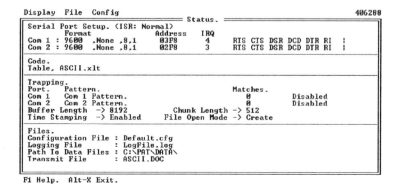

Figure 7.12
Screenshot of the protocol analysis tool

7.4.3   Detailed troubleshooting

7.4.3.1   Mis-wired communication cabling

EIA-232 wiring for 3-wire point-to-point mode

There are various wiring scenarios such as point-to-point, multi-drop two-wire, multi-drop 4-wire, etc. In a point-to-point configuration, the physical standard used is usually EIA-232 on a 25-pin D connector, which means a minimum of three wires, are used. They are: on the DTE (master) side, transmit (TxD – pin 2), receive (RxD – pin 3) and signal ground (common – pin 7). On the DCE (slave) side they are; receive (RxD – pin 2), transmit (TxD – pin 3) and signal ground (common – pin 7).

The other pins were primarily used for handshaking between the two devices for data flow control. Nowadays, these pins are rarely used as the flow control is controlled via software handshaking protocols. Details of these protocols can be obtained from the *Practical Industrial Data Networks: Design, Installation and Troubleshooting* book released by IDC.

With the advent of VLSI technology, the footprint of the various devices have a tendency to shrink; even the physical connections of communications have been now standardized on 9-pin D connectors. The pin assignment that has been adopted is the IBM standard in which the pin configurations are as follows. On the DTE (master) side, transmit (TxD – pin 3), receive (RxD – pin 2) and signal ground (common – pin 5). On the DCE (slave) side receive (RxD – pin 3), transmit (TxD – pin 2) and signal ground (common – pin 5).

It follows that the cabling between the two devices must be straight through pin-for-pin. In this manner, the transmit pin of the master is directly connected to the receive pin of the slave and *vice versa*. Such cable is referred to as a straight-through cable. These cables are standard off-the-shelf products available in standard pre-determined lengths. Alternatively they can be fabricated to custom lengths, with ease.

Master devices usually are the present-day IBM compatible computers, with a Modbus application installed on it, and therefore have the standard IBM EIA-232 port provided. The slave devices usually have a user selectable option to have either an EIA-232 or EIA-485 port for communication. Unfortunately some manufacturers, in order to force the customers to return to them time-and-again, employ a strategy of modifying these standards to their own advantage.

Illustrated below are a couple of these combinations:

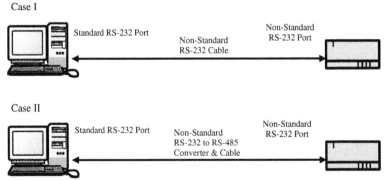

Figure 7.13
Typical customizations in EIA-232 cabling

In case I, the manufacturer has modified the standard pin-out of the EIA-232 connection or has totally replaced the standard IBM EIA-232 9-pin DIN connector with his own standard. In situations like these, it is then imperative to possess the manufacturer's manuals of the device being used.

With the aid of a continuity checker and an EIA-232 breakout box, you can determine the non-standard pin-out of the cable, and fabricate an alternative cable to the required length without having to revert to the supplier.

In case II, the manufacturer has embedded an EIA-232 to EIA-485 converter into the cable itself and this would mimic a standard off-the-shelf EIA-232 serial cable. This obviously would not be obvious if you were not aware of the modification. With the aid of a voltage tester, you can determine the operating voltages and therefore be in a position to decipher the type of standard being employed.

In implementations where multidropped EIA-485 is used, the installations usually cater for both two-wire and four-wire communications. In the case of single master configurations, either of the two wiring modes could be used. In the case of multi-master configurations, only two-wire communication can be used.

EIA-485/EIA-422 wiring for 4-wire, repeat or multidrop modes

The four-wire configuration is, in actual fact, the EIA-422 standard where there is one line driver connected to multiple receivers, or multiple line drivers connected to one line receiver. When using 4-wire EIA-422 communications, messages are transmitted on one pair of shielded, twisted wires and received on another (e.g. Belden part number 9729, 9829, or equivalent). Both multidrop and repeat configurations are possible when EIA-422 is used.

This is the classic case where the solitary line driver belongs to the master and the remaining receivers belong to the multiple slaves. The slaves receive their commands via this link and respond via their line drivers that are all connected to the master's receiver via the bus.

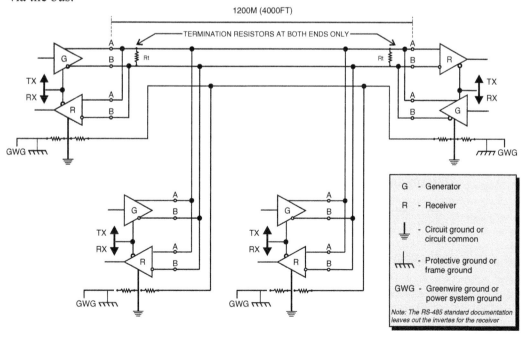

Figure 7.14
Typical two-wire EIA-422/485 cabling

EIA-485 wiring for 2-wire, multidrop mode only

The two-wire configuration is the actual EIA-485 standard where all line drivers and receivers are connected to the same bus. When using 2-wire EIA-485 communications, messages are transmitted and received on the same pair of shielded, twisted wires (e.g. Belden part number 9841, 9341, or equivalent). Care must be taken to turn individual line drivers on only when sending messages. For the remainder of the time they must be switched off for the purpose of receiving messages. This can be accomplished using software or hardware. Only multidrop wiring configuration is possible when EIA-485 is used.

Typically, the maximum number of physical EIA-485 nodes on a network is 32. If more physical nodes are placed on the network an EIA-485 repeater must be used.

This system works inherently similar to the four-wire system except that the master transmits requests to the various slaves and the slaves respond to the master over the same pair of wires.

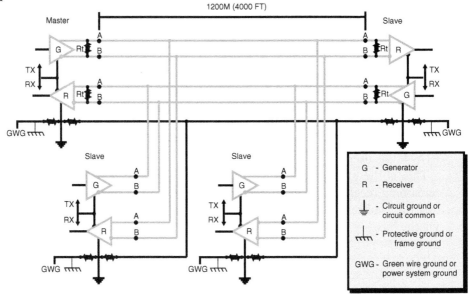

Figure 7.15
Typical four-wire EIA-485 multidrop cabling

Note: The shield grounds should be connected to system ground at one point only. This will eliminate the possibility of ground loops causing erroneous operation.

Grounding

The logic ground of all controllers and any other communication device on the same network must reference the same ground. If one power supply is used for all controllers, then they will be referenced to the same ground. When multiple supplies are used, there are a couple of options to consider:

- Connect the power supply output ground to a solid earth ground on all controllers. If the grounding system is good, all controllers will be referenced to the same ground. (Note that the typical PC follows this practice). In some older buildings, the earth ground could vary by several volts within the building. In this case, the following option should be considered.

- Use the GND position on each controller to tie all controllers to the same potential. This will necessitate a third conductor, preferably with a 100-ohm ½ watt resistor in series at both ends, as per the sketch. It is not recommended that the shield of a cable be used for this purpose, use a separate conductor.
- Where it is not possible to get all controllers referenced to the same ground, use an isolated EIA-485 repeater between controllers that are located at different ground potentials. Isolated repeaters are also an excellent way to clean up signals and extend distances in noisy environments.

7.4.3.2   Faulty communication interfaces

EIA-232 driver failed

The EIA-232 driver may be tested with a good high-impedance voltmeter. Place the meter in the DC voltage range. Place the RED probe on the transmit pin (TxD) and the BLACK probe on the signal ground (common). While the node is not transmitting (TX light off), there should be a voltage between –5 V DC and –25 V DC across these pins. This is referred to as the idle state of the EIA-232 driver. When the node is transmitting, the voltage will oscillate through between ±5 V DC and ±25 V DC. It may be difficult to see the deflection at the higher baud rates and an oscilloscope is suggested for advanced troubleshooting. If the voltage is fixed at anywhere between –25 V DC and +25 V DC and doesn't oscillate as the TX light blinks, then the transmitter is probably damaged.

EIA-232 receiver failed

This requires the use of an EIA-232 loop-back plug or an EIA-232 breakout box with a jumper installed between the transmit and receive pins. After connecting this to the communication port of the node to be tested, the node is made to transmit data, and when the TxD light of the node flashes, the RxD light must also flash, then it can be said that the EIA-232 receiver of that node is good. If the RxD light does not flash at the same time as the TxD, then the EIA-232 receiver may be bad. Needless to say, the baud rate has to be reduced to 1 or 2 baud in order to observe the LED flashing.

Alternatively, two nodes may be connected to each other with a tested and working communication cable and both their EIA-232 drivers working. If one of the nodes is transmitting to the other, the second node indicates a good reception by flashing its RxD light as the first node transmits, then the EIA-232 receiver on the second node is good. The same is true when the second node transmits and the first flashes its RxD light.

EIA-485 driver failed

The EIA-485 driver too may be tested with a good high-impedance voltmeter. Place the meter in the DC V voltage range. Place the RED probe on the non-inverted transmit terminal (TX+) and the BLACK probe on the inverted transmit terminal (TX–). While the node is not transmitting (TX light off), there should be approximately –4 V DC across these wires. When the node is transmitting, the voltage will oscillate through ±4 V DC. It may be difficult to see the deflection at the higher baud rates and an oscilloscope is suggested for advanced troubleshooting. The minimum voltage obtained during a full 1200 baud to 19 200 baud sweep will be around –1.6 V DC. If the voltage is fixed at around +4, 0, or –4 volts, and doesn't oscillate as the TX light blinks, then the transmitter is probably damaged.

EIA-485 receiver failed

This requires two nodes connected to each other with a tested and working communication cable and both their EIA-485 drivers working. If one of the nodes is transmitting to the other, the second node indicates a good reception by flashing its RX light as the first node transmits, then the EIA-485 receiver on the second node is good. If this does not happen first check that the TX+ and TX– wires are not connected the wrong way around before assuming that the second node's receiver is damaged.

7.4.3.3 Software related problems

See the suggestions under Modbus Plus. The important issues to consider are:

No response to message from master to slave

This could mean that either the slave does not exist or there are CRC errors in the transmitted message due to noise (or an incorrectly formatted message).

Exception responses

See the list of potential problems reported by the exception responses. This could vary from slave address problems to I/O addresses being illegal.

7.4.4 Conclusion

We have seen that the master computer sends commands to the various slave units to determine the status of its various process inputs or to change the status of its outputs using the Modbus protocol. The commands are transmitted over a single pair of twisted wires (EIA-485) or two pairs of twisted wires (EIA-422) at speeds of up to 10 Mbaud. The addressed slave decodes the commands and returns the appropriate response. If the master computer is an IBM PC or compatible, inexpensive interface driver software is available. This software dramatically simplifies sending and receiving these messages. If you prefer, you can use one of the off-the-shelf graphics-based data acquisition and control software packages. Many of these packages offer a Modbus compatible driver.

If you were to follow the recommended installation and startup procedure and took extra precaution with the following items, then you would achieve trouble-free startup.

- Setting the serial communications parameters (baud rate etc)
- Installing the communications wiring
- Terminating the communications bus

8

Modbus Plus protocol overview

Objectives

When you have completed study of this chapter, you will be able to:
- Describe the main features of Modbus Plus
- Fix problems with:
  - Cabling
  - Grounding/earthing
  - Shielding
  - Terminations

8.1 General overview

Besides the standard Modbus protocol, there exist two other Modbus protocol structures:
- Modbus Plus
- Modbus II

The more popular one is the Modbus Plus. It is not an open standard as the classical Modbus has become. Modbus II is not used much due to additional cabling requirements and other difficulties.

Modbus Plus protocol was developed to overcome the 'single-master' limitation prevalent in the Modbus Protocol. As described earlier, in order to share information across various Modbus networks, a designer would have had to employ a complex hierarchical network structure in order to achieve system wide networking.

Modbus Plus is a local area network system for industrial control applications. Networked devices can exchange messages for the control and monitoring of processes at remote locations in the industrial plant. The Modbus Plus network was one of the first token-passing protocol networks that pioneered the development of other more advanced deterministic protocols today.

The Modbus Plus protocol topography allows for up to 64 devices on a network segment. Each device on a network segment must be assigned a unique network address within the range 01 through 64 inclusive. Multiple networks segments may be bridged together to form large systems. Messages are passed from one device to another by the appropriate use of a route within the message. The route will include the network address of the network address of the target device, and any inter-network addresses required. The route field of a message may be up to 5 layers deep, is terminated with the value 00, and all unused drops are also set to 00.

Preamble	Opening Flag	Broadcast Address	MAC / LLC Data	Error Check Field	Closing Flag
1 Byte	1 Byte	1 Byte	Variable	2 Bytes	1 Byte

Destination Address	Source Address	MAC Function	Byte Count	LLC Field (including Modbus Command)
1 Byte	1 Byte	1 Byte	2 Bytes	Variable

Master Output Path	Router Counter	Transaction Sequence No.	Routing Path	Modbus Frame (without CRC / /LRC)
1 Byte	1 Byte	1 Byte	5 Bytes	Variable

Figure 8.1
Format of Modbus Plus message frame

Each network supports up to 64 addressable node devices. Up to 32 nodes can connect directly to the network cable over a length of 1500 ft (450 meters). Repeaters can extend the cable distance to its maximum of 6000 ft (1800 meters) and the node count to its maximum of 64. Fiber optic repeaters are available for longer distances.

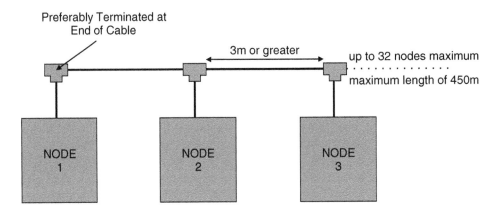

Figure 8.2
Typical Modbus Plus network with termination

Multiple networks can be inter-networked through bridge plus devices. Messages originating on one network are routed through one or more bridges to a destination on another network. Bridges do not permit deterministic timing of I/O processes. In networks requiring deterministic I/O timing, messages remain on that network only, and they do not pass through bridges.

Network nodes are identified by addresses assigned by the user. Each node's address is independent of its physical site location. Addresses are within the range of 1 to 64 decimal, and they do not have to be sequential. Duplicate addresses are not allowed.

Network nodes function as peer members of a logical ring, gaining access to the network upon receipt of a token frame. The token is a grouping of bits that is passed in a rotating address sequence from one node to another. Each network maintains its own token rotation sequence independent of the other networks. Where bridges join multiple networks, the token is not passed through the bridge device.

While holding the token, a node initiates message transactions with other nodes. Each message contains routing fields that define its source and destination, including its routing path through bridges to a node on a remote network.

When passing the token, a node can write into a global database that is broadcast to all nodes on the network. Global data is transmitted as a field within the token frame. Other nodes monitor the token pass and can extract the global data if they have been programmed to do so. Use of the global database allows rapid updating of alarms, setpoints, and other data. Each network maintains its own global database, as the token is not passed through a bridge to another network.

The network bus consists of twisted pair shielded cable that is run in a direct path between successive nodes. The two data lines in the cable are not sensitive to polarity, however a standard wiring convention is followed in this guide to facilitate maintenance.

On dual-cable networks, the cables are known as cable A and cable B. Each cable can be up to 1500 ft (450 m) long, measured between the two extreme end devices on a cable section. The difference in length between cables A and B must not exceed 500 ft (150 m), measured between any pair of nodes on the cable section.

SYMMETRY IN PLACEMENT OF REPEATERS

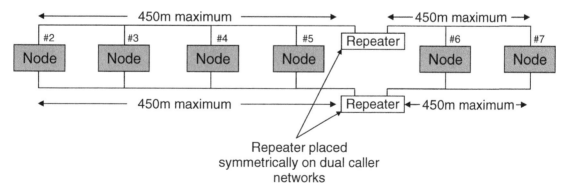

Figure 8.3
Typical Modbus Plus network with dual cabling

The token passing sequence is determined by the node addresses. Token rotation begins at the network's lowest-addressed active node, proceeding consecutively through each higher-addressed node, until the highest-addressed active node receives the token. That node thcn passes the token to the lowest one to begin a new rotation.

If a node leaves the network, a new token-passing sequence will be established to bypass it, typically within 100 milliseconds. If a new node joins, it will be included in the address sequence, typically within 5 seconds (worst case time is 15 seconds). The process of deleting and adding nodes is transparent to the user application.

Where multiple networks are interconnected with bridges, tokens are not passed through a bridge device from one network to another. Each network performs its token passing process independently of the other networks.

While a node holds the token, it sends its application messages if it has any to transmit. The other nodes monitor the network for incoming messages. When a node receives a message, it sends an immediate acknowledgment to the originating node. If the message is a request for data, the receiving node will begin assembling the requested data into a reply message.

When the message is ready, it will be transmitted to the requestor when the node receives a subsequent token granting it access to transmit. After a node sends all of its messages, it passes the token on to the next node. Protocols for token passing and messaging are transparent to the user application.

8.2 Troubleshooting

8.2.1 Common problems and faults

The Modbus Plus network is a 3-wire (one pair and a shield) twisted pair cable with the nodes connected in a daisy-chained configuration. There is no polarity requirement at the node's transceiver, so the data cable pair may be connected either way at a node. A 220–ohm terminator is required at each end of the network cable. There are limits upon the maximum number of nodes per segment, the number of repeaters, and the lengths of cable segments on the Modbus Plus network. For more information, refer to the Modicon Modbus Plus network planning and installation guide (GM–MBPL–001).

The most common hardware-related problems have already been discussed in the previous section. Please refer to the hardware-related troubleshooting list for the Modbus protocol. You should configure the node address of the Modbus Plus port on the nodes before connecting them to the network. This should avoid possible duplicate address problems with other units on the network.

Following are a few of the software-related issues that could arise in Modbus Plus networks. While most issues arise from the use of invalid slave port addressing, illegal slave memory addressing, illegal data formats and even perhaps use of unrecognized function codes. Other issues are related to the actual configuration of the communication hardware itself; for example, the SA85 Modbus Plus network interface card, that is used to provide an IBM compatible personal computer access to the Modbus Plus network.

- Remote PC not listening for call
- Slave xx reported an illegal data address
- Slave xx reported an illegal data value
- Slave xx reported an illegal function
- Card SA85:0 not found, bad NCB_LANA_NUM – device not found
- Troubleshooting the SA85 adapter card under Windows NT

The connection to the Modbus Plus network is made just like with any Modbus Plus device. Many a time, most problems could be easily identified by referring to the manufacturer supplied device user manual, and looking up the meanings of the various status indicators provided on the front panel of the Modbus Plus device. For example, there usually is a Modbus Active LED provided on the node.

The table below is an example of such a Modbus Plus network status indicator.

Modbus active LED

The Modbus Act LED on the node flashes repetitive patterns to display node status. The table below lists typical Modbus active LED blinking patterns and their description.

Blinking patterns	Description
Flash every 160 ms normal operation	This node is successfully receiving and passing the token. All nodes should blink at this rate when working properly.
Flash every 1 sec	This node is in the MONITOR_OFFLINE state. It must remain in this state for 5 seconds and is not allowed to transmit messages onto the network. While in this state, the node is listening to all other nodes and building a table of the known nodes.
2 Flashes, off 2 sec	This node is in the MAC_IDLE state. It is hearing other nodes get the token but it is never receiving the token itself. This node may have a bad transmitter.
3 Flashes, off 1.7 sec	This node is not hearing any other nodes and it is periodically claiming the token and finds no other nodes to pass it to. This node may be the only active node on the network or it may have a bad receiver.
4 Flashes, off 1.4 sec	This node has heard a duplicate address packet. There is another node on the network with the same address as this node. This node is now in the DUPLICATE_OFFLINE state where it will passively monitor the link until it does not hear the duplicate node for 5 seconds.

Table 8.1
Typical list of Modbus active LED blinking patterns

8.2.2   Description of tools used

Apart from the standard set of tools that have been listed in the previous section while discussing Modbus Protocols to troubleshoot hardware-related problems, there are only the manufacturer supplied application packages that report any software errors that occur in the Modbus Plus communication networks.

The following section discusses the cause of some of the most common Modicon Modbus and Modbus Plus error messages that you may encounter when attempting to establish communications with one of the various off the shelf Modbus Plus-based applications packages. For example, the Wonderware® Modicon Modbus or Modicon Modbus Plus I/O server and how to resolve them. (Source http://www.wonderware.com)

8.2.3   Detailed troubleshooting

Remote PC not listening for call

This error means that the Modbus or Modbus Plus I/O server is unable to 'see' a slave device. That is, it is unable to get to the specified node(s), which has been defined in the topic slave path. Either the slave path does not have the correct path defined or one or more of the node addresses in the slave path do not match the addresses of the PLCs.

If you are unable to establish communications through more than one Modbus Bridge plus (BP), then start with the first bridge plus listed in the slave path and access a slave device on that Modbus Plus network. For example, if your slave path is defined as

'24.32.16.10' and you are getting the 'Remote PC not listening for call' error, then begin by establishing communications to a node through BP 24.

If you are addressing a slave device with an address of 25 on the first network, then set the slave path to '24.25.0.0.' Then, establish communications to node 25 on the first network. (That is, advise a point on node 25 and verify that the data is being retrieved.) If you are successful, then move onto the next BP and communicate to a node within that network. Then, modify the slave path to '24.32.x.0' where x represents the address of a node within the next network. Repeat this step until the target PLC is reached. By following this technique, it will help you to verify two things:

1) The correct node addresses are being used.
2) The slave path is defined correctly.

Slave xx reported an illegal data address

This error message means that a tagname's DDE item name has the correct format and was accepted by the Modbus or Modbus Plus I/O Server, but the PLC (slave device) rejected the DDE item name because the register has not been defined. 'xx' refers to the node address of the slave device. That is, the referenced address, or DDE item name, was not a defined address for the slave device. For example, say we had the following configuration:

- The I/O table for a Modbus 984 PLC is defined up to register 45 999
- A DDE topic in the I/O server is defined with a slave type of '584/984'
- A DDE tagname is defined with an item of 49 999

Since register 49 999 is not defined in the I/O table when the I/O server attempts to advise this register, the error 'slave xx reported an illegal data address' is logged in the Wonderware Logger (WWLogger).

To resolve this error, verify the DDE items to be advised fall within the range of the defined I/O table for the PLC.

Note that this error does not mean the same thing as getting an 'advised failed' error. The 'slave xx reported an illegal data address' error means that the DDE item of a tagname is not addressing a register defined in the PLC's I/O map.

Slave xx reported an illegal data value

(The xx refers to the node address of the slave device.) This error message means the data which is being read from or written to a register in the slave device does not match the expected data type for that register ('xx' refers to the node address of the slave). That is, the data is not the same data type that the Modbus or Modbus Plus I/O server expects for that register.

For example, if an I/O server topic is defined with a slave type of '584/984' and the I/O server tries to write to coil 9999, but the data from the tagname is something other than a discrete data type, then the error 'slave xx reported an illegal data value' occurs. This error could also occur if the slave type selected is not correct for the type of PLC that is being advised. For example, if you try to write data to a Modbus 984 PLC but the slave type selected was '484', then this error will occur.

Slave xx reported an illegal function

This error message occurs when the slave device cannot perform a particular Modbus function that was issued by the master node because either the slave device does not understand the function or it is unable to execute the function. The computers that have either the Modbus or Modbus Plus I/O server installed are considered to be a 'master' node. 'xx' refers to the node address of the slave device.

When receiving this type of error, verify that the slave device supports the Modbus function that it is being requested to execute. You can verify the Modbus function code that the I/O server issued and the response that was received from the slave device by enabling the I/O server's 'show send' and 'show receive' menu options.

To enable the 'show send and 'show receive' menu options, follow these steps:

1. Edit the Win.ini file in the C:\Windows\System directory and depending on the server you are using, that is, the Modbus or Modbus Plus I/O server, locate either [Modbus] or [MBPLUS] file section. Look for a 'DebugMenu=x' entry in the file section. If it exists, modify the entry as follows:

   DebugMenu = 1

2. If it does not exist, add this entry to the end of the file section.
3. Restart the Modbus or Modbus Plus I/O server and click the window's control menu box. You should now see the 'show send' and 'show receive' menu options. Single-click each option and the option 'all topics' (or a list of the current active topics defined in the I/O server) will display to the right of the 'show send' or 'show receive' menu option. Enable either 'all topics' or one or more of the topics for which the problem is occurring.
4. Initiate communications to the slave device again.

With the 'show send' and 'show receive' menu options enabled, the I/O server will log all Modbus protocol messages in WWLogger when it attempts to communicate 's' to the slave device, such as the following send request examples (those issued by the I/O server to the slave device):

    S: :0301005601683D\r\n

    S: :0101001101E20A\r\n

The Modbus and Modbus Plus I/O servers use the standard Modbus protocol and here are the function codes that the I/O server will issue (which are tied to the protocol messages):

Control code	Function code
Read coil	01
Read contact	02
Read holding register	03
Read input register	04
Force coil	05
Load register	06
Force multiple coils	15
Load multiple registers	16
Read general reference	20
Write general reference	21

Table 8.2
Function codes issued by Modbus Plus I/O server

For information on how to correctly read the Modbus send requests in WWLogger and decipher the Modbus function code that is listed as part of the message, see the document titled 'Interpreting the Modbus Protocol,' which is part of the 'Wonderware Knowledge Base.'

Or, you may go to the Modicon Web site at http://www.modicon.com/.

Note: The 'show send' and 'show receive' menu options will generate frequent writes to WWLogger, and therefore, both should be enabled during troubleshooting only. If 'show send' and 'show receive' are left activated during the normal operation of the I/O server, then the performance of your computer will slow considerably.

Card SA85:0 not found, bad NCB_LANA_NUM – device not found

This error message means the Modbus SA85 adapter card is not being acknowledged by the Modbus Plus I/O server. This error may result when the SA85 adapter card does not initialize during the startup of the computer. If this card does not initialize properly, then the I/O server will not be able to communicate with any slave device on the Modbus Plus network.

To resolve this error, if you are running Windows, Windows for Workgroups or Windows 95, then make sure the following device statement is in the config.sys file which is located in the root (C:\) directory:

$$\text{Device} = \text{MBPHOST.SYS /mXXXX /nY /r2 /sZZ}$$

Where XXXX represents the 'memory window' setting that is set with the switches on the SA85 adapter card; Y represents the number of the adapter card that is installed (examples: n0, n1); and ZZ is the hexadecimal address of the software interrupt that is to be used by the device driver. Note the spaces between the parameters in the device statement. Once this SA85 device statement is defined correctly, the SA85 adapter card should initialize properly. (See tech note number 5, solving Modbus Plus channel allocation errors for specific instructions on how to define this device statement.)

Note that release 5.2a and later of the Modbus Plus I/O server will automatically add this device statement to the config.sys file (earlier releases of the I/O server require that you enter the device statement in the config.sys file, as well as edit the Modicon.ini file). If you are running release 5.2a or later, then make sure that the Mbphost.sys file is located in the root directory.

Note: Wonderware technical support suggests that you install version 3.4 or later of the Mbphost.sys file instead of the SA85.sys file in the root directory. To obtain the latest Mbphost.sys file, contact your Group Schneider/Modicon representative or go to the Modicon web site.

Verify that the Mbphost.sys file is being loaded to the SA85 adapter card during startup of the computer by manually stepping through the startup procedure of the operating system. For Windows and Windows for Workgroups, you do this by pressing the <F8> key when the operating system proceeds from its self-diagnostic test to the execution of the config.sys file. For Windows 95, you do this by pressing the <F8> key at the beginning of startup and select 'step by step confirmation' from the Windows 95 startup menu screen. This will allow you to step through the execution of each line of the config.sys and autoexec.bat files. (For Windows NT 3.51 and 4.0, see the section 'Troubleshooting the SA85 adapter card under Windows NT' below.)

After you enter the manual step through execution mode, look for information that is displayed after the 'Device=' statement in which the Mbphost.sys file is executed. It should indicate that the SA85 device driver was loaded correctly, as shown in this example:

```
SA85 Device Driver Version 3.4

<<PG-MBPL-500>>

Copyright (c) 1989-1994 Modicon In. All Rights Reserved.

set for polled mode, SW int:0x5D Memory at 0xD000. LAN
adapter #0

node address = 31

This example shows that the SA85 adapter card is:

Set for polled mode (which is required for the SA85
Adapter card);

Using interrupt 5D;

Using memory location D000;
```

Set to address 31 on the Modbus Plus network. (Note that for both 'c' and 'd', the memory and node address numbers must always match the dip switch settings on the SA85 adapter card.)

Following the procedure in this section, it would be possible to help determine why the SA85 adapter card did not initialize properly, and thus, it would be possible to resolve the error, 'Card SA85:0 not found, bad NCB_LANA_NUM – device not found.'

Troubleshooting the SA85 adapter card under Windows NT

Troubleshooting the cause of the error, 'Card SA85:0 not found, bad NCB_LANA_NUM – device not found' under Windows NT 3.5 or 4.0 is performed differently than under Windows or WFW.

Note: When installing or troubleshooting any communication problems with the Modbus or Modbus Plus I/O server under Windows NT 3.5 or 4.0, make sure that you log on as 'administrator' to ensure that all files are copied and updated as necessary.

During the installation of the Modbus Plus I/O server, updates are added to the Windows NT registry. If you did not log on as 'administrator,' then the registry will not be updated, and thus, the I/O server will not work properly. The I/O server under Windows NT does not rely on the config.sys file during startup of the computer. Instead, during the installation of the I/O server, the file Mbplus.sys is installed into the '\%SystemRoot%\System32\Drivers' directory. Then, once you configure the I/O server's adapter card settings and reboot the computer, the parameters for the Mbplus.sys file are automatically defined and the registry will be updated.

Start up the Windows NT Event Viewer, which is located under the Administrative Tools program group. Verify that Windows NT acknowledged the SA85 adapter card with no errors. (That is, there are not 'STOP' errors for 'MBPLUS'.)

Further, verify that Windows NT acknowledged the SA85 adapter card by double-clicking the Devices program group from the Control Panel window. Then, scroll down the Device list and locate the 'MBPLUS' device. Verify that the status of the device is listed as 'Started' and the startup is defined as 'Automatic'. If these statuses are not listed and the Modbus Plus I/O server attempts to communicate with a slave device, then the error, 'Card SA85:0 not found, bad NCB_LANA_NUM – device not found' will result.

8.2.4   Conclusion

As we have seen, Modbus Plus is a local area network system for industrial control applications. Networked devices can exchange messages for the control and monitoring of processes at remote locations in the industrial plant. Products supporting Modbus Plus communication include programmable controllers and network adapters. The network is also supported by a variety of products from other manufacturers. Each controller can connect to Modbus Plus directly from a port on its front panel.

Additional networks can be accessed through network option modules (NOMs) installed in the common backplane. The network also provides an efficient means for servicing input/output subsystems. Modbus Plus distributed I/O (DIO) drop adapters and terminal block I/O (TIO) modules can be placed at remote I/O sites to allow the application to control field devices over the network link.

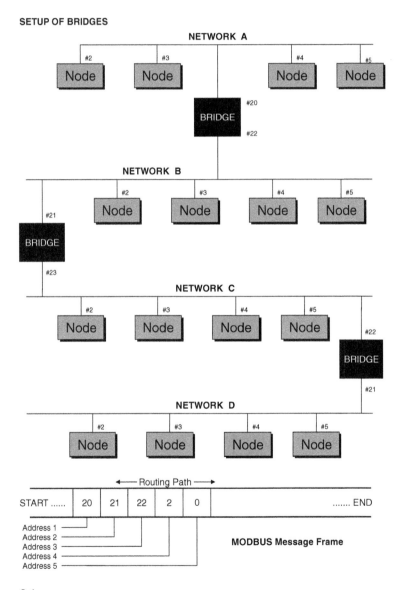

Figure 8.4
Modbus plus routing

With a little care and proper documentation of all related parameters, it would be possible to successfully setup a fully functional Modbus Plus network with ease, as illustrated in the figure above.

9

Data Highway Plus/DH485 overview

Objectives

When you have completed study of this chapter, you will be able to:

- Describe the main features of Data Highway Plus
- Fix problems with:
  - Cabling
  - Grounding/earthing
  - Shielding
  - Terminations
  - Token passing timing

9.1 Allen Bradley Data Highway (Plus) protocol

9.1.1 Overview of Allen Bradley protocol

There are three main protocol standards used in Allen Bradley data communications:

The Data Highway protocol

This is a local area network (LAN) that allows peer-to-peer communications up to 64 nodes. It uses a half-duplex (polled) protocol and rotation of link mastership. It operates at 57.6 kbaud.

The Data Highway Plus protocol

This is similar to the Data Highway network although designed for fewer PCs and operates at a data rate of 57.6 kbaud. This has peer-to-peer communications with a token passing scheme to rotate link mastership among the nodes connected to that link.

Note that both protocol standards implement peer-to-peer communications through a modified token passing system called the floating master. This is a fairly efficient mechanism as each node has an opportunity to become a master at which time it can immediately transmit without checking with each mode for the requisite permission to commence transmission.

The Allen Bradley Data Highway Plus uses the three layers of the OSI layer model:

- Hardware (a physical layer)
- Data link layer protocol
- Application layer

DH-485

This is used by the SLC range of Allen Bradley controllers and is based on RS-485.

9.1.2 Physical layer (hardware layer)

This is based on twin axial cable with three conductors essentially in line with the EIA-485 specifications. The troubleshooting and specifications for RS-485 are covered in an earlier chapter.

9.1.3 Full-duplex data link layer

Note that the asynchronous link can use either a full-duplex (unpolled) protocol or a master–slave communication through a half-duplex (unpolled) protocol. Although both types of protocols are available, the tendency today is to use the full-duplex protocol as this explains the high performance nature of the link. Hence this protocol will be examined in more detail in the following sections.

Full-duplex protocol is character oriented. It uses the ASCII control characters listed in the following table, extended to eight bits by adding a zero for bit number seven (i.e. the eighth bit).

The following ASCII characters are used:

Abbreviation	Hex value
STX	02
ETX	03
ENQ	05
ACK	06
DLE	10
NAK	11

Table 9.1
ASCII control characters

Full-duplex protocol combines these characters into control and data symbols. The following table lists the symbols used for full-duplex implementation.

Symbol	Type	Description
DLE STX	Control symbol	Sender symbol that indicates the start of a message.
DLE ETX BCC/CRC	Control symbol	Sender signal that terminates a message.
DLE ACK	Control symbol	Responses symbol which signals that a message has been successfully received.
DLE NAK	Control symbol	Response symbol which signals that a message was not received successfully.
DLE ENQ	Control symbol	Sender symbol that requests retransmission of a response symbol from the receiver.
APP DATA	Data symbol	Single character data values between 00–OF and 11–FF includes data from the application layer including user programs and common application routines.
DLE DLE	Data symbol	Symbol that represent the data value 10 Hex.

Table 9.2
Symbols used for full-duplex mode

Format of a message

Note that response symbols transmitted within a message packet are referred to as embedded responses.

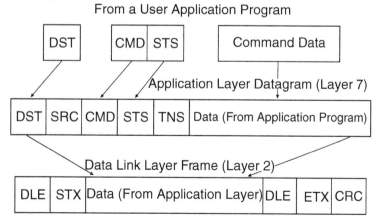

Figure 9.1
Protocol structure

The CRC-16 calculation is done using the value of the application layer data bytes and the ETX byte. The CRC-16 result consists of two bytes.

Note that to transmit the data value of 10H, the sequence of data symbols DLE must be used. Only one of these DLE bytes and no embedded responses are included in the CRC value.

Message limitations

- Minimum size of a valid message is six bytes.
- Duplicate message detection algorithm – receiver compares the second, third, fifth and sixth bytes of a message with the same bytes in the previous message.

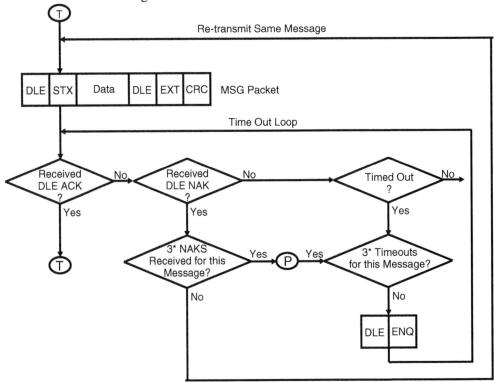

Figure 9.2
Software logic for transmitter

P = Recovery procedure
T = Ready to transmit next message
* = Default values used by module

Depending on the highway traffic and saturation level, there may be a wait for a reply from the remote node before transmitting the next message.

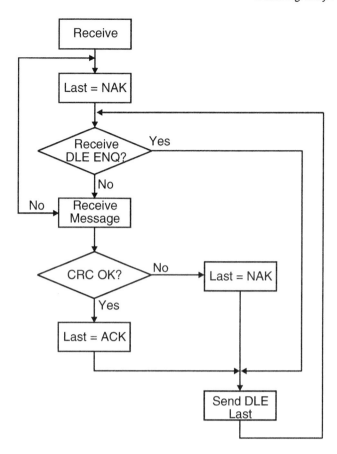

Figure 9.3
Software logic for receivers

Software logic for receivers

The following diagrams show typical events that occur in the communications process.

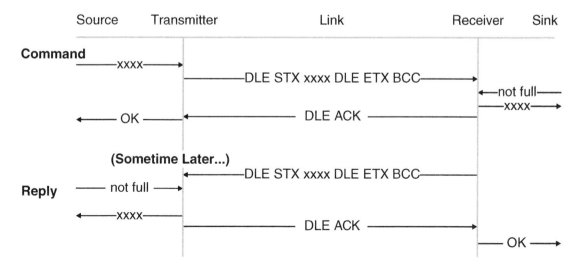

Figure 9.4
Normal message transfer

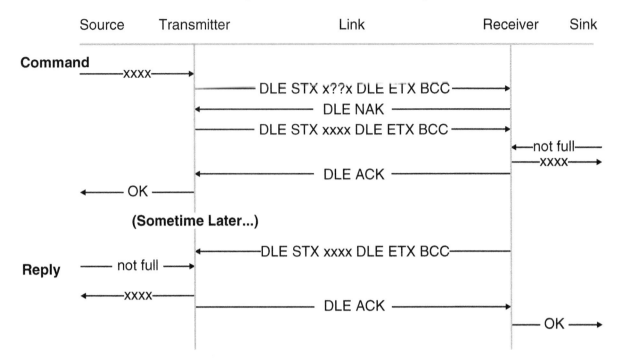

Figure 9.5
Message transfer with NAK

There are two types of application programs:

- Command initiators
- Command executors

Command initiators specify which command function to execute at a particular remote node.

The command executor must also issue a reply message for each command it receives. If the executor cannot execute a command, it must send the appropriate error code.

The reply message may contain an error. The command initiator must check for this condition and depending on the type of error, retransmit the message or notify the user.

If the command executor reply is lost due to noise, the command initiator should maintain a timer for each outstanding command message. If the timer expires, the command initiator should take appropriate action (notify the user or retransmit to executor).

If the application layer software cannot deliver a command message, it should generate a reply message with the appropriate error code and send the reply to the initiator. If it cannot deliver a reply message, the application layer should destroy the reply without notification to the command executor.

Command

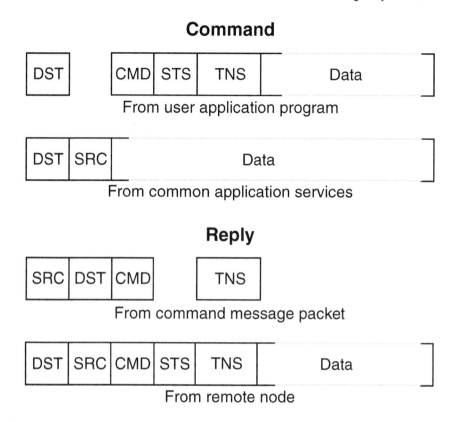

From user application program

From common application services

Reply

From command message packet

From remote node

Figure 9.6

Note that not all command messages have FNC, ADR, SIZE or DATA bytes. Not all reply messages have DATA or ETX STS bytes.

Explanation of bytes

- DST destination byte. This contains the ultimate destination of the node
- SRC source node of the message
- CMD command byte
- FNC function byte

These together define the activity to be performed by the command message at the destination node. Note that bit five of the command byte shall always be zero (normal priority).

STS and ETX SYS – Status and extended status bytes

In command messages, the STS byte is set to zero. In reply messages, the STS byte may contain a status code. If the four high bits of the STS byte are ones, there is extended status information in an ETX STS byte.

TNS – Transaction bytes (two bytes)

The application level software must assign a unique 16-bit transaction number (generated via a counter). When the command initiator receives reply to one of its command messages, it can use the TNS value to associate the reply message with its corresponding command.

Whenever the command executor receives a command from another node, it should copy the TNS field of the command message into the same field of the corresponding reply message.

ADDR

Address field contains the address of a memory location in the command executor where the command is to begin executing. The ADDR field specifies a byte address (not a word address as in PLC programming).

Size

The size byte specifies the number of data bytes to be transferred by a message.

Data

The data field contains binary data from the application programs.

PLC-5 command set message packet fields

- **Packet offset**

    This field contains the offset between the DATA field of the current message packet and the DATA field of the first packet in the transmission.

- **Total trans**

    This field contains the total number of PLC-5 data elements transferred in all message packets initiated by a command.

Basic command set

The asynchronous link message packet formats to be used are delivered below:

In the lists below, privileged commands are initiated by computer and executed by PLCs. A PLC or a computer initiates non-privileged commands. The CMD values listed are for non-priority command message packets.

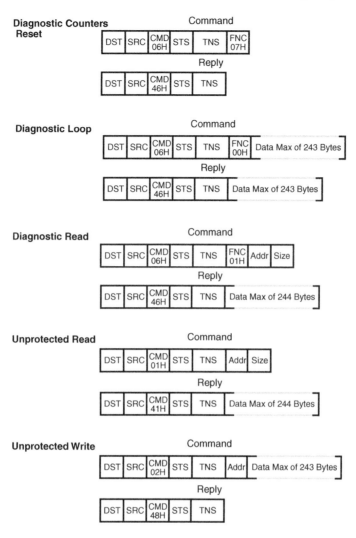

Figure 9.7
Basic command set for PLC-5

Synchronous link status code (STS, ETX STS)

The TS bytes provide information about the execution or failure of the corresponding command that was transmitted from the computer. If the reply returns a code of 00, the command was executed at the remote node. All other codes are divided into two types:

- Local error – local node is unable to transmit a message to the remote node.
- Remote error – remote node is unable to execute the command.

Local STS error code

Code	Description
00	Success – no error
01	Destination node out of buffer space
02	Remote node does not ACK command message
03	Duplicate token holder detected
04	Local port is disconnected

Remote STS error codes

Code	Description
00	Success – no errors
10	Illegal command or format
20	Host has a problem and will not communicate
30	Remote node is missing, disconnected
40	Host could not complete function due to hardware fault
50	Addressing problem or memory protect rungs
60	Function disallowed due to command protection selection
70	Processor is in program mode
80	Compatibility mode file is missing or communication zone problem
A0	Not used
B0	Remote node problems due to download
F0	Error in the ETX STS byte
C0 to E0	Not used

ETX STS Byte

There is only an ETX STS byte if the STS code is F0. If the command code is 00 to 08, there is not an ETX STS byte. Commands used in this implementation are in this range, hence the ETX STS byte is not being used.

Diagnostic counter for each module

Diagnostic counters are bytes of information stored in RAM in each Data Highway and Data Highway Plus module. When using the 'diagnostic read' command, a dummy value should be used for the address. The reply contains the entire counter block.

9.2 Troubleshooting

9.2.1 Introduction

This section is broken down into the following sections:
- Data Highway Plus wiring troubleshooting
- Data Highway Plus network diagnostics

Note that the rules for troubleshooting the physical side of these two cables are very similar to that of RS-485. In fact, DH-485 is identical to RS-485 whilst Data Highway Plus is essentially a transformer isolated version.

The difficult part in diagnosing problems with the Data Highway Plus is in the operation of the protocol.

9.2.2 Data Highway Plus wiring troubleshooting

One should inspect the cable closely for wiring problems if the operation of the network appears intermittent. Typical problems that should be examined are:

- Damage to the cable
- No termination resistance (or removed) at the end of the line
- Screen removed or damaged

A few suggestions to look for on the wiring are described below.

Wiring recommendations

- Ensure that twin axial cable has been used for trunk lines and drop lines (Belden 9463 or equivalent)
- Connect ground wire from station connector to the earth ground
- Route cables away from sources of electrical interference

Station connector wiring

- Connect trunk lines to the upper side of the station connector
- Connect drop lines and earth to the lower connections

Cable preparation

- Cut the drop line to required length
- Remove one inch of insulation, foil, braid and filler cord
- Strip about one eighth of an inch of insulation off each of the insulated wires
- Insert the cable through hood of D-shell connector
- Fit about half an inch of heatshrink tubing over each of the conductors

Solder wires to connector

Solder as following:

- Blue wire to pin 6
- Drain wire to pin 7
- Clear wire to pin 8
- Slide heatshrink over soldered connections and shrink with a heat gun
- Assemble the 15-pin connector to the D-shell hood

Prepare other end of the drop line cable

- Remove about 3 inches of insulation, foil, braid and filler
- Free drain wire, clear and blue insulated wires
- Strip about ¼ inch of insulation off each conductor

Trunk line preparation

Remove cover and terminal block from enclosure and mount enclosure where you want to add the drop line. Cut trunk line cable. Feed incoming and outgoing cables through the respective cable clamps at the top of the enclosure.

Wiring enclosure

- Twist the wires of the same color together from each trunk line cable
- Secure the wires under their respective screw clamp terminals

Note: It is easier to do this with the terminal block out of the enclosure.

Terminating stations

- Install a 150-ohm resistor at the first and last station connectors on the trunk line.
- Use a one inch length of heatshrink to protect resistor and connect between terminals 1 and 3.

9.2.3 Data Highway Plus network diagnostics

There are literally hundreds of errors; but there are a few important ones worth highlighting in this section.

Many of them are the result of excessive noise on the network and can be corrected by examining the actual wiring and removing the source of noise, if possible. If not (for example, due to the data highway running parallel to a power cable in a cable tray), consideration will have to be given to the use of fiber cabling as a replacement for the copper-based cable.

A few errors (which are identified in the diagnostics registers on the interface module) worth mentioning are:

1. ACK Time out. This indicates where a sender has been transmitting packets and has not received any acknowledgement responses. This is often due to long cable or the presence of reflections and low level noise.
2. Contention. This occurs on noisy or over length cables. A similar remedy to the one above.
3. False poll. This indicates the number of times this station has tried to relinquish bus mastership and the station that was expected to take over didn't do so. This is often the case on a noisy highway.
4. Transmitted messages and received messages. While these are not errors, they are worth examining (and recording) to see the level of activity (number of packets sent and received) against the level of errors.
5. Data pin allocation. Not all Data Highway Plus equipment use the same pin configuration on the 15-pin connector. With certain types of Allen Bradley PLC the allocation of pins 6 and 8 are transposed.

10

HART overview

Objectives

When you have completed study of this chapter, you will be able to:

- Describe the fundamental operation of HART
- Fix problems with:
  - Cabling
  - Configuration

10.1 Introduction to HART and smart instrumentation

Smart (or intelligent) instrumentation protocols are designed for applications where actual data is collected from instruments, sensors, and actuators by digital communication techniques. These components are linked directly to programmable logic controllers (PLCs) and computers.

The HART (highway addressable remote transducer) protocol is a typical smart instrumentation fieldbus that can operate in a hybrid 4–20 mA digital fashion.

HART is, by no means, the only protocol in this sphere. There are hundreds of smart implementations produced by various manufacturers – for example, Honeywell – that compete with HART. This chapter deals specifically with HART.

At a basic level, most smart instruments provide core functions such as:

- Control of range/zero/span adjustments
- Diagnostics to verify functionality
- Memory to store configuration and status information (such as tag numbers etc)

Accessing these functions allows major gains in the speed and efficiency of the installation and maintenance process. For example, the time consuming 4–20 mA loop check phase can be achieved in minutes, and the device can be readied for use in the process by zeroing and adjustment for any other controllable aspects such as the damping value.

10.2   HART protocol

This protocol was originally developed by Rosemount and is regarded as an open standard, available to all manufacturers. Its main advantage is that it enables an instrumentation engineer to keep the existing 4–20 mA instrumentation cabling and to use, simultaneously, the same wires to carry digital information superimposed on the analog signal. This enables most companies to capitalize on their existing investment in 4–20 mA instrumentation cabling and associated systems and to add further capability of HART without incurring major costs.

HART is a hybrid analog and digital protocol, as opposed to most fieldbus systems, which are purely digital.

The HART protocol uses the frequency shift keying (FSK) technique based on the Bell 202 communications standard. Two individual frequencies of 1200 and 2200 Hz, representing digits 1 and 0 respectively, are used. The average value of the sine wave (at the 1200 and 2200 Hz frequencies), which is superimposed on the 4–20 mA signal, is zero. Hence, the 4–20 mA analog information is not affected.

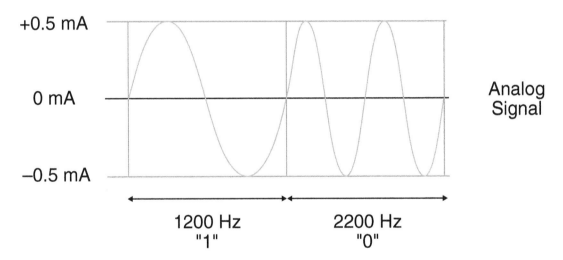

Figure 10.1
Frequency allocation of HART protocol

The HART protocol can be used in three ways:

- In conjunction with the 4–20 mA current signal in point-to-point mode
- In conjunction with other field devices in multidrop mode
- In point-to-point mode with only one field device broadcasting in burst mode

Traditional point-to-point loops use zero for the smart device polling address. Setting the smart device polling address to a number greater than zero creates a multidrop loop. The smart device then sets its analog output to a constant 4 mA and communicates only digitally.

The HART protocol has two formats for digital transmission of data:

- Poll/response mode
- Burst (or broadcast) mode

In the poll/response mode the master polls each of the smart devices on the highway and requests the relevant information. In burst mode the field device continuously transmits process data without the need for the host to send request messages. Although this mode is fairly fast (up to 3.7 times/second), it cannot be used in multidrop networks.

The protocol is implemented with the OSI model using layers 1, 2 and 7. The actual implementation is covered in this chapter.

10.3   Physical layer

The physical layer of the HART protocol is based on two methods of communication.
* Analog 4–20 mA
* Digital frequency shift keying (FSK)

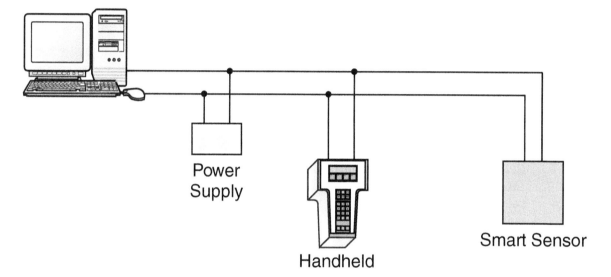

Figure 10.2
HART point-to-point communications

The basic communication of the HART protocol is the 4–20 mA current system. This analog system is used by the sensor to transmit an analog value to the HART PLC or HART card in a PC. In a 4–20 mA system, the sensor outputs a current value somewhere between 4 and 20 mA that represents the analog value of the sensor. For example, a water tank that is half full – say 3400 kiloliters – would put out 12 mA. The receiver would interpret this 12 mA as 3400 kiloliters. This communication is always point-to-point, i.e. from one device to another. It is not possible to do multidrop communication using this method alone. If two or more devices put some current on the line at the same time, the resulting current value would not be valid for either device.

Digital multidrop communications

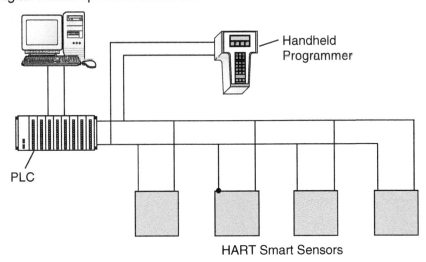

Figure 10.3
HART multi-point communications

For multidrop communications, the HART protocol uses a digital/analog modulation technique known as frequency shift keying (FSK). This technique is based on the Bell 202 communication standard. Data transfer rate is 1200 baud with a digital '0' frequency (2200 Hz) and a digital '1' frequency (1200 Hz). Category 5 shielded, twisted pair wire is recommended by most manufacturers. Devices can be powered by the bus or individually. If the bus powers the devices, only 15 devices can be connected. As the average DC current of an AC frequency is zero, it is possible to place a 1200 Hz or 2200 Hz tone on top of a 4–20 mA signal. The HART protocol does this to allow simultaneous communications on a multidrop system.

The HART handheld communicator

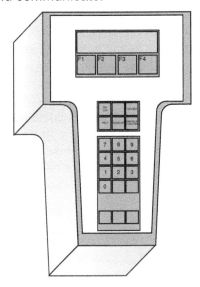

Figure 10.4
HART handheld controller

The HART system includes a handheld control device. This device can be a second master on the system. It is used to read, write, range and calibrate devices on the bus. It can be taken into the field and used for temporary communications. The battery operated handheld has a display and key input for specific commands.

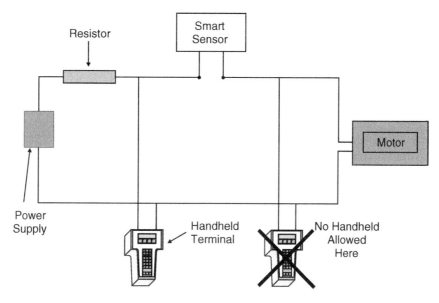

Figure 10.5
HART handheld connection method

The HART field controller in Figure 10.5 is wired in series with the field device (valve positioner or other actuator). In some cases, a bypass capacitor may be required across the terminals of the valve positioner to keep the positioner's series impedance below the 100 ohm level required by HART specifications. Communications with the field controller requires the communicating device (handheld terminal or PC) to be connected across a loop impedance of at least 230 ohm. Communications is not possible across the terminals of the valve positioner because of its low impedance (100 ohm). Instead, the communicating device must be connected across the transmitter or the current sense resistor.

10.4   Data link layer

The data link frame format is shown in Figure 10.7.

	Layer	Description	HART™
7	Application	Serves up formatted data	Hart Commands
6	Presentation	Translates Data	
5	Session	Controls Dialogue	
4	Transport	Ensures Message Integrity	
3	Network	Routes Information	
2	Data Link	Detects Errors	Protocol Rules
1	Physical	Connects Device	Bell 202

Figure 10.6
HART protocol implementation of OSI layer model

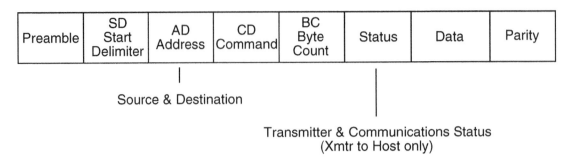

Preamble	SD Start Delimiter	AD Address	CD Command	BC Byte Count	Status	Data	Parity

Source & Destination

Transmitter & Communications Status
(Xmtr to Host only)

Figure 10.7
HART data link frame format

Two-dimensional error checking, including vertical and longitudinal parity checks, is implemented in each frame. Each character or frame of information has the following parameters:

- 1 start bit
- 8 data bits
- 1 odd parity bit
- 1 stop bit

10.5   Application layer

The application layer allows the host device to obtain and interpret field device data. There are three classes of commands:

- Universal commands
- Common practice commands
- Device specific commands

Examples of these commands are listed below.

Universal commands

- Read manufacturer and device type
- Read primary variable (PV) and units
- Read current output and per cent of range
- Read up to 4 predefined dynamic variables
- Read or write 8-character tag, 16-character descriptor, date
- Read or write 32 character message
- Read device range, units and damping time constant
- Read or write final assembly number
- Write polling address

Common practice commands

- Read selection of up to 4 dynamic variables
- Write damping time constant
- Write device range
- Calibrate (set zero, set span)
- Set fixed output current
- Perform self-test
- Perform master reset

- Trim pv zero
- Write PV units
- Trim DAC zero and gain
- Write transfer function (square root/linear)
- Write sensor serial number
- Read or write dynamic variable assignments

Instrument specific commands

- Read or write low flow cut-off value
- Start, stop or clear totalizer
- Read or write density calibration factor
- Choose PV (mass flow or density)
- Read or write materials or construction information
- Trim sensor calibration

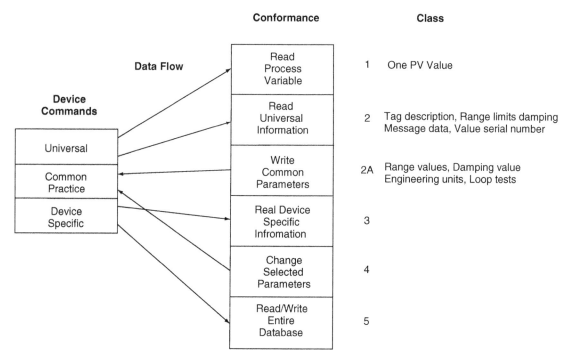

Figure 10.8
HART application layer implementation

Summary of HART benefits

- Simultaneous analog and digital communications
- Allows other analog devices on the highway
- Allows multiple masters to control same smart instrument
- Multiple smart devices on the same highway
- Long distance communications over telephone lines
- Two alternative transmission modes
- Flexible messaging structure for new features
- Up to 256 process variables in any smart field device

Hardware recommendations

Minimum cable size:	24 AWG, (0.51-mm diameter)
Cable type:	Single-pair shielded or multiple-pair with overall shield

10.6   Troubleshooting

Beside the actual instruments that require calibration, the only major problem that can occur with HART is the cable length calculation.

The HART protocol is designed to work over existing analog signal cables but the reliable length of cable depends on:

- Loop load resistance
- Cable resistance
- Cable capacitance
- Number and capacitance of field devices
- Resistance and position of other devices in the loop

The main reason for this is that network must pass the HART signal frequencies (1200 Hz and 2200 Hz) without excessive loss or distortion. A software package such as H-Sim can be used to calculate whether you are operating with the correct signal level. In addition, you should ensure that you have the correct bandwidth of at least 2500 Hz. You can do this by ensuring that the product of the cable resistance and capacitance is less than 65 microseconds.

11

AS-interface (AS-i) overview

Objectives

When you have completed study of this chapter, you will be able to:

- Describe the main features of AS-i
- Fix problems with:
  - Cabling
  - Connections
  - Gateways to other standards

11.1 Introduction

The actuator sensor interface is an open system network developed by eleven manufacturers. These manufacturers created the AS-i association to develop the AS-i specifications. Some of the more widely known members of the association include Pepperl-Fuchs, Allen-Bradley, Banner Engineering, Datalogic Products, Siemens, Telemecanique, Turck, Omron, Eaton and Festo. The governing body is ATO, the AS-i Trade Organization. The number of ATO members currently exceeds fifty and continues to grow. The ATO also certifies that products under development for the network meet the AS-i specifications. This will assure compatibility between products from different vendors.

AS-i is a bit-oriented communication link designed to connect binary sensors and actuators. Most of these devices do not require multiple bytes to adequately convey the necessary information about the device status, so the AS-i communication interface is designed for bit-oriented messages in order to increase message efficiency for these types of devices.

The AS-i interface is just that, an interface for binary sensors and actuators, designed to interface binary sensors and actuators to microprocessor-based controllers using bit-length 'messages.' It was not developed to connect intelligent controllers together since this would be far beyond the limited capability of bit-length message streams.

Modular components form the central design of AS-i. Connection to the network is made with unique connecting modules that require minimal, or in some cases no, tools d provide for rapid, positive device attachment to the AS-i flat cable. Provision is made in the communications system to make 'live' connections, permitting the removal or addition of nodes with minimum network interruption.

Connection to higher level networks (e.g. ProfiBus) is made possible through plug-in PC and PLC cards or serial interface converter modules.

The following sections examine these features of the AS-i network in more detail.

11.2   Layer 1 – the physical layer

AS-i uses a two-wire untwisted, unshielded cable that serves as both communication link and power supply for up to thirty-one slaves. A single master module controls communication over the AS-i network, which can be connected in various configurations such as bus, ring, or tree. The AS-i flat cable has a unique cross-section that permits only properly polarized connections when making field connections to the modules. Alternatively, ordinary 2-wire cable (#16 AWG, 1.5 mm) can be used. A special shielded cable is also available for high noise environments.

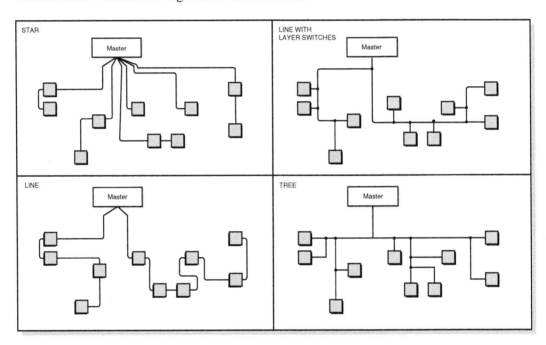

Figure 11.1
Various AS-i configurations

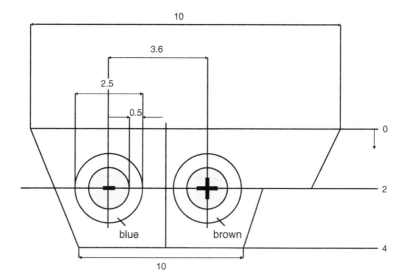

Figure 11.2
Cross-section of AS-i cable (mm)

Each slave is permitted to draw a maximum of 65mA from the 30 V DC-power supply. If devices require more than this, separate supplies must be provided for each device. With a total of 31 slaves drawing 65 mA, a total limit of 2 A has been established to prevent excessive voltage drop over the 100 m permitted network length. A 16 AWG cable is specified to insure this condition. If this limitation on power drawn from the (yellow) signal cable is a problem, then a second (black) cable, identical in dimensions to the yellow cable, can be used in parallel for power distribution only.

The slave (or field) modules are available in four configurations:

- Input modules for 2- and 3-wire DC sensors or contact closure
- Output modules for actuators
- Input/output (I/O) modules for dual purpose applications
- Field connection modules for direct connection to AS-i compatible devices
- 12-bit analog to digital converter

The original AS-i specification (V2) allowed for 31 devices per segment of cable, with a total of 124 digital inputs and 124 digital outputs that is, a total of 248 I/O points. The latest specification, V2.1, allows for 62 devices, resulting in 248 inputs and 186 outputs, a total of 434 I/O points. With the latest specification, even 12-bit A to D converters can be read over 5 cycles.

A unique design allows the field modules to be connected directly into the bus while maintaining network integrity. The field module is composed of an upper and lower section, secured together once the cable is inserted. Specially designed contact points pierce the self-sealing cable providing bus access to the I/O points and/or continuation of the network. True to the modular design concept, two types of lower sections and three types of upper sections are available to permit 'mix-and-match' combinations to accommodate various connection schemes and device types. Plug connectors are utilized to interface the I/O devices to the slave (or with the correct choice of modular section screw terminals) and the entire module is sealed from the environment with special seals provided where the cable enters the module. The seals conveniently store away within the module when not in use.

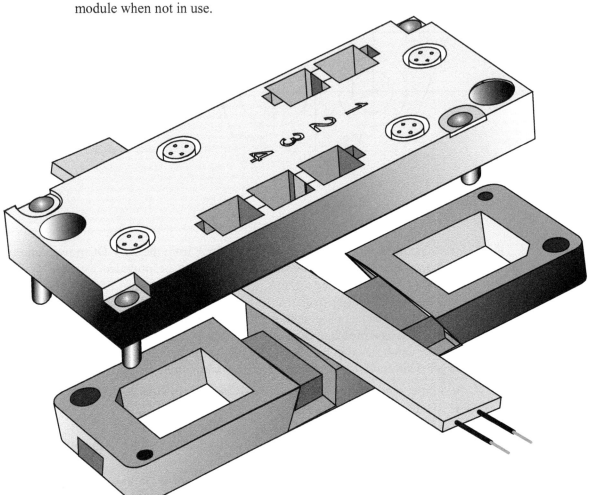

Figure 11.3
Connection to the cable

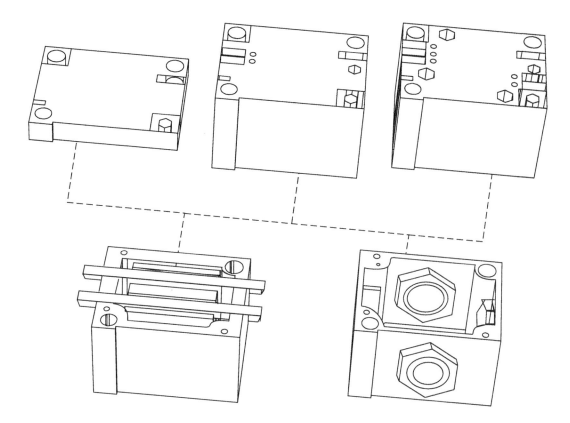

Figure 11.4
Various connection schemes and device types

The AS-i network is capable of a transfer rate of 167 kbps. Using an access procedure known as 'master–slave access with cyclic polling', the master continually polls all the slave devices during a given cycle to ensure rapid update times. For example, with 31 slaves and 124 I/O points connected, the AS-i network can ensure a 5 ms-cycle time, making the AS-i network one of the fastest available.

A modulation technique called 'alternating pulse modulation' provides this high transfer rate capability as well as high data integrity. This technique will be described in the following section.

11.3   Layer 2 – the data link layer

The data link layer of the AS-i network consists of a master call-up and slave response. The master call-up is exactly fourteen bits in length while the slave response is 7 bits. A pause between each transmission is used for synchronization. Refer to the following figure, for example, call-up and answer frames.

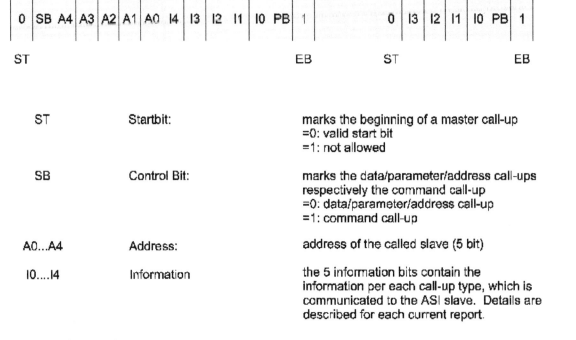

Figure 11.5
Example call-up and response frames

Various code combinations are possible in the information portion of the call-up frame and it is precisely these various code combinations that are used to read and write information to the slave devices. Examples of some of the master call-ups are listed in the following figure. A detailed explanation of these call-ups is available from the ATO literature and is only included here to illustrate the basic means of information transfer on the AS-i network.

The modulation technique used by AS-i is known as 'alternating pulse modulation' (APM). Since the information frame is of a limited size, providing conventional error checking was not possible and therefore the AS-i developers chose a different technique to insure a high level of data integrity.

Referring to the following figure, the coding of the information is similar to Manchester II coding but utilizing a 'sine squared' waveform for each pulse. This waveform has several unique electrical properties, which reduce the bandwidth required of the transmission medium (permitting faster transfer rates) and reduce the end of line reflections common in networks using square wave pulse techniques. Also, notice that each bit has an associated pulse during the second half of the bit period. This property is utilized as a bit level of error checking by all AS-i devices. The similarity to Manchester II coding is no accident since this technique has been used for many years to pass synchronizing information to a receiver along with the actual data.

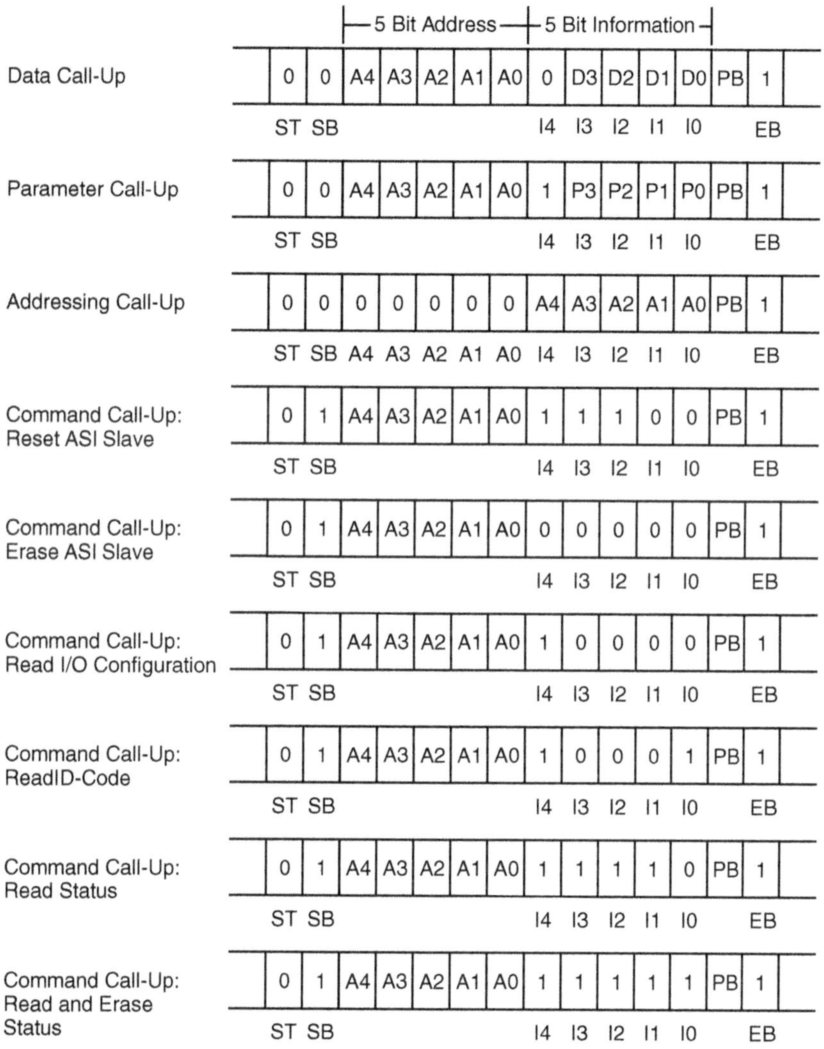

	5 Bit Address	5 Bit Information	

Data Call-Up

0	0	A4	A3	A2	A1	A0	0	D3	D2	D1	D0	PB	1

ST SB   I4 I3 I2 I1 I0   EB

Parameter Call-Up

0	0	A4	A3	A2	A1	A0	1	P3	P2	P1	P0	PB	1

ST SB   I4 I3 I2 I1 I0   EB

Addressing Call-Up

0	0	0	0	0	0	0	A4	A3	A2	A1	A0	PB	1

ST SB A4 A3 A2 A1 A0 I4 I3 I2 I1 I0   EB

Command Call-Up:
Reset ASI Slave

0	1	A4	A3	A2	A1	A0	1	1	1	0	0	PB	1

ST SB   I4 I3 I2 I1 I0   EB

Command Call-Up:
Erase ASI Slave

0	1	A4	A3	A2	A1	A0	0	0	0	0	0	PB	1

ST SB   I4 I3 I2 I1 I0   EB

Command Call-Up:
Read I/O Configuration

0	1	A4	A3	A2	A1	A0	1	0	0	0	0	PB	1

ST SB   I4 I3 I2 I1 I0   EB

Command Call-Up:
ReadID-Code

0	1	A4	A3	A2	A1	A0	1	0	0	0	1	PB	1

ST SB   I4 I3 I2 I1 I0   EB

Command Call-Up:
Read Status

0	1	A4	A3	A2	A1	A0	1	1	1	1	0	PB	1

ST SB   I4 I3 I2 I1 I0   EB

Command Call-Up:
Read and Erase
Status

0	1	A4	A3	A2	A1	A0	1	1	1	1	1	PB	1

ST SB   I4 I3 I2 I1 I0   EB

Figure 11.6
Some AS-i call-ups

In addition, AS-i developers also established a set of rules for the APM coded signal that is used to further enhance data integrity. For example, the start bit or first bit in the AS-i telegram must be a negative impulse and the stop bit a positive impulse. Two subsequent impulses must be of opposite polarity and the pause between two consecutive impulses should be 3 microseconds. Even parity and a prescribed frame length are also incorporated at the frame level. As a result the 'odd' looking waveform, in combination with the rules of the frame formatting, the set of APM coding rules and parity checking, work together to provide timing information and high level data integrity for the AS-i network.

11.4   Operating characteristics

AS-i node addresses are stored in non-volatile memory and can be assigned either by the master or one of the addressing or service units. Should a node fail, AS-i has the ability to automatically reassign the replaced node's address and, in some cases, reprogram the node itself allowing rapid response and repair times.

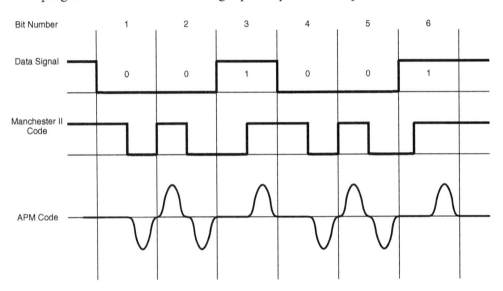

Figure 11.7
Sine squared wave form

Since AS-i was designed to be an interface between lower level devices, connection to higher level systems enables the capability to transfer data and diagnostic information. Plug-in PC cards and PLC cards are currently available. The PLC cards allow direct connection with various Siemens PLCs. Serial communication converters are also available to enable AS-i connection to conventional EIA-232, 422, and 485 communication links. Direct connection to a Profibus field network is also possible with the Profibus coupler, enabling several AS-i networks access to a high-level digital network.

Handheld and PC-based configuration tools are available, which allow initial startup programming and also serve as diagnostic tools after the network is commissioned. With these devices, on-line monitoring is possible to aid in determining the health of the network and locating possible error sources.

11.5   Troubleshooting

11.5.1   Introduction

The AS-i system has been designed with a high degree of 'maintenance friendliness' in mind and has a high level of built-in auto-diagnosis. The system is continuously monitoring itself against faults such as:

- Operational slave errors (permanent or intermittent slave failure, faulty configuration data such as addresses, I/O configuration, and ID codes)
- Operational master errors (permanent or intermittent master failure, faulty configuration data such as addresses, I/O configuration, and ID codes)
- Operational cable errors (short circuits, cable breakage, corrupted telegrams due to electrical interference and voltage outside of the permissible range)
- Maintenance related slave errors (false addresses entered, false I/O configuration, false ID codes)
- Maintenance related master errors (faulty projected data such as I/O configuration, ID codes, parameters etc)
- Maintenance related cable errors (counter poling the AS-i cable)

The fault diagnosis is displayed by means of LEDs on the master.

Where possible, the system will protect itself. During a short-circuit, for example, the power supply to the slaves is interrupted, which causes all actuators to revert to a safe state. Another example is the jabber control on the AS-i chips, whereby a built-in fuse blows if too much current is drawn by a chip, disconnecting it from the bus.

The following tools can be used to assist in fault finding.

11.5.2   Tools of the trade

11.5.2.1   Addressing handheld

Before an AS-i system can operate, all the operating addresses must be assigned to the connected slaves, which store this on their internal non-volatile memory (EEPROM). Although this can theoretically be done on-line, it requires that a master device with this addressing capability be available.

In the absence of such a master, a specialized battery powered addressing handheld (for example, one manufactured by Pepperl and Fuchs) can be used. The device is capable of reading the current slave address (from 0 to 31) as well as reprogramming the slave to a new address entered via the keyboard.

The slaves are attached to the handheld device, one at a time, by means of a special short cable. They are only powered via the device while the addressing operation takes place (about 1 second) with the result that several hundred slaves can be configured in this way before a battery change is necessary.

11.5.2.2   Monitor

A monitor is essentially a protocol analyzer, which allows a user to capture and analyze the telegrams on the AS-i bus. A good monitor should have triggering and filtering capabilities, as well as the ability to store, retrieve and analyze captured data. Monitors are usually implemented as PC-based systems.

11.5.2.3   Service device

An example of such a device is the SIE 93 handheld manufactured by Siemens. It can perform the following:

- Slave addressing, as described above.
- Monitoring, i.e. the capturing, analysis and display of telegrams.
- Slave simulation, in which case it behaves like a supplementary slave, the user can select its operating address.

- Master simulation, in which case the entire cycle of master requests can be issued to test the parameters, configuration and address of a specific slave device (one at a time).

11.5.2.4   Service book

A 'service book' is a commissioning and servicing tool based on a notebook computer. It is capable of monitoring an operating network, recording telegrams, detecting errors, addressing slaves off-line, testing slaves off-line, maintaining a database of sensor/ actuator data and supplying help functions for user support.

Bus data for use by the software on the notepad is captured, preprocessed and forwarded to the laptop by a specialized network interface, a so-called 'hardware checker'. The hardware checker is based on an 80C535 single chip micro-controller and connects to the notepad via an EIA-232 interface.

11.5.2.5   Slave simulator

Slave simulators are PC-based systems used by software developers to evaluate the performance of a slave (under development) in a complete AS-i network. They can simulate the characteristics of up to 32 slaves concurrently and can introduce errors that would be difficult to setup in real situations.

12

DeviceNet overview

Objectives

When you have completed study of this chapter you will be able to:

- List the main features of DeviceNet
- Identify and correct problems with:
  - Cable topology
  - Power and earthing
  - Signal voltage levels
  - Common mode voltages
  - Terminations
  - Cabling
  - Noise
  - Node communications problems

12.1 Introduction

DeviceNet, developed by Allen Bradley, is a low-level device oriented network based on the CAN (controller area network) developed by Bosch (GmbH) for the automobile industry. It is designed to interconnect lower level devices (sensors and actuators) with higher level devices (controllers).

The variable, multi-byte format of the CAN message frame is well suited to this task as more information can be communicated per message than with bit-type systems. The Open DeviceNet Vendor Association, Inc. (ODVA) has been formed to issue DeviceNet specifications, ensure compliance with the specifications and offer technical assistance for manufacturers wishing to implement DeviceNet. The DeviceNet specification is an open specification and available through the ODVA.

DeviceNet can support up to 64 nodes, which can be removed individually under power and without severing the trunk line. A single, four-conductor cable (round or flat) provides both power and data communications. It supports a bus (trunk line drop line) topology, with branching allowed on the drops. Reverse wiring protection is built into all nodes, protecting them against damage in the case of inadvertent wiring errors.

The data rates supported are 125, 250 and 500 k baud (i.e. bits per second in this case), although a specific installation does not have to support all as data rates can be traded for distance.

As DeviceNet was designed to interface lower level devices with higher level controllers, a unique adaptation of the basic CAN protocol was developed. This is similar to the familiar poll/response or master/slave technique but still utilizes the speed benefits of the original CAN.

Figure 12.1 below illustrates the positioning of DeviceNet and CANBUS within the OSI model. Note that DeviceNet only implements layers 1, 2 and 7 of the OSI model. Layers 1 and 2 provide the basic networking infrastructure, whilst layer 7 provides an interface for the application software. Due to the absence of layers 3 and 4, no routing and end-to-end control is possible.

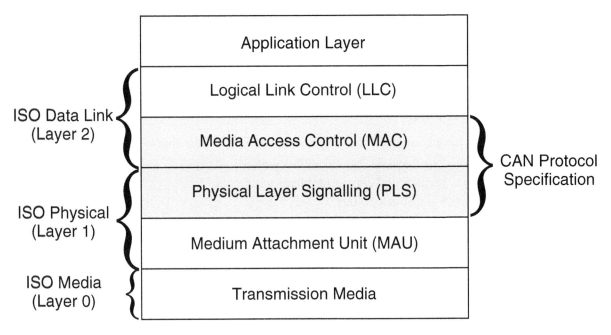

Figure 12.1
DeviceNet vs the OSI model

12.2   Physical layer

12.2.1   Topology

The DeviceNet media consists of a physical bus topology. The bus or 'trunk' (white and blue wires) is the backbone of the network and must be terminated at either end by a 120 ohm ¼ W resistor.

Drop lines of up to 6 meters (20 feet) in length enable the connection of nodes (devices) to the main trunk line, but care is to be taken not to exceed the total drop line budget for a specific speed. Branching to multiple nodes is allowed only on drop lines.

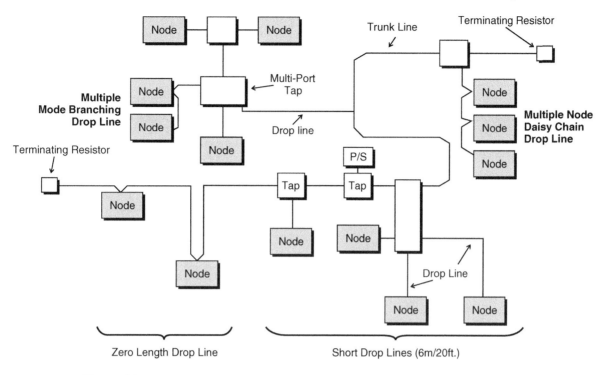

Figure 12.2
DeviceNet topology

Three types of cable are available, all of which can be used as the trunk. They are thick, thin and flat wire.

12.3 Connectors

DeviceNet has adopted a range of open and closed connectors that are considered suitable for connecting equipment onto the bus and drop lines. This range of recommended connectors is described in this section. DeviceNet users can connect to the system using other proprietary connectors, the only restrictions placed on the user regarding the types of connectors used are as follows:

- All nodes (devices), whether using sealed or unsealed connections, supplying or consuming power, must have male connectors.
- Whatever connector is chosen, it must be possible for the related device to be connected or disconnected from the DeviceNet bus without compromising the system's operation.
- Connectors must be rated to carry high levels (8 amps or more at 24 volts, or 200 VA) of current.
- A minimum of 5 isolated connector pins is required, with the possible requirement of a 6th, or metal body shield connection for safety ground use.

There are two basic styles of DeviceNet connectors that are used for bus and drop line connections in normal, harsh, and hazardous conditions. These are:

- An open style connector (pluggable or hardwired), and
- A closed style connector (mini or micro style)

12.3.1 Pluggable (unsealed) connector

This is a 5-pin, unsealed open connector utilizing direct soldering, crimping, screw terminals, barrier strips or screw-type terminations. This type of connector entails removing system power for connection.

Important: DeviceNet requires that connectors on *devices* must have male contacts

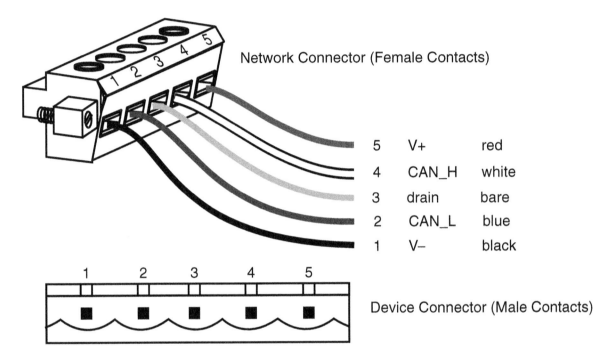

5	V+	red
4	CAN_H	white
3	drain	bare
2	CAN_L	blue
1	V–	black

Network Connector (Female Contacts)

Device Connector (Male Contacts)

Figure 12.3
Unsealed screw connector

12.3.2 Hardwired (unsealed) connection

Loose wire connections can be used to make direct attachment to a node or a bus tap without the presence of a connector, although this is not a preferred method. It is only a viable option if the node can be removed from the trunk line without severing the trunk.

The ends of the cable are 'live' if the cable has been removed from the node in question and are still connected as part of the bus infrastructure. As such, care MUST be exercised to insulate the exposed ends of the cable.

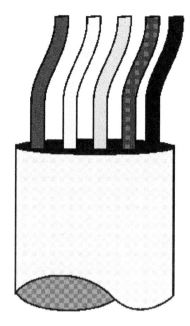

Important: This hard-wired solution is an option, provided that the node can be removed from the network without severing the trunk.

Important: Wires should not be installed while the network is active. This will prevent problems such as shorting the network supply or disrupting communications.

Figure 12.4
Open wire connection

12.3.3   Mini (sealed) connector

This 18mm round connector is recommended for harsh environments (field connections). This connection must meet ANSI/B93.55M-1981. The female connector (attached to the bus cable) must have rotational locking. This connector requires a minimum voltage rating of 25 volts, and for trunk use a current rating of 8 amps is required. Additional options can include oil and water resistance.

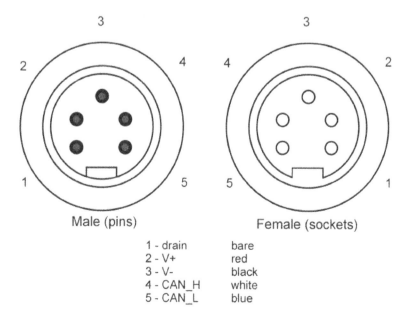

Figure 12.5
Sealed mini-type connector (face views)

12.3.4 Micro (sealed) connector

This connector is effectively a 12 mm diameter miniature version of the mini style connector, except its suitability is for thin wire drop connections requiring reduction in both physical and current carrying capacity.

It has 5 pins, 4 in a circular periphery pattern and the fifth pin in the center. This connector should have a minimum voltage rating of 25 volts, and for drop connections a current rating of 3 amps is required. The male component must mate with Lumberg style RST5-56/xm or equivalent, the female component part must also conform to Lumberg style RST5-56/xm or equivalent. Additional options can include oil and water resistance.

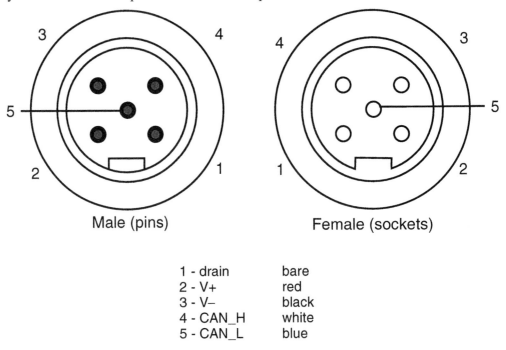

Male (pins)   Female (sockets)

1 - drain   bare
2 - V+   red
3 - V–   black
4 - CAN_H   white
5 - CAN_L   blue

Figure 12.6
Sealed microstyle connector (face views)

12.4 Cable budgets

DeviceNet's transmission media can be constructed of either DeviceNet thick, thin or flat cable or a combination thereof. Thick or flat cable is used for long distances and is stronger and more resilient than the thin cable, which is mainly used as a local drop line connecting nodes to the main trunk line.

The trunk line supports only tap or multiport taps that connect drop lines into the associated node. Branching structures are allowed only on drop lines and not on the main trunk line.

The following tables show the distance vs length trade-off for the different types of cable.

DATA RATES	125 Kbaud	250 Kbaud	500 Kbaud
Trunk Distance	500m (1640ft)	250m (830ft)	100m (328ft)
Max. Drop Length	20ft	20ft	20ft
Cumulative Drop	512ft	256ft	128ft
Number of Nodes	64	64	64

Figure 12.7
Constraints: Thick wire

DATA RATES	125 Kbaud	250 Kbaud	500 Kbaud
Trunk Distance	100m (328ft)	100m (328ft)	100m (328ft)
Max. Drop Length	20ft	20ft	20ft
Cumulative Drop	512ft	256ft	128ft
Number of Nodes	64	64	64

Figure 12.8
Constraints: Thin wire

DATA RATES	125 Kbaud	250 Kbaud	500 Kbaud
Trunk Distance	420m (1378ft)	200m (656ft)	75m (246ft)
Max. Drop Length	20ft	20ft	20ft
Cumulative Drop	512ft	256ft	128ft
Number of Nodes	64	64	64

Figure 12.9
Constraints: Flat wire

12.5 Device taps

12.5.1 Sealed taps

Sealed taps are available in single port (T type) and multiport configurations. Regardless of whether the connectors are mini or micro style, DeviceNet requires that male connectors must have external threads while female connectors must have internal threads. In either case, the direction of rotation is optional.

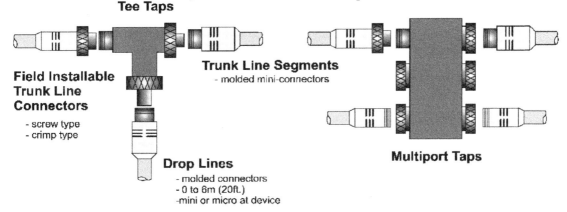

Figure 12.10
Sealed taps

12.5.2   IDC taps

IDCs (insulation displacement connectors) are used for KwikLink flat cable. They are modular, relatively inexpensive and compact. They are compatible with existing media and require little installation effort. The enclosure conforms to NEMA 6P and 13, and IP 67.

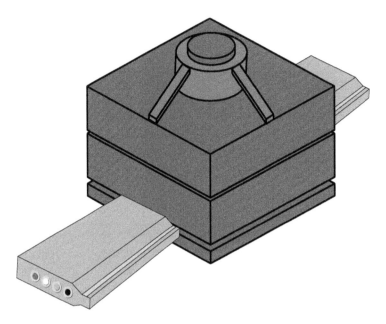

Figure 12.11
Insulation displacement connector

12.5.3   Open style taps

DeviceNet has three basic forms of open taps. They are:

- Zero length drop line, suitable for daisy-chain applications.
- Open tap, able to connect a 6 meter (20 foot) drop line onto the trunk.
- An open style connector, supporting 'temporary' attachment of a node to a drop line.

The temporary connector is suitable for connection both to and from the system when the system is powered. It is of similar construction to a standard telephone wall plug, being of molded construction and equipped with finger grips to assist removal, and is styled as a male pin-out. The side cheeks are polarized to prevent reversed insertion into the drop line open tap connector.

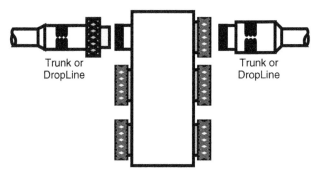

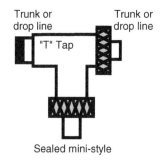

Sealed mini-style

Sealed multi-port tap
with connectors for four drop lines

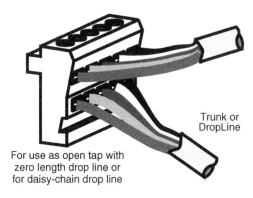

For use as open tap with
zero length drop line or
for daisy-chain drop line

Trunk or
DropLine

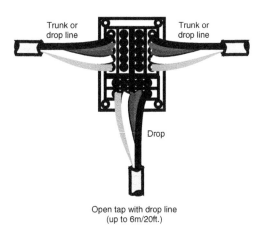

Open tap with drop line
(up to 6m/20ft.)

Figure 12.12
Open and temporary DeviceNet taps

12.5.4 Multiport open taps

If a number of nodes or devices are located within a close proximity of each other, e.g. within a control cabinet or similar enclosure, an open tap can be used.

Alternatively, devices can be wired into a DeviceBox multiport tap. The drops from the individual devices are not attached to the box via sealed connectors but are fed in via cable grips and connected to a terminal strip.

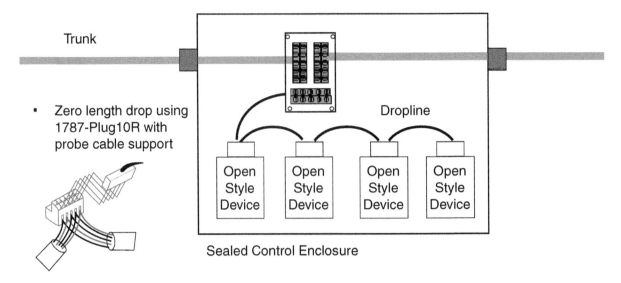

Figure 12.13
Multiport open taps

12.5.5   Power taps

Power taps differ from device taps in that they have to perform four essential functions that are not specifically required by the device taps. These include:

- Two protection devices in the V+ supply
- Connection from the positive output of the supply to the V+ bus line via a Schottky diode
- Provision of a continuous connection for the signaling pair, drain and negative wires through the tap
- Provision of current limiting in both directions from the tap

The following figure illustrates the criteria listed above.

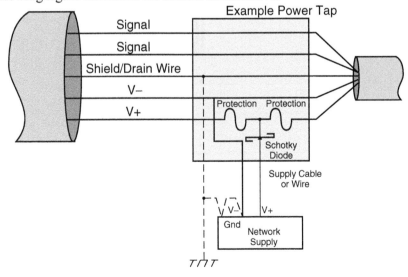

Figure 12.14
Principle of a DeviceNet power tap

12.6   Cable description

The 'original' (round) DeviceNet cable has two shielded twisted pairs. These are twisted on a common axis with a drain wire in the center, and are equipped with an overall braid.

12.6.1   Thick cable

This cable is used as the trunk line when length is important. Overall diameter is 0.480 inches (10.8 mm) and it comprises of:

- A signal pair, consisting of one twisted pair (3 twists per foot) coded blue/white with a wire size of #18 (19 × 30 AWG) copper and individually tinned; the impedance is 120 ohms ± 10% at 1 MHz, the capacitance between conductors is 12 pF/foot and the propagation delay is 1.36 nS/foot maximum
- A power pair, consisting of one twisted pair (3 twists per foot) coded black/red with a wire size of #15 (19 × 28 AWG) copper and individually tinned

This is completed by separate aluminized Mylar shields around each pair and an overall foil/braided shield with an #18 (19 × 30 AWG) bare drain wire. The power pair has an 8 amp power capacity and is PVC/nylon insulated. It is also flame resistant and UL oil resistant II.

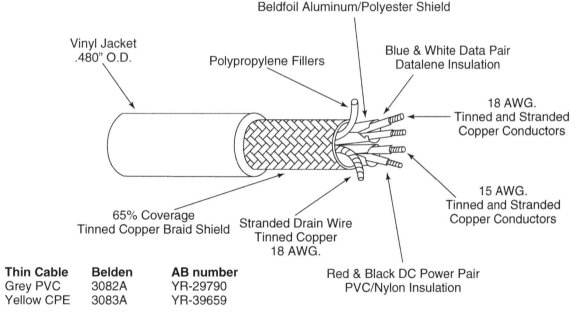

Thin Cable	Belden	AB number
Grey PVC	3082A	YR-29790
Yellow CPE	3083A	YR-39659

Figure 12.15
DeviceNet thick cable

12.6.2   Thin cable specification

This cable is used for both drop lines as well as short trunk lines. Its overall diameter is 0.27 inches (6.13 mm) and it comprises of:

- A signal pair, consisting of a twisted pair (4.8 twists per foot) coded blue/white with a wire size of #24 (19 × 36 AWG) copper and individually tinned; the impedance is 120 ohms ± 10% at 1 MHz, the capacitance

between conductors is 12 pF/Foot and the propagation delay is 1.36 nS /foot maximum
- A power pair consisting of one twisted pair (4.8 twists per foot) coded black/red with a wire size of #22 (19 × 34 AWG) copper and individually tinned

This is completed by separate aluminized Mylar shields around each pair and an overall foil/braided shield with an #22 (19 × 34 AWG) bare drain wire. The power pair has a 3 amp power capacity and is PVC insulated.

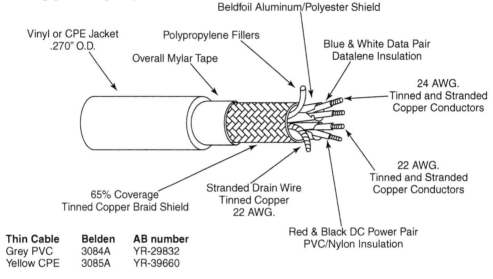

Thin Cable	Belden	AB number
Grey PVC	3084A	YR-29832
Yellow CPE	3085A	YR-39660

Figure 12.16
DeviceNet thin cable

12.6.3   Flat cable

DeviceNet flat cable is a highly flexible cable that works with existing devices. It has the following specifications:
- 600 V 8 A rating
- A physical key
- Fitting into 1 inch (25 mm) conduit
- The jacket made of TPE/Santoprene

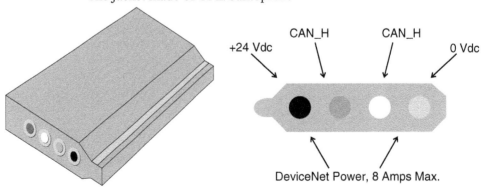

Figure 12.17
DeviceNet thin cable

12.7 Network power

12.7.1 General approach

One or more 24 V power supplies can be used to power the devices on the DeviceNet network, provided that the 8 A current limit on thick/flat wire and the 3 A limit on thin wire is not exceeded. The power supplies used should be dedicated to the DeviceNet cable power ONLY!

Although, technically speaking, any suitable power supply can be used, supplies such as the Rockwell Automation 1787-DNPS 5.25A supply are certified specifically for DeviceNet.

The power calculations can be done by hand, but it is easier to use a design spreadsheet such as the Rockwell Automation/Allen Bradley DeviceNet Power Supply Configuration toolkit running under Microsoft Excel.

The network can be constructed of both thick and thin cable as long as only one type of cable is used per section of network, comprising a section between power taps or between a power tap and the end of the network.

Using the steps illustrated below, a quick initial evaluation can be achieved as to the power requirements needed for a particular network.

Sum the total current requirements of all network devices, then evaluate the total permissible network length (be conservative here) using the following table:

Thick cable network current distribution and allowable current loading												
Network length in meters	0	25	50	100	150	200	250	300	350	400	450	500
Network length in feet	0	83.3	167	333	500	666	833	999	1166	1332	1499	1665
Maximum current in amps	8	8	5.42	2.93	2.01	1.53	1.23	1.03	0.89	0.78	0.69	0.63

Thin cable network current distribution and allowable current loading											
Network length in meters	0	10	20	30	40	50	60	70	80	90	100
Network length in feet	0	33	66	99	132	165	198	231	264	297	330
Maximum current in amps	8	8	5.42	2.93	2.01	1.53	1.23	1.03	0.89	0.78	0.69

Table 12.1
Thick and thin cable length and power capacity

Depending on the final power requirement cost and network complexity, either a single supply end or center connected can be used.

12.7.2 Single supply – end connected

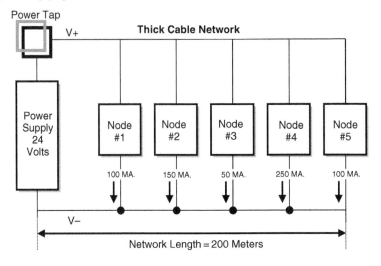

Figure 12.18
Single supply – end connected

Total network length = 200 meters (656 feet)
Total current = Sum of node 1, 2, 3, 4 and 5 currents = 0.65 amps
Referring to Table 12.1 the current limit for 200 meters = 1.53 amps
Configurations are correct as long as THICK cable is used.

12.7.3 Single supply – center connected

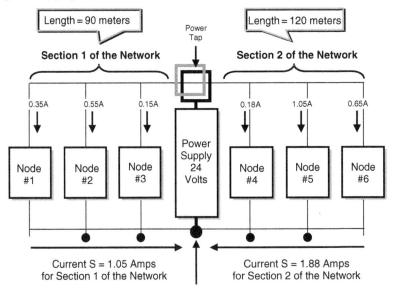

Figure 12.19
Single supply – center connected

Current in section 1 = 1.05 amps over a length of 90 meters (300 feet)

Current in section 2 = 1.88 amps over a length of 120 meters (400 feet)

Current limits for a distance of 90 meters is 3.3 amps and for 120 meters is 2.63 amps. Power for both sections is correct, and a 3 amp (minimum) power supply is required.

The following table indicates parameters that control load limits and allowable tolerances as related to DeviceNet power.

System power load limits	
Max. voltage drop on both the −Ve and +Ve power lines	5.0 volts on each line
Maximum thick cable trunk line current	8.0 amps in any section
Maximum thin cable trunk line current	3.0 amps in any section
Maximum drop line current	0.75 to 3.0 amps
Voltage range at each node	11.0 to 25.0 volts
Operating current on each product	Specified by the product manufacturer

Table 12.2
System power load limits

Maximum drop line currents	
Current limits are calculated by the following equations, where I = allowable drop line current and L = distance. In Meters: I = 4.57/L In Feet I = 15/L	
Drop length	**Maximum allowable current**
1.00	3.00
3.00	3.00
5.00	3.00
7.50	2.00
10.00	1.50
15.00	1.00
20.00	0.75

Table 12.3
Maximum drop line currents

12.7.4 Suggestions for avoiding errors and power supply options

The following steps can be used to minimize errors when configuring power on a network.

- Ensure the calculations made for current and distances are correct (be conservative)
- Conduct a network survey to verify correct voltages, remembering that a minimum of 11 volts at a node is required and that a maximum voltage drop of 10 volts across each node is allowed
- Allow for a good margin to have reserves of power to correct problems if needed
- If using multiple supplies, it is essential that they be all turned on simultaneously to prevent both power supply and cable overloading occurring
- Power supplies MUST be capable of supporting linear and switching regulators
- Supply MUST be isolated from both the AC supply and chassis

12.8 System grounding

Grounding of the system must be done at one point only, preferably as close to the physical center of the network as possible. This connection should be done at a power tap where a terminal exists for this purpose. A main ground connection should be made from this point to a good earth or building ground via a copper braid or at least an 8 AWG copper conductor not more than 3 meters (10 feet) in length.

At this point of connection, the following conductors and circuits should be connected together in the form of a 'star' configuration.

- The drain wire of the main trunk cable
- The shield of the main trunk cable
- The negative power conductor
- The main ground connection as described above

If the network is already connected to ground at some other point, do NOT connect the ground terminal of a power tap to a second ground connection. This can result in unwanted ground loop currents occurring in the system. It is essential that a single ground connection is established for the network bus and failure to ground the bus –ve supply at ONE POINT only will result in a low signal to noise ratio appearing in the system.

Care must be exercised when connecting the drain/shield of the bus or drop line cable at nodes, which are already grounded. This can happen when the case or enclosure of the equipment, comprising the node, is connected to ground for electrical safety and/or a signaling connection to other self-powered equipment. In cases where this condition exists, the drain/shield should be connected to the node ground through a 0.01uF/500 volt capacitor wired in parallel with a 1 megohm ¼ watt resistor. If the node has no facility for grounding, the drain and shield must be left UNCONNECTED.

12.9 Signaling

DeviceNet is a two wire differential network. Communication is achieved by switching the CAN–H wire (white) and the CAN–L wire (blue) relative to the V– wire (black). CAN–H swings between 2.5 V DC (recessive state) and 4.0 V DC (dominant state) while CAN–L swings between 2.5 V DC (recessive state) and 1.5 V DC (dominant state).

With no network master connected, the CAN–H and CAN–L lines should be in the recessive state and should read (with a voltmeter set to DC mode) between 2.5 V and 3.0 V relative to V– at the point where the power supply is connected to the network. With a network master connected AND polling the network, the CAN–H to V– voltage will be around 3.2 V DC and the CAN–L to V– voltage will be around 2.4 V DC. This is because the signals are switching, which affects the DC value read by the meter.

The voltage values given here assume that no common mode voltages are present. Should they be present, voltages measured closer to the power supply will be higher than those measured furthest from the power supply. However, the differential voltages (CAN–H minus CAN–L) will not be affected.

DeviceNet uses a differential signaling system. A logical '1' is represented by CAN–H being low (recessive) and CAN–L being high (recessive). Conversely, a logical '0' is represented by CAN–H being high (dominant) and CAN–L being low (dominant). Figure 12.22 depicts this graphically.

The nodes are all attached to the bus in parallel, resulting in a wired-AND configuration. This means that as long as ANY one node imposes a low signal (logical 0) on the bus, the resulting signal on the bus will also be low. Only when ALL nodes output a high signal (logical 1), will the signal on the bus be high as well.

12.10 Data link layer

12.10.1 Frame format

The format of a DeviceNet frame is shown here. Note that the data field is rather small (8 bytes) and that any messages larger than this need to be fragmented.

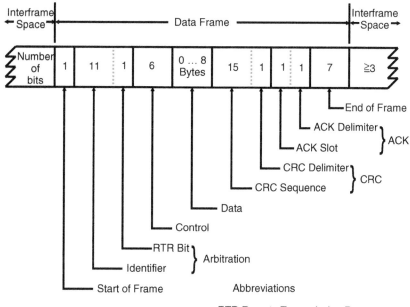

Figure 12.20
DeviceNet frame

This frame will be placed on the bus as sequential 1s and 0s, by changing the levels of the CAN–H and CAN–L in a differential fashion.

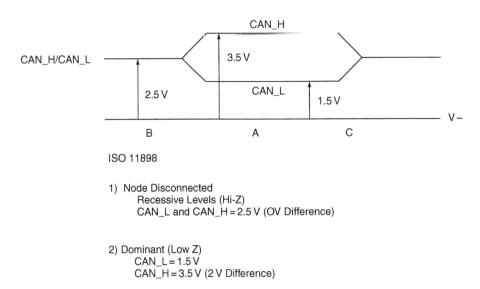

ISO 11898

1) Node Disconnected
    Recessive Levels (Hi-Z)
    CAN_L and CAN_H = 2.5 V (OV Difference)

2) Dominant (Low Z)
    CAN_L = 1.5 V
    CAN_H = 3.5 V (2 V Difference)

Figure 12.21
DeviceNet transmission

In this figure, A represents the CAN–H signal in its dominant (high) state (+3.5 V DC to +4.0 V DC), C represents the CAN–L signal in its dominant (low) state (+1.5 V DC to 2.5 V DC) and B represents both the CAN–H signal recessive (low) and the CAN–L signal recessive (high) states of +2.5 V–3.0 V DC.

12.10.2   Medium access

The medium access control method could be described as 'carrier sense multiple access with bit-wise arbitration', where the arbitration takes place on a bit-by-bit basis on the first field in the frame (the 11-bit identifier field). If a node wishes to transmit, it has to defer to any existing transmission. Once that transmission has ended, the node wishing to transmit has to wait for 3-bit times before transmitting. This is called the interframe space.

Despite this precaution, it is possible for two nodes to start transmitting concurrently. In the following example nodes 1 and 2 start transmitting concurrently, with both nodes monitoring their own transmissions. All goes well for the first few bits since the bit sequences are the same. Then the situation arises where the bits are different. Since the '0' state is dominant, the output of node 2 overrides that of node 1. Node 1 loses the arbitration and stops transmitting. It does, however, still ACK the message by means of the ACK field in the frame.

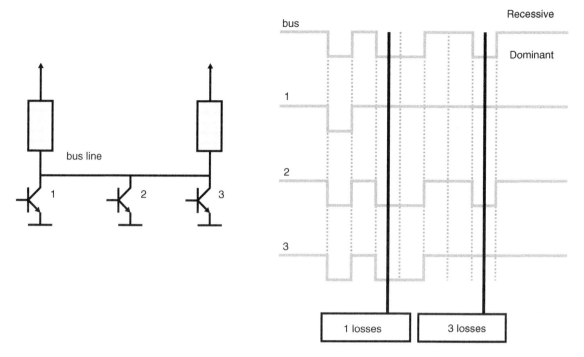

Figure 12.22
DeviceNet arbitration

Because of this method of arbitration, the node with the lowest number (i.e. the most significant '0's in its identifier field) will win the arbitration.

12.10.3   Fragmentation

Any device that needs more than 8 bytes of data sent in any direction will cause fragmentation to occur. This happens since a frame can only contain 8 bytes of data. When fragmentation occurs, only 7 bytes of data can be sent at a time since the first byte is used to facilitate the reassembly of fragments. It is used as follows:

First byte	**Significance**
00	First fragment (number 0)
41–7F	Intermediate fragment (lower 6 bits of the byte is the fragment number)
80–FF	Last fragment (lower 6 bits of the byte is the fragment number)

Example:

Data packet	**Description**
00 12 34 56 78 90 12 34	First fragment, number 0
41 56 78 90 12 34 56 78	Intermediate fragment number 1
42 90 12 34 56 78 90 12	Intermediate fragment number 2
83 34 56	Last fragment number 3

12.11   The application layer

The CAN specification does not dictate how information within the CAN message frame fields are to be interpreted – that was left up to the developers of the DeviceNet application software.

Through the use of special identifier codes (bit patterns) in the identifier field, master is differentiated from slave. Also, sections of this field tell the slaves how to respond to the master's message. For example, slaves can be requested to respond with information simultaneously in which case the CAN bus arbitration scheme assures the timeliest consecutive response from all slaves in decreasing order of priority. Or, slaves can be polled individually, all through the selection of different identifier field codes. This technique allows the system implementers more flexibility when establishing node priorities and device addresses.

12.12   Troubleshooting

12.12.1   Introduction

Networks, in general, exhibit the following types of problems from time to time.

The first type of problem is of an electronic nature, where a specific node (e.g. a network interface card) malfunctions. This can be due to a component failure or to an incorrect configuration of the device.

The second type is related to the medium that interconnects the nodes. Here, the problems are more often of an electromechanical nature and include open and short circuits, electrical noise, signal distortion and attenuation. Open and short circuits in the signal path are caused by faulty connectors or cables. Electrical interference (noise) is caused by incorrect grounding, broken shields or external sources of electromagnetic or radio frequency interference. Signal distortion and attenuation can be caused by incorrect termination, failure to adhere to topology guidelines (e.g. drop cables too long), or faulty connectors.

Whereas these are general network-related problems, the following ones are very specific to DeviceNet:

- Missing terminators
- Excessive common mode voltage, caused by faulty connectors or excessive cable length
- Low power supply voltage caused by faulty connectors or excessive cable length
- Excessive signal propagation delays caused by excessive cable length

These problems will be discussed in more detail.

12.12.2   Tools of the trade

The following list is by no means complete, but is intended to give an overview of the types of tools available for commissioning and troubleshooting DeviceNet networks. Whereas some tools are sophisticated and expensive, many DeviceNet problems can be sorted out with common sense and a multimeter.

Multimeter

A multimeter capable of measuring DC volts, resistance, and current is an indispensable troubleshooting tool. On the current scale, it should be capable of measuring several amperes.

Oscilloscope

An inexpensive 20 MHz dual-trace oscilloscope comes in quite handy. It can be used for all the voltage tests as well as observing noise on the lines, but caution should be exercised when interpreting traces.

Firstly, signal lines must be observed in differential mode (with probes connected to CAN_H and CAN_L. If they are observed one at a time with reference to ground, they may seem unacceptable due to the common mode noise (which is not a problem since it is rejected by the differential mode receivers on the nodes).

Handheld analyzers

Handheld DeviceNet analyzers such as the NetAlert NetMeter or DeviceNet Detective can be used for several purposes. Depending on the capabilities of the specific device, they can configure node addresses and baud rates, monitor power and signal levels, log errors and network events of periods ranging from a few minutes to several days, indicate short circuits and poorly wired connections, and obtain configuration states as well as firmware versions and serial numbers from devices.

Figure 12.23
NetAlert NetMeter

Intelligent wiring components

Examples of these are the NetAlert traffic monitor and NetAlert power monitor. These are 'intelligent' tee-pieces that are wired into the system. The first device monitors and displays network traffic by means of LEDs and gives a visual warning if traffic levels exceed 90%. The second device monitors voltages and visually indicates whether they are OK, too high, too low, or totally out of range.

The NetMeter can be attached to the above-mentioned tees for more detailed diagnostics.

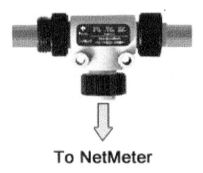

To NetMeter

Figure 12.24
PowerMonitor tee

Controller software

Many DeviceNet controllers have associated software, running under various operating systems such as Windows 2000 and NT4 that can display sophisticated views of the network for diagnostic purposes. The example given here is one of many generated by the ApplicomIO software and displays data obtained from a device, down to bit level (in hexadecimal).

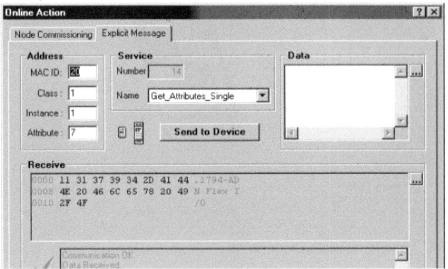

Figure 12.25
ApplicomIO display

CAN bus monitor

Since DeviceNet is based on the controller area network (CAN), CAN protocol analysis software can be used on DeviceNet networks to capture and analyze packets (frames). An example is Synergetic's CAN Explorer for Windows 95/98 and NT, running on a PC. The same vendor also supplies the ISA, PC/104, PCI and parallel port network interfaces for connecting the PC to the DeviceNet network.

A PC with this software cannot only function as a protocol analyzer, but also as a data logger.

12.12.3 Fault finding procedures

In general, the system should not be operating i.e. there should be no communication on the bus, and all devices should be installed.

A low-tech approach to troubleshooting could involve disconnecting parts of the network, and observing the effect on the problem. This does, unfortunately, not work well for problems such as excessive common mode voltage and ground loops since disconnecting part of the network often solves the problem.

12.12.3.1 Incorrect cable lengths

If the network exhibits problems during the commissioning phase or after modifications/additions have been made to the network, the cable lengths should be double-checked against the DeviceNet topology restrictions. The maximum cable lengths are as follows:

	125 Kb/s	**250 Kb/s**	**500 Kb/s**
Thick trunk length	500m	250m	100m
Thin trunk length	100m	100m	100m
Single drop	6m	6m	6m
Cumulative all drops	156m	78m	39m

For simplicity, only the metric sizes are given here.

They following symptoms are indicative of a topology problem.

- If drop lines too long, i.e. the total amount of drops exceeds the permitted length, variations on CAN signal amplitude will occur throughout the network
- If a trunk line is too long it will cause 'transmission line' effects in which reflections in the network cause faulty reception of messages; this will result in CAN frame errors

12.12.3.2 Power and earthing problems

Shielding

Shielding is checked in the following way.

Connect a 16A DC ammeter from DC common to shield at the end of the network furthest from the power supply. If the power supply is in the middle, then this test must be performed at both ends. In either case, there should be significant current flow. If practical, this test can also be performed at the end of each drop.

If there is no current flowing, then the shield is broken or the network is improperly grounded.

Grounding

In general, the following rules should be observed:

- Physically connect the DC power supply ground wire and the shield together to earth ground at the location of the power supply

- In the case of multiple power supplies, connect this ground only at the power supply closest to the middle of the network
- Ensure that all nodes on the network connect to the shield, the signal and power lines

Note: CAN frame errors are a symptom of grounding problems. CAN error messages can be monitored with a handheld DeviceNet analyzer or CAN bus analyzer.

Break the shield at a few points in the network, and insert a DC ammeter in the shield.

If there is a current flow, then the shield is connected to DC common or ground in more than one place, and a ground loop exists.

Power

This can be measured with a voltmeter or handheld DeviceNet analyzer and is measured between V+ (red) and V– (black).

Measure the network voltage at various points across the network, especially at the ends and at each device. The measured voltage should ideally be 24 V, but no more than 25 V and not less than 11 V DC.

If devices draw a lot of current, then voltages on the bus can fluctuate hence bus voltages should be monitored over time.

If the voltages are not within specification, then:

- Check for faulty or loose connectors
- Check the power system design calculations by measuring the current flow in each section of cable

On some DeviceNet analyzers, one can set a supply alarm voltage below which a warning should be generated. Plug the analyzer in at locations far from the power supply and leave it running over time. If the network voltage falls below this level at any time, this low voltage event will be logged by the analyzer.

Note that 'THIN' cable, which has a higher DC resistance, will have greater voltage drop across distance.

12.12.3.4 Incorrect signal voltage levels

The following signal levels should be observed with a voltmeter, oscilloscope or DeviceNet analyzer. Readings that differ by more than 0.5 V with the following values are most likely indicative of a problem.

CAN_H can NEVER be lower than CAN_L and if this is observed, it means that the two wires have probably been transposed.

(a) If bus communications are OFF (idle) the following values can be observed with any measuring device.

CAN_H (white)	2.5 V DC
CAN_L (blue)	2.5 V DC

(b) If bus communications are ON, the following can be observed with a voltmeter:

CAN_H (white)	3.0 V DC
CAN_L (blue)	2.0 V DC

Alternatively, the voltages can be observed with an oscilloscope or DeviceNet analyzer, in which case both minimum and maximum values can be observed.

These are:

CAN_H (white)	2.5 V min, 4.0 V max
CAN_L (blue)	1.0 V min, 2.5 V max

12.12.3.5 Common mode voltage problems

This test assumes that the shield had already been checked for continuity and current flow and can be done with a voltmeter or an oscilloscope:

- Turn all network power supplies on
- Configure all nodes to draw the maximum amount of power from the network
- Turn on all outputs that draw current

Now measure the DC voltage between V– and the shield. The difference should, technically speaking, be less than 4.65 V. For a reasonable safety margin, this value should be kept below 3 V.

These measurements should be taken at the two furthest ends (terminator position), at the DeviceNet master(s) and at each power supply. Should a problem be observed here, a solution could be to relocate the power supply to the middle of the network or to add additional power supplies.

In general, one can design a network using any number of power supplies, providing that:

- The voltage drop in the cable between a power supply and each station it supplies does not exceed 5 V DC
- The current does not exceed the cable/connector ratings
- The power supply common ground voltage level does not vary by more than 5 V between any two points in the network.

12.12.3.6 Incorrect termination

These tests can be performed with a MultiMate. They must be done with all bus communications off (bus off) and the meter set to measure resistance.

Check the resistance from CAN_H to CAN_L at each device. If the values are larger than 60 ohms (120 ohms in parallel with 120 ohms) there could be a break in one of the signal wires or there could be a missing terminator or terminators somewhere. If, on the other hand, the measured values are less than 50 ohms, this could indicate a short between the network wires, (an) extra terminating resistor(s), one or more faulty transceivers or unpowered nodes.

12.12.3.7 Noise

Noise can be observed with a loudspeaker or with an oscilloscope. However, more important than the noise itself, is the way in which the noise affects the actual transmissions taking place on the medium. The most common effect of EMI/RFI problems are CAN frame errors, which can be monitored with a CAN analyzer or DeviceNet analyzer.

The occurrence of frame errors must be related to specific nodes and to the occurrence of specific events e.g. a state change on a nearby variable frequency drive.

12.12.3.8 Node communication problems

Node presence

One method to isolate defective nodes is to use the master configuration software or a DeviceNet analyzer to create a 'live list' to see which nodes are active on the network, and to compare this with the list of nodes that are supposed to be on the network.

Excessive traffic

The master configuration software or a DeviceNet analyzer can measure the percentage traffic on the network. A figure of 30–70% is normal and anything over 90% is excessive.

Loads over 90% indicate problems. High bus loads can indicate any of the following:

- Some nodes could be having difficulty making connections with other nodes and have to retransmit repeatedly to get messages through. Check termination, bus length, topology, physical connections and grounding
- Defective nodes can 'chatter' and put garbage on the network
- Nodes supplied with corrupt or noisy power may chatter
- Change of state (COS) devices may be excessively busy with rapidly changing data and cause high percentage bus load
- Large quantities of explicit messages (configuration and diagnostic data) being sent can cause high percentage bus load
- Diagnostic instruments such as DeviceNet analyzers add traffic of their own; if this appears to be excessive, the settings on the device can be altered to reduce the additional traffic

MACID/baud rate settings

A network status LED is built into many devices. This LED should always be flashing GREEN. A solid RED indicates a communication fault, possibly an incorrect baud rate or a duplicate station address (MACID).

Network configuration software can be used to perform a 'network who' to verify that all stations are connected and communicating correctly.

In the absence of indicator LEDs, a DeviceNet analyzer will only be able to indicate that one or more devices have wrong baud rates. The devices will have to be found by inspection, and the baud rate settings corrected. First disconnect the device with the wrong baud rate, correct the setting, and then reconnect the device.

In the absence of an indicator LED, there is no explicit way of checking duplicate MACIDs either. If two nodes have the same address, one will just passively remain off-line. One solution is to look for nodes that should appear in the live list, but do not.

13

ProfiBus PA/DP/FMS overview

Objectives

When you have completed study of this chapter, you will be able to:

- List the main features of Profibus PA/DP/FMS
- Fix problems with:
  - Cabling
  - Fiber
  - Shielding
  - Grounding/earthing
  - Segmentation
  - Color coding
  - Addressing
  - Token bus operation
  - Unsolicited messages
  - Fine tuning of impedance terminations
  - Drop-line lengths
  - GSD files usage
  - Intrinsic safety concerns

13.1   Introduction

ProfiBus (PROcess FIeld BUS) is a widely accepted international networking standard, commonly found in process control and in large assembly and material handling machines. It supports single-cable wiring of multi-input sensor blocks, pneumatic valves, complex intelligent devices, smaller sub-networks (such as AS-i), and operator interfaces.

ProfiBus is nearly universal in Europe and also popular in North America, South America, and parts of Africa and Asia. It is an open, vendor independent standard. It adheres to the OSI model and ensures that devices from a variety of different vendors can communicate together easily and effectively. It has been standardized under the German

National standard as DIN 19 245 Parts 1 and 2 and, in addition, has also been ratified under the European National standard EN 50170 Volume 2.

The development of ProfiBus was initiated by the BMFT (German Federal Ministry of Research and Technology) in cooperation with several automation manufacturers in 1989. The bus interfacing hardware is implemented on ASIC (application specific integrated circuit) chips produced by multiple vendors, and is based on the EIA-485 standard as well as the European EN50170 electrical specification. The standard is supported by the ProfiBus Trade Organization, whose web site can be found at *www.profibus.com*.

ProfiBus uses 9-pin D-type connectors (impedance terminated) or 12 mm quick-disconnect connectors. The number of nodes is limited to 127. The distance supported is up to 24 km (with repeaters and fiber optic transmission), with speeds varying from 9600 bps to 12 Mbps. The message size can be up to 244 bytes of data per node per message (12 bytes of overhead for a maximum message length of 256 bytes), while the medium access control mechanisms are polling and token passing. ProfiBus supports two main types of devices, namely, masters and slaves.

- Master devices control the bus and when they have the right to access the bus, they may transfer messages without any remote request. These are referred to as active stations.
- Slave devices are typically peripheral devices i.e. transmitters/sensors and actuators. They may only acknowledge received messages or, at the request of a master, transmit messages to that master. These are also referred to as passive stations.

There are several versions of the standard, namely, ProfiBus DP (master/slave), ProfiBus FMS (multi-master/peer-to-peer), and ProfiBus PA (intrinsically safe).

- ProfiBus DP (distributed peripheral) allows the use of multiple master devices, in which case each slave device is assigned to one master. This means that multiple masters can read inputs from the device but only one master can write outputs to that device. ProfiBus DP is designed for high speed data transfer at the sensor/actuator level (as opposed to ProfiBus FMS which tends to focus on the higher automation levels) and is based around DIN 19 245 parts 1 and 2 since 1993. It is suitable as a replacement for the costly wiring of 24 V and 4–20 mA measurement signals.

    The data exchange for ProfiBus DP is generally cyclic in nature. The central controller, which acts as the master, reads the input data from the slave and sends the output data back to the slave. The bus cycle time is much shorter than the program cycle time of the controller (less than 10 mS).
- ProfiBus FMS (Fieldbus message specification) is a peer-to-peer messaging format, which allows masters to communicate with one another. Just as in ProfiBus DP, up to 126 nodes are available and all can be masters if desired. FMS messages consume more overhead than DP messages.
- 'COMBI mode' is when FMS and DP are used simultaneously in the same network, and some devices (such as Synergetic's DP/FMS masters) support this. This is most commonly used in situations where a PLC is being used in conjunction with a PC, and the primary master communicates with the secondary master via FMS. DP messages are sent via the same network to I/O devices.
- The ProfiBus PA protocol is the same as the latest ProfiBus DP with V1 diagnostic extensions, except that voltage and current levels are reduced to meet the requirements of intrinsic safety (class I, division II) for the process

industry. Many DP/FMS master cards support ProfiBus PA, but barriers are required to convert between DP and PA. PA devices are normally powered by the network at intrinsically safe voltage and current levels, utilizing the transmission technique specified in IEC 61158-2. (which Foundation Fieldbus H1 uses as well).

13.2 ProfiBus protocol stack

The architecture of the ProfiBus protocol stack is summarized in the figure below. Note the addition of an eighth layer, the so-called 'user' layer, on top of the 7-layer OSI model.

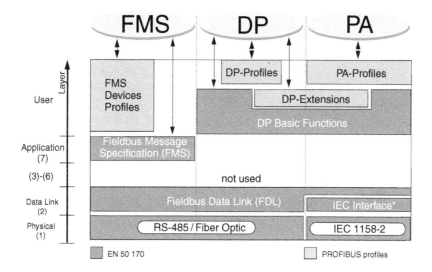

Figure 13.1
ProfiBus protocol stack

All three ProfiBus variations namely FMS, DP and PA use the same data link layer protocol (layer 2). The DP and PA versions use the same physical layer (layer 1) implementation, namely EIA-485, while PA uses a variation thereof (as per IEC 61158-2) in order to accommodate intrinsic safety requirements.

13.2.1 Physical layer (layer 1)

The physical layer of the ProfiBus DP standard is based on EIA-485 and has the following features:

- The network topology is a linear bus, terminated at both ends.
- Stubs are possible.
- The medium is a twisted pair cable, with shielding conditionally omitted depending on the application. Type A cable is preferred for transmission speeds greater than 500 kbaud. Type B should only be used for low baud rates and short distances. These are very specific cable types of which the details are given below.
- The data rate can vary between 9.6 kbps and 12 Mbps, depending on the cable length. The values are:

9.6 kbps	1200 m
19.2 kbps	1200 m

93.75 kbps	1200 m
187.5 kbps	600 m
500 kbps	200 m
1.5 Mbps	200 m
12 Mbps	100 m

The specifications for the two types of cable are as follows:

Type A cable

Impedance:	135 up to 165 ohm (for frequency of 3 to 20 MHz)
Cable capacity:	<30 pF per meter
Core diameter:	>0.34 mm^2 (AWG 22)
Cable type:	twisted pair cable. 1×2 or 2×2 or 1×4
Resistance:	<110 ohm per km
Signal attenuation:	max. 9 dB over total length of line section
Shielding:	Cu shielding braid or shielding braid and shielding foil

Type B cable

Impedance:	135 up to 165 ohm (for frequency >100 kHz)
Cable capacity:	<60 pF per meter
Core diameter:	>0.22 mm^2 (AWG 24)
Cable type:	twisted pair cable. 1×2 or 2×2 or 1×4
Resistance:	<110 ohm per km
Signal attenuation:	max. 9 dB over total length of line section
Shielding:	Cu shielding braid or shielding braid and shielding foil

For a more detailed discussion of EIA-485, refer to the chapter on EIA-485 installation and troubleshooting.

13.2.2   Data link layer (layer 2)

The second layer of the OSI model implements the functions of medium access control as well as that of the logical link control i.e. the transmission and reception of the actual frames. The latter includes the data integrity function i.e. the generation and checking of checksums.

The medium access control determines when a station may transmit on the bus and ProfiBus supports two mechanisms, namely, token passing and polling.

Token passing is used for communication between multiple masters on the bus. It involves the passing of software tokens between masters, in a sequence of ascending addresses. Thus, a logical ring is formed (despite the physical topology being a bus). The polling method (or master–slave method), on the other hand, is used by a master that currently has the token to communicate with its associated slave devices (passive stations).

ProfiBus can be setup either as a pure master–master system (token passing), or as a polling system (master–slave), or as a hybrid system using both techniques.

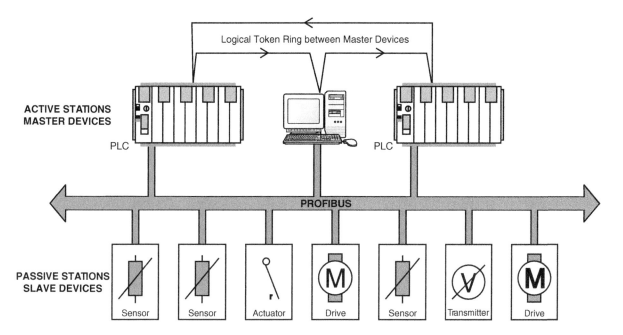

Figure 13.2
Hybrid medium access control

The following is a more detailed description of the token-passing mechanism.

- The token is passed from master station to master station in ascending order.
- When a master station receives the token from a previous station, it may then transfer messages to slave devices as well as to other masters.
- If the token transmitter does not recognize any bus activity within the slot time, it repeats the token and waits for another slot time. It retires if it recognizes bus activity. If there is no bus activity, it will repeat the token frame for the last time. If there is still no activity, it will try to pass the token to the next but one master station. It continues repeating the procedure until it identifies a station that is alive.
- Each master station is responsible for the addition or removal of stations in the address range from its own station address to the next station. Whenever a station receives the token, it examines one address in the address range between itself and its current successor. It does this maintenance whenever its currently queued message cycles have been completed. Whenever a station replies saying that it is ready to enter the token ring it is then passed the token. The current token holder also updates its new successor.
- After a power up and after a master station has waited a predefined period, it claims the token if it does not see any bus activity. The master station with the lowest station address commences initialization. It transmits two token frames addressed to itself. This then informs the other master stations that it is now the only station on the logical token ring. It then transmits a 'request field data link status' to each station in an increasing address order. The first master station that responds is then passed the token. The slave stations and 'master not ready' stations are recorded in an address list called the GAP list.

- When the token is lost, it is not necessary to re-initialize the system. The lowest address master station creates a new token after its token timer has timed out. It then proceeds with its own messages and then passes the token onto its successor.
- The real token rotation time is calculated by each master station on each cycle of the token. The system reaction time is the maximum time interval between two consecutive high priority message cycles of a master station at maximum bus load. From this, a target token rotation time is defined. The real token rotation time must be less than the target token rotation time for low priority messages to be sent out.
- There are two priorities that can be selected by the application layer, namely 'low' and 'high'. The high priority messages are always dispatched first. Independent of the token rotation time, a master station can always transmit one high priority message. The system's target token rotation time depends on the number of stations, the number of high priority messages and the duration of each of these messages. Hence it is important only to set very important and critical messages to high priority. The predefined target token rotation time should contain sufficient time for low priority message cycles with some safety margin built in for retries and loss of messages.

Basically the ProfiBus layer 2 operates in a connectionless fashion, i.e. it transmits frames without prior checking as to whether the intended recipient is able or willing to receive the frame. In most cases, the frames are 'unicast,' i.e. they are intended for a specific device, but broadcast and multicast communication is also possible. Broadcast communication means that an active station sends an unconfirmed message to all other stations (masters and slaves). Multicast communication means that a device sends out an unconfirmed message to a group of stations (masters or slaves).

Layer 2 provides data transmission services to layer 7. These services are as defined in DIN 19241-2, IEC 955, ISO 8802-2 and ISO/IEC JTC 1/SC 6N 4960 (LLC Type 1 and LLC Type 3) and comprise three acyclic data services as well as one cyclic data service.

The following data transmission services are defined:

- Send-data-with-acknowledge (SDA) – acyclic.
- Send-data-with-no-acknowledge (SDN) – acyclic.
- Send-and-request-data-with-reply (SRD) – acyclic.
- Cyclic-send-and-request-data-with-reply (CSRD) – cyclic.

All layer 2 services are accessed by layer 7 in software through so-called service access points or SAPs. On both active and passive stations, multiple SAPs (service access points) are allowed simultaneously.

- 32 Stations are allowed without repeaters, but with repeaters this number may be increased to 127.
- The maximum bus length is 1200 meters. This may be increased to 4800 m with repeaters.
- Transmission is half-duplex, using NRZ (non-return to zero) coding.
- The data rate can vary between 9.6 kbps and 12 Mbps, with values of 9.6, 19.2, 93.75, 187.5, 500, 1500 kbps or 12 Mbps.
- The frame formats are according to IEC 870-5-1, and are constructed with a Hamming distance of 4. This means that despite up to 4 consecutive faulty bits in a frame (and despite a correct checksum), a corrupted message will still be detected.
- There are two levels of message priority.

13.2.3   Application layer

Layer 7 of the OSI model provides the application services to the user. These services make an efficient and open (as well as vendor independent) data transfer possible between the application programs and layer 2. The ProfiBus application layer is specified in DIN 19 245 part 2 and consists of:

- The Fieldbus message specification (FMS)
- The lower layer interface (LLI)
- The FieldBus management services – layer 7 (FMA 7)

13.2.4   Fieldbus message specification (FMS)

From the viewpoint of an application process (at layer 8), the communication system is a service provider offering communication services, known as the FMS services. These are basically classified as either confirmed or unconfirmed services.

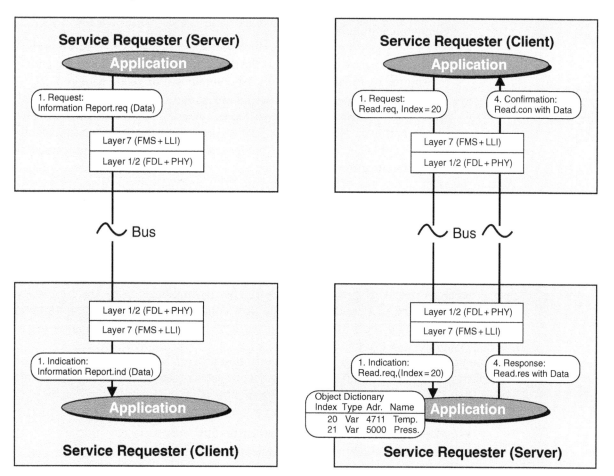

Figure 13.3
Execution of confirmed and unconfirmed services

Confirmed services are only permitted on connection-oriented communication relationships while unconfirmed services may also be used on connectionless relationships. Unconfirmed services may be transferred with either a high or a low priority.

In the ProfiBus standard, the interaction between requester and responder, as implemented by the appropriate service is described by a service primitive.

The ProfiBus FMS services can be divided into the following groups:

- Context management services allow establishment and release of logical connections, as well as the rejection of inadmissible services
- Variable access services permit access (read and write) to simple variables, records, arrays and variable lists
- The domain management services enable the transmission (upload or download) of contiguous memory blocks. The application process splits the data into smaller segments (fragments) for transmission purposes
- The program invocation services allow the control (start, stop etc) of program execution
- The event management services are unconfirmed services, which make the transmission of alarm messages possible. They may be used with high or low priority, and messages may be transmitted on broadcast or multicast communication relationships
- The VFD support messages permit device identification and status reports. These reports may be initiated at the discretion of individual devices, and transmitted on broadcast or multicast communication relationships
- The OD management services permit object dictionaries to be read and written. Process objects must be listed as communication objects in an object dictionary (OD). The application process on the device must make its objects visible and available before these can be addressed and processed by the communication services

As can be seen, there are a large amount of ProfiBus-FMS application services to satisfy the various requirements of field devices. Only a few of these (5, in fact) are mandatory for implementation in all ProfiBus devices. The selection of further services depends on the specific application and is specified in the so-called profiles.

13.2.5   Lower layer interface (LLI)

Layer 7 needs a special adaptation to layer 2. This is implemented by the LLI in the ProfiBus protocol. The LLI conducts the data flow control and connection monitoring as well as the mapping of the FMS services onto layer 2, with due consideration of the various types of devices (master or slave).

Communications relationships between application processes with the specific purpose of transferring data must be defined before a data transfer is started. These definitions are listed in layer 7 in the communications relationship list (CRL).

The main tasks of the LLI are:

- Mapping of FMS services onto the data link layer services
- Connection establishment and release
- Supervision of the connection
- Flow control

The following types of communication relationships are supported by:

- Connectionless communication, which can be either
- Broadcast, or
- Multicast, and
- Connection oriented communication, which can be either
- Master/master (cyclic or acyclic), or

- Master/slave – with or without slave initiative – (cyclic or acyclic)

Connection oriented communication relationships represent a logical peer-to-peer connection between two application processes. Before any data can be sent over this connection, it has to be established with an initiate service, one of the context management services. This comprises the connection establishment phase. After successful establishment, the connection is protected against third party access and can then be used for data communication between the two parties involved. This comprises the data transfer phase. In this phase, both confirmed and unconfirmed services can be used. When the connection is no longer needed, it can be released with yet another context management service, the Abort service. This comprises the connection release phase.

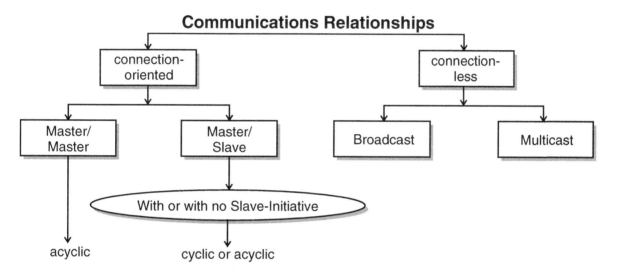

Figure 13.4
Supported communication relationships

13.2.6   Fieldbus management layer (FMA 7)

This describes object and management services. The objects are manipulated locally or remotely using management services. There are three groups here:

- **Context management**

   This provides a service for opening and closing a management connection

- **Configuration management**

   This provides services for the identification of communication components of a station, for loading and reading the communication relationship list (CRL) and for accessing variables, counters and the parameters of the lower layers

- **Fault Management**

   This provides services for recognizing and eliminating errors

13.3   The ProfiBus communication model

From a communication point of view, an application process includes all programs, resources and tasks that are not assigned to one of the communication layers. The ProfiBus communication model permits the combination of distributed application processes into a common process, using communications relationships. This acts to unify distributed application processes to a common process. That part of an application process in a field device that is reachable for communication is called a virtual field device (VFD).

All objects of a real device that can be communicated with (such as variables, programs, data ranges) are called communication objects. The VFD contains the communication objects that may be manipulated by the services of the application layer via ProfiBus.

13.4   Relationship between application process and communication

Between two application processes, one or more communication relationships may exist; each one having a unique communication end point as shown in the following diagram:

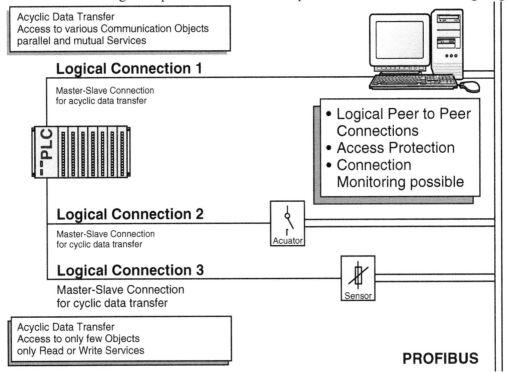

Figure 13.5
Assignment of communication relationships to application process

Mapping of the functions of the VFD onto the real device is provided by the application layer interface. The diagram below shows the relationship between the real field device and the virtual field device.

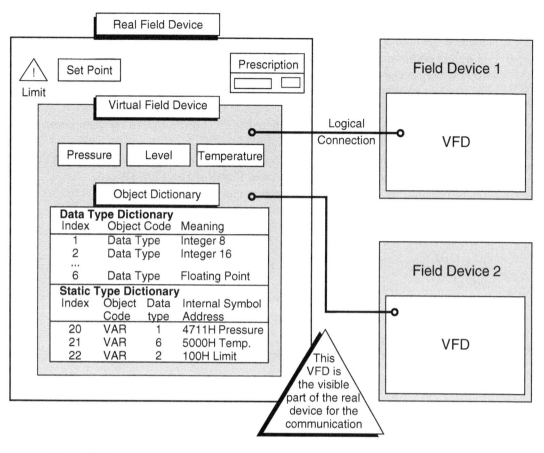

Figure 13.6
Virtual field device (VFD) with object dictionary (OD)

In this example, only the variables pressure, fill level and temperature may be read or written via two communication relationships.

13.5   Communication objects

All communication objects of a ProfiBus station are entered into a local object dictionary. This object dictionary may be predefined at simple devices; however on more complex devices it is configured and locally or remotely downloaded into the device.

The object dictionary (OD) structure contains:

- A header, which contains information about the structure of the OD
- A static list of types, containing the list of the supported data types and data structures
- A static object dictionary, containing a list of static communication objects
- A dynamic list of variable lists, containing the actual list of the known variable lists, and a dynamic list of program invocations, which contains a list of the known programs

Defined static communication objects include simple variable, array (a sequence of simple variables of the same type), record (a list of simple variables not necessarily of the same type), domain (a data range) and event.

Dynamic communication objects are entered into the dynamic part of the OD. They include program invocation and variable list (a sequence of simple variables, arrays or records). These can be predefined at configuration time, dynamically defined, deleted or changed with the application services in the operational phase.

Logical addressing is the preferred way of addressing of communication objects. They are normally accessed with a short address called an index (unsigned 16-bit). This makes for efficient messaging and keeps the protocol overhead down. There are, however, two other optional addressing methods:

- Addressing by name, where the symbolic name of the communication objects is transferred via the bus.
- Physical addressing. Any physical memory location in the field device may be accessed with the services PhysRead and PhysWrite.

It is possible to implement password protection on certain objects and also to make them read-access only, for example.

13.6 Performance

A short reaction time is one of the main advantages of ProfiBus DP. The figures are typical.

512 Inputs and outputs distributed over 32 stations can be accessed:

- in 6 mS at 1.5 Mbps and
- in 2 mS at 12 Mbps.

Figure 13.7 gives a visual indication of ProfiBus performance.

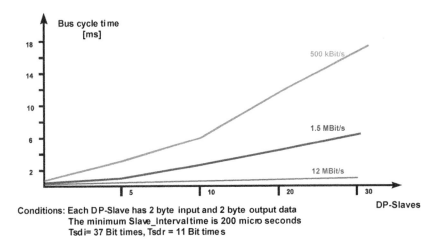

Conditions: Each DP-Slave has 2 byte input and 2 byte output data
The minimum Slave_Interval time is 200 micro seconds
Tsdi= 37 Bit times, Tsdr = 11 Bit times

Figure 13.7
Bus cycle time of a ProfiBus DP mono-master system

The main service used to achieve these results is the send and receive data service of layer 2. This allows for the transmission of the input and output data in a single message cycle. Obviously, the other reason for increased performance is the higher transmission speed of 12 Mbps.

13.7 System operation

13.7.1 Configuration

The choice is up to user as to whether the system should be a mono-master or multi-master system. Up to 126 stations (masters or slaves) can be accommodated.

There are different device types:

- DP – master class 1 (DPM1). This is typically a PLC (programmable logical controller).
- DP – master class 2 (DPM2). These devices are used for programming, configuration or diagnostics.
- DP – slave A. This is typically a sensor or actuator. The amount of I/O data is limited to 246 bytes.

The two configurations possible are shown in the diagrams below.

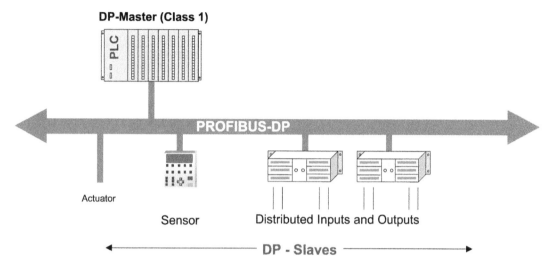

Figure 13.8
ProfiBus DP mono-master system

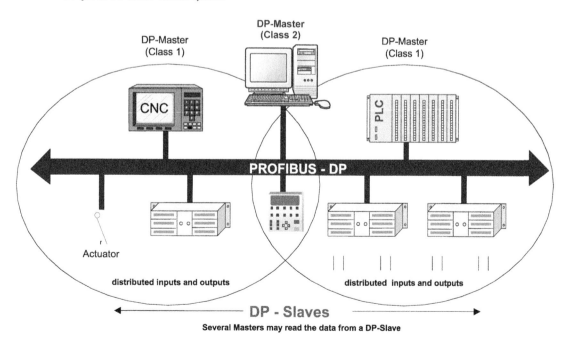

Figure 13.9
ProfiBus DP multi-master system

The following states can occur with DPM1:

- Stop. In this state, no data transfer occurs between the DPM1 and DP-slaves.
- Clear. The DPM1 puts the outputs into a fail-safe mode and reads the input data from the DP-slaves.
- Operate. The DPM1 is in the data transfer state with a cyclic message sequence where input data is read and output data is written down to the slave.

13.7.2 Data transfer between DPM1 and the DP-slaves

During configuration of the system, the user defines the assignment of a DP slave to a DPM1 and which of the DP-slaves are included in the message cycle. In the so-called parameterization and configuration phases, each slave device compares its real configuration with that received from the DPM1. This configuration information has to be identical. This safeguards the user from any configuration faults. Once this has been successfully checked, the slave device will enter into the data transfer phase as indicated in the figure below.

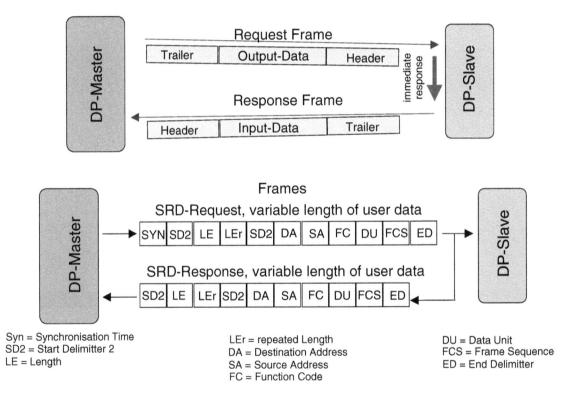

Figure 13.10
User data exchange for ProfiBus DP

13.7.3 Synchronization and freeze modes

In addition to the standard cyclic data transfer mechanisms automatically executed by the DPM1, it is possible to send control commands from a master to an individual or group of slaves.

If the 'sync' command is transmitted to the appropriate slaves, they enter this state and freeze the outputs. They then store the output data during the next cyclic data exchange. When they receive the next 'sync' command, the stored output data is issued to the field.

If a 'freeze' command is transmitted to the appropriate slaves, the inputs are frozen in the present state. The input data is only updated on receiving the next 'freeze' command.

13.7.4   Safety and protection of stations

At the DPM1 station, the user data transfer to each slave is monitored with a watchdog timer. If this timer expires indicating that no successful transfer has taken place, the user is informed and the DPM1 leaves the OPERATE state and switches the outputs of all the assigned slave devices to the fail-safe state. The master changes to the CLEAR state. Note that the master ignores the timer if the automatic error reaction has been enabled (Auto_Clear = True).

At the slave devices, the watchdog timer is again used to monitor any failures of the master device or the bus. The slave switches its outputs autonomously to the fail-safe state if it detects a failure.

13.7.5   Mixed operation of FMS and DP stations

Where lower reaction times are acceptable, it is possible to operate FMS and DP devices together on the same bus. It is also possible to use a composite device, which supports both FMS and DP protocols simultaneously. This can make sense if the configuration is done using FMS and the higher speed cyclic operations are done for user data transfer. The only difference between the FMS and DP protocols are of course the application layers.

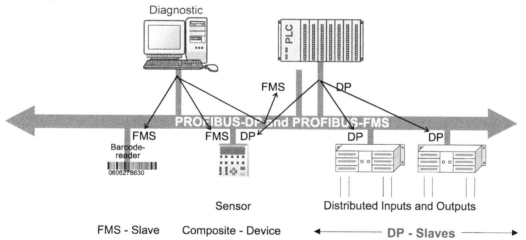

Figure 13.11
Mixed operation of ProfiBus FMS and DP

13.8   Troubleshooting

13.8.1   Introduction

ProfiBus DP and FMS use EIA-485 at the physical layer (layer 1) and therefore all the EIA-485 installation and troubleshooting guidelines apply. Refer to the appropriate chapter in this manual. Profibus PA uses the same physical layers as the 61158-2 standard (which is the same as the Foundation Fieldbus H1 standard). This section will discuss some additional specialized tools.

13.8.2   Troubleshooting tools

13.8.2.1   Handheld testing device

These are similar to the ones available for DeviceNet, and can be used to check the copper infrastructure before connecting any devices to the cable. A typical example is the unit made by Synergetic. They can indicate:

- A switch (i.e. reversal) of the A and B lines
- Wire breaks in the A and B lines as well as in the shield
- Short circuits between the A and B lines and the shield
- Incorrect or missing terminations

The error is indicated via text shown in the display of the device.

These devices can also be used to check the EIA-485 interfaces of ProfiBus devices, after they have been connected to the network. Typical functions include:

- Creating a list with the addresses of all stations connected to the bus (useful for identifying missing devices)
- Testing individual stations (e.g. identifying duplicate addresses)
- Measuring distance (checking whether the installed segment lengths comply with the Profibus requirements)
- Measuring reflections (e.g. locating an interruption of the bus line)

13.8.2.2   D-type connectors with built-in terminators

For further location of cable break errors reported by a handheld tester, 9-pin D connectors with integrated terminations are very helpful. When the termination is switched to 'on' at the connector, the cable leading out of the connector is disconnected. This feature can be used to identify the location of the error, as follows:

If, for example, the handheld is connected at the beginning of the network and a wire break of the A line is reported, plug the D connector somewhere in the middle of the network and switch the termination to 'on'. If the problem is still reported by the tester, it means that the introduced termination is still not 'seen' by the tester and thus the cable break must be between the beginning of the network and the D connector.

13.8.2.3   Configuration utilities

Each ProfiBus network must be configured and various products are commercially available to perform this task. Examples include the ProfiBus DP configuration tool by SST, the Allen Bradley Plug & Play software and the Siemens COM package. In many cases, the decision on the tool to be used for configuration is made automatically by choosing the controlling device for the bus. The choice of configuration tool should not be treated lightly because the easier the tool is to use, the less likely a configuration error will be made.

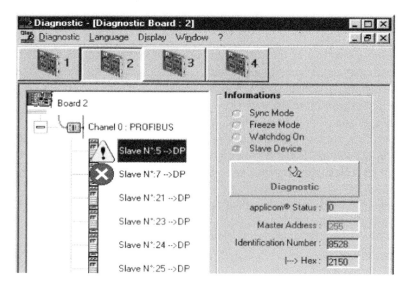

Figure 13.12
Applicom configuration tool

With ProfiBus, all parameters of a device (including text to provide a good explanation of the parameters and of the possible choices, values and ranges) are specified in a so-called GSD file, which is the electronics data sheet of the device. Therefore the configuration software has all the information it needs to create a friendly user interface with no need for the interpretation of hexadecimal values.

13.8.2.4 Built-in diagnostics

Several diagnostic functions were designed into ProfiBus. The standard defines different timers and counters used by the physical layer to maintain operation and monitor the quality of the network. One counter, for example, counts the number of invalid start delimiters received as an indication of installation problems or an interface not working properly. These timers and counters can be used by the ProfiBus device or by its configuration tool to identify a problem and to indicate it to the user.

For ProfiBus DP, a special diagnostic message is defined, which can be indicated by a ProfiBus DP slave or requested by a ProfiBus DP master. The first 6 bytes are implemented for all ProfiBus DP devices. This information is used to indicate various problems to the user and could include:

- Configuration of the specific device incorrect
- Required features not supported on the device or
- Device does not answer

The user normally gets access to all this information through the configuration tool. The user selects a device, the tool reads the diagnostic information from the device, and provides high-level text information.

During operation, a DP device automatically reports problems to the ProfiBus DP master. The master stores the diagnostic information and provides it to the user. This can, for example, be done by a PC-based system that utilizes diagnostic flow charts to evaluate the information and then make it available to the operator.

The definition of additional diagnostics enables each manufacturer to simplify matters for end-users. The additional information differentiates between a device-related, an identifier-related, and a channel-related part. The device-related part provides the opportunity to encode manufacturer specific details. This can be used to report that the module place in slot #4 is not the same as the one configured. Module-related diagnostics provide an overview of the status of all modules; it identifies whether a module supports diagnostics or not. The channel-related part offers the possibility to report problems down to the bit level. This means a DP slave can indicate that channel #3 of the module in slot #5 has a short-circuit to ground.

With the additional diagnostics, a ProfiBus DP device can send very detailed error reports to the controlling device. As a result, the master device is able to provide details to the user such as 'ERROR oven control: lower temperature limit exceeded' or 'station address 23 (conveyor control): wire break at module 2, channel 5'. This feature provides not only the flexibility to report any kind of error at a device but also often how to correct it. Because the protocol for ProfiBus PA is identical to that for DP, the diagnostic mechanism is the same.

13.8.2.5 Bus monitors

A bus monitor (protocol analyzer) is an additional tool for troubleshooting a ProfiBus network, enabling the user to perform packet content and timing verification. It is capable of monitoring and capturing all ProfiBus network activity (messages, ACKs etc) on the bus, and then saving the captured data to disk. Each captured message is time-stamped with sub-millisecond resolution A monitor does not have a ProfiBus station address nor does it affect the speed or efficiency of the network.

A monitor provides a wide range of trigger and filter functions that allow capturing messages between two stations only or triggering on a special event like diagnostic requests. Such a tool can be used for the indication of problems with individual devices (e.g. wrong configuration) and also to visualize physical problems.

Bus monitors are typically PCs with a special ProfiBus interface card and the appropriate data capturing software. An example is the ProfiBus DP capture utility by SST. Bus monitors are sophisticated tools and are recommended only for people with a reasonable knowledge of ProfiBus and its protocol.

13.8.3 Tips

ProfiBus (especially DP) is straightforward from a user's point of view, as the messaging format is fairly simple. However, the following notes will be helpful in identifying common problems:

- ProfiBus has a relatively high (12 Mbps) maximum data rate but it can also be operated at speeds as low as 9600 baud. If ProfiBus is to be used at high speeds, it might be necessary to use a scope or analyzer to check and fine-tune impedance terminations and drop-line lengths. Such problems are magnified at higher speeds. Users who initially intend to run their network at maximum speed often find that a lower speed setting performs just as well and is easier to get working.
- One of the most common problems encountered in configuring a ProfiBus network is selecting the wrong GSD (device description) file for a particular node. As GSD files reside in a separate disk and are not embedded in the product itself, files are sometimes paired with the wrong devices.
- When installing a new network, follow the ProfiBus installation guidelines:
- Use connectors suitable for an industrial environment and according to the defined standard
- Use only the specified (blue) cable

- Make sure the cable has no wire break and none of the wires causes a short circuit condition
- Do not crisscross the wires; always use the green wire for A and the red wire as B throughout the whole network
- Make sure the segment length is according to the chosen transmission rate (use repeaters to extend the network)
- Make sure the number of devices/EIA-485 drivers per segment does not exceed 32 (use repeaters where necessary)
- Check proper termination of all copper segments (an EIA-485 segment must be terminated at both ends)
- If so-called 'activated terminations' are used, they must be powered at all times
- Avoid drop lines or make sure the overall length does not exceed the specified maximum. In case T-drops are needed, use repeaters or active bus terminals
- In case the network connects buildings or runs in a hazardous environment, consider the use of fiber optics
- Check whether the station addresses are set to the correct value
- Check if the network configurations match the physical setup
- For EIA-485 implementations (ProfiBus FMS and ProfiBus DP), type A cable is preferred for transmission speeds greater than 500 kbaud; type B should only be used for low baud rates and short distances.

The specifications for the two types of cable are as follows:

Type A Cable		
Impedance:	135 upto 165 Ohm (for frequency of 3 to 20 MHz)	
Cable Capacity:	<30 pF per Meter	
Core Diameter:	>0.34 mm2 (AWG 22)	
Cable Type:	twisted pair cable. 1×2 or 2×2 or 1×4.	
Resistance:	<110 Ohm per km	
Signal Attenuation:	Max. 9dB over total length of line section	
Shielding:	Cu shielding braid or shielding braid and shielding foil	

Type B Cable		
Impedance:	135 upto 165 Ohm (for frequency >100 kHz)	
Cable Capacity:	<60 pF per Meter	
Core Diameter:	>0.22 mm2 (AWG 24)	
Cable Type:	twisted pair cable. 1×2 or 2×2 or 1×4.	
Resistance:	<110 Ohm per km	
Signal Attenuation:	Max. 9dB over total length of line section	
Shielding:	Cu shielding braid or shielding braid and shielding foil	

The connection between shield and protective ground is made via the metal cases and screw tops of the D-type connectors. Should this not be possible, then the connection should be made via pin 1 of the connectors. This is not an optimum solution and it is probably better to bare the cable shield at the appropriate point and to ground it with a cable as short as possible to the metallic structure of the cabinet.

14

Foundation Fieldbus overview

Objectives

When you have completed study of this chapter, you will be able to:
- Describe how Foundation Fieldbus operates
- Remedy problems with:
    - wiring
    - earths/grounds
    - shielding
    - wiring polarity
    - power
    - terminations
    - intrinsic safety
    - voltage drop
    - power conditioning
    - surge protection
    - configuration

14.1   Introduction to Foundation Fieldbus

Foundation Fieldbus (FF) takes full advantage of the emerging 'smart' field devices and modern digital communications technology allowing end user benefits such as:
- Reduced wiring
- Communications of multiple process variables from a single instrument
- Advanced diagnostics
- Interoperability between devices of different manufacturers
- Enhanced field level control
- Reduced startup time
- Simpler integration

The concept behind Foundation Fieldbus is to preserve the desirable features of the present 4–20 mA standard (such as a standardized interface to the communications link, bus power derived from the link and intrinsic safety options) while taking advantage of the new digital technologies. This will provide the features noted above because of:

- Reduced wiring due to the multidrop capability
- Flexibility of supplier choices due to interoperability
- Reduced control room equipment due to distribution of control functions to the device level
- Increased data integrity and reliability due to the application of digital communications.

Foundation Fieldbus consists of four layers. Three of them correspond to OSI layers 1, 2 and 7. The fourth is the so-called 'user layer' that sits on top of layer 7 and is often said to represent OSI 'layer 8', although the OSI model does not include such a layer. The user layer provides a standardized interface between the application software and the actual field devices.

14.2   The physical layer and wiring rules

The physical layer standard has been approved and is detailed in the IEC 61158-2 and the ISA standard S50.02-1992. It supports communication rates of 31.25 kbps and uses the Manchester Bi-phase L encoding scheme with four encoding states as shown in Figure 14.2. Devices can be optionally powered from the bus under certain conditions. The 31.25 kbps (or H1, or low-speed bus) can support from 2 to 32 devices that are not bus powered, two to twelve devices that are bus powered or two to six devices that are bus powered in an intrinsically safe area. Repeaters are allowed and will increase the length and number of devices that can be put on the bus. The H2 or high-speed bus option was not implemented as originally planned, but was superseded by the high-speed Ethernet (HSE) standard. This is discussed later in this section.

The low-speed (H1) bus is intended to utilize existing plant wiring and uses #22 AWG type B wiring (shielded twisted pair) for segments up to 1200 m (3936 feet) and #18 AWG type A wiring (shielded twisted pair) up to 1900 meters (6232 feet). Two additional types of cabling are specified and are referred to as type C (multi-pair twisted without shield) and type D (multi-core, no shield). Type C using #26 AWG cable is limited to 400 meters (1312 feet) per segment and type D with #16 AWG is restricted to segments less than 200 meters (660 feet).

- Type A   #18 AWG   1900 m (6232 feet)
- Type B   #22 AWG   1200 m (3936 feet)
- Type C   #26 AWG   400 m (1312 feet)
- Type D   #16 AWG multi-core   200 m (660 feet)

The Foundation Fieldbus wiring is floating/balanced and equipped with a termination resistor (RC combination) connected across each end of the transmission line. Neither of the wires should ever be connected to ground. The terminator consists of a 100 ohm quarter watt resistor and a capacitor sized to pass 31.25 kHz. As an option, one of the terminators can be center-tapped and grounded to prevent voltage buildup on the bus. Power supplies must be impedance matched. Off-the-shelf power supplies must be conditioned by fitting a series inductor. If a 'normal power supply' is placed across the line, it will load down the line due to its low impedance. This will cause the transmitters to stop transmitting.

Fast response times for the bus are one of the FF goals. For example, at 31.25 kbps on the H1 bus, response times as low as 32 microseconds are possible. This will vary, based on the loading of the system, but will average between 32 microseconds and 2.2 ms with an average of approximately 1 ms.

Spurs can be connected to the 'home run'. The length of the spurs depends on the type of wire used and the number of spurs connected. The maximum length is the total length of the spurs and the home run.

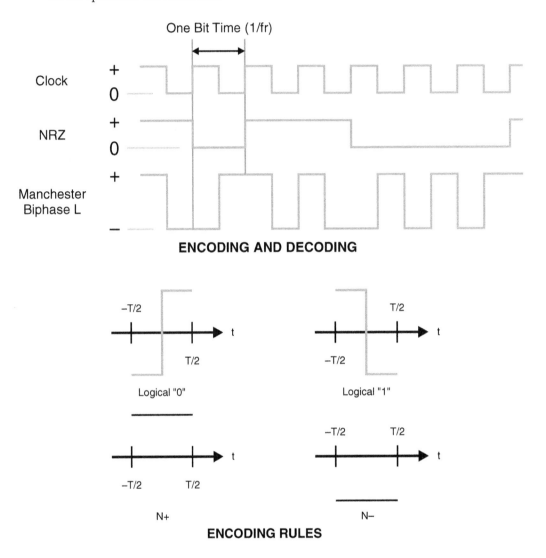

ENCODING AND DECODING

ENCODING RULES

Symbols		Encoding
1	(ONE)	Hi-Lo transition (mid-bit)
0	(ZERO)	Lo-Hi transition (mid-bit)
N+	(NON-DATA PLUS)	Hi (No transition)
N-	(NON-DATA MINUS)	Lo (No transition)

ENCODING RULES

Figure 14.1
Foundation Fieldbus physical layer

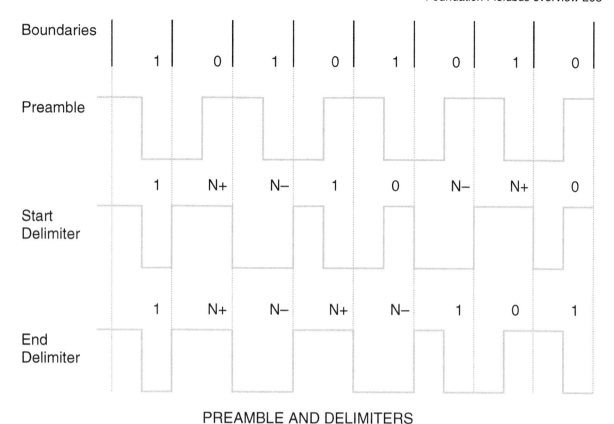

PREAMBLE AND DELIMITERS

Figure 14.2
Use of N+ and N– encoding states

The physical layer standard has been out for some time. Most of the recent work has been focused on these upper layers and are defined by the FF as the 'communications stack' and the 'user layer'. The following sections will explore these upper layers:

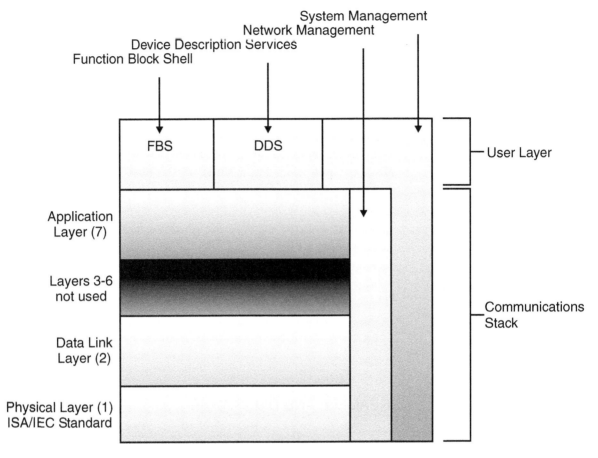

Figure 14.3
The OSI model of the FF protocol stack

14.3   The data link layer

The communications stack as defined by the FF corresponds to OSI layers two and seven, the data link and applications layers. The DLL (data link layer) controls access to the bus through a centralized bus scheduler called the link active scheduler (LAS). The DLL packet format is shown below:

GENERAL PACKET LAYOUT

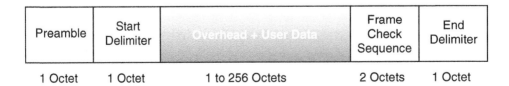

Figure 14.4
Data link layer packet format

The link active scheduler (LAS) controls access to the bus by granting permission to each device according to predefined 'schedules'. No device may access the bus without LAS permission. There are two types of schedules implemented: cyclic (scheduled) and acyclic (unscheduled). It may seem odd that one could have an unscheduled 'schedule', but these terms actually refer to messages that have a periodic or non-periodic routine, or 'schedule'.

The cyclic messages are used for information (process and control variables) that requires regular, periodic updating between devices on the bus. The technique used for information transfer on the bus is known as the publisher–subscriber method. Based on the user predefined (programmed) schedule, the LAS grants permission for each device in turn access to the bus. Once the device receives permission to access the bus, it 'publishes' its available information. All other devices can then listen to the 'published' information and read it into memory (subscribe) if it requires it for its own use. Devices not requiring specific data simply ignore the 'published' information.

The acyclic messages are used for special cases that may not occur on a regular basis. These may be alarm acknowledgment or special commands such as retrieving diagnostic information from a specific device on the bus. The LAS detects time slots available between cyclic messages and uses these to send the acyclic messages.

14.4 The application layer

The application layer in the FF specification is divided into two sub layers: the Foundation Fieldbus access sub layer (FAS) and the Foundation Fieldbus messaging specification (FMS).

The capability to pre-program the 'schedule' in the LAS provides a powerful configuration tool for the end user since the time of rotation between devices can be established and critical devices can be 'scheduled' more frequently to provide a form of prioritization of specific I/O points. This is the responsibility and capability of the FAS. Programming the schedule via the FAS allows the option of implementing (actually, simulating) various 'services' between the LAS and the devices on the bus.

Three such 'services' are readily apparent such as:

- Client/server with a dedicated client (the LAS) and several servers (the bus devices)
- Publisher/subscriber as described above, and
- Event distribution with devices reporting only in response to a 'trigger' event, or by exception, or other predefined criteria.

These variations, of course, depend on the actual application and one scheme need not necessarily be 'right' for all applications, but the flexibility of the Foundation Fieldbus is easily understood from this example.

The second sub layer, the Foundation Fieldbus messaging specification (FMS), contains an 'object dictionary' that is a type of database that allows access to Foundation Fieldbus data by tag name or an index number. The object dictionary contains complete listings of all data types, data type descriptions, and communication objects used by the application. The services allow the object dictionary (application database) to be accessed and manipulated.

Information can be read from or written to the object dictionary allowing manipulation of the application and the services provided.

14.5   The user layer

The FF specifies an eighth layer called the user layer that resides 'above' the application layer of the OSI model, this layer is usually referred to as layer 8. In the Foundation Fieldbus, this layer is responsible for three main tasks viz. network management, system management and function block/device description services. Figure 14.5 illustrates how all the layer's information packets are passed to the physical layer.

The network management service provides access to the other layers for performance monitoring and managing communications between the layers and between remote objects (objects on the bus). The system management takes care of device address assignment, application clock synchronization, and function block scheduling. This is essentially the time coordination between devices and the software and ensures correct time stamping of events throughout the bus.

MESSAGE ENCODING/DECODING EXAMPLE

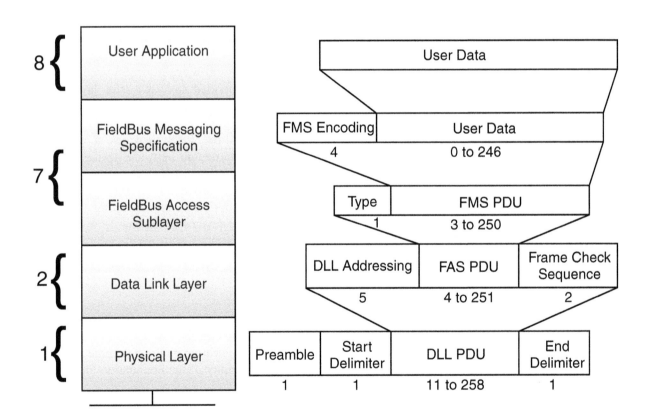

Figure 14.5
The passage of information packets to the physical layer

Function blocks and device description services provide pre-programmed 'blocks', which can be used by the end user to eliminate redundant and time-consuming configuration. The block concept allows selection of generic functions, algorithms, and even generic devices from a library of objects during system configuration and programming. This process can dramatically reduce configuration time since large 'blocks' are already configured and simply need to be selected. The goal is to provide an open system that supports interoperability and a device description language (DDL), which will enable multiple vendors and devices to be described as 'blocks' or 'symbols'. The user would select generic devices then refine this selection by selecting a DDL object to specify a specific vendor's product. Entering a control loop 'block' with the appropriate parameters would nearly complete the initial configuration for the loop. Advanced control functions and mathematics 'blocks' are also available for more advanced control applications.

14.6   Error detection and diagnostics

FF has been developed as a purely digital communications bus for the process industry and incorporates error detection and diagnostic information. It uses multiple vendors' components and has extensive diagnostics across the stack from the physical link up through the network and system management layers by design.

The signaling method used by the physical layer timing and synchronization is monitored constantly as part of the communications. Repeated messages and the reason for the repetition can be logged and displayed for interpretation.

In the upper layer, network and system management is an integral feature of the diagnostic routines. This allows the system manager to analyze the network 'on-line' and maintain traffic loading information. As devices are added and removed, optimization of the link active scheduler (LAS) routine allows communications optimization dynamically without requiring a complete network shutdown. This ensures optimal timing and device reporting, giving more time to higher priority devices and removing, or minimizing, redundant or low priority messaging.

With the device description (DD) library for each device stored in the host controller (a requirement for true interoperability between vendors), all the diagnostic capability of each vendors' products can be accurately reported and logged and/or alarmed to provide continuous monitoring of each device.

14.7   High-speed Ethernet (HSE)

High-speed Ethernet (HSE) is the Fieldbus Foundation's backbone network running at 100 Mbits/second. HSE field devices are connected to the backbone via HSE linking devices. A HSE linking device is a device used to interconnect H1 Fieldbus segments to HSE to create a larger network. A HSE switch is an Ethernet device used to interconnect multiple HSE devices such as HSE linking devices and HSE field devices to form an even larger HSE network. HSE hosts are used to configure and monitor the linking devices and H1 devices. Each H1 segment has its own link active scheduler (LAS) located in a linking device. This feature enables the H1 segments to continue operating even if the hosts are disconnected from the HSE backbone. Multiple H1 (31.25 kbps) Fieldbus segments can be connected to the HSE backbone via linking devices.

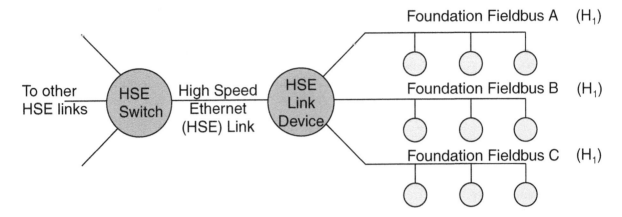

Figure 14.6
High-speed Ethernet and Foundation Fieldbus

14.8 Good wiring and installation practice with Fieldbus

14.8.1 Termination preparation

If care is taken in the preparation of the wiring, there will be fewer problems later and minimal maintenance required.

A few points to be noted here are:

- Strip 50 mm of the cable sheathing from the cable and remove the cable foil.
- Strip 6 mm of the insulation from the ends. Watch out to avoid wire nicking or cutting off strands of wire. Use a decent cable-stripping tool.
- Crimp a ferrule on the wire ends and on the end of the shield wire. Crimp ferrules are preferable as they provide a gas tight connection between the wire and the ferrule that is corrosion resistant. It is the same metal as the terminal in the wiring block.
- An alternative strategy is to twist the wires together and to tin them with solder. Wires can be put directly into the wire terminal but make sure all strands are in and they are not touching each other. Make sure that the strands are not stretched to breaking point.
- Do not attach shield wires together in a field junction box. This can be the cause of ground loops.
- Do not ground the shield in more than one place (inadvertently).
- Use good wire strippers to avoid damaging the wire.

14.8.2 Installation of the complete system

Other system components can be installed soon after the cable is installed. This includes the terminators, power supply, power conditioners, spurs and in some cases, the intrinsic safety barriers. Some devices already have terminator built-in. In that case, be careful that you are not doubling up with terminators.

Check whether the grounding is correct. There should only be one point for the shield ground point.

Once these checks have been performed; switch on the power supply and check the wiring system. The Fieldbus tester (or an alternative simpler device) can be used to indicate:

- Polarities are correct
- Power carrying capability of the wire system is ok.
- The attenuation and distortion parameters are within specification.

A few additional wiring tips and suggestions with reference to the diagram below.

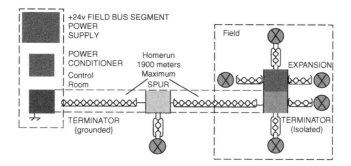

Figure 14.7
Overall diagram of FieldBus wiring configuration (courtesy Relcom Inc.)

1. It is not possible to run two homerun cables in parallel for redundancy under the H1 standard. H1 Fieldbus is a balanced transmission line that must be terminated at each end. In some cases, it is a good idea to run a parallel cable for future use. In case of physical damage, you need to disconnect the damaged cable and put in the undamaged one. Ensure that, if this is the philosophy, you do not route both cables in the same cable tray.
2. Do not ground the shield of the cable at each Fieldbus device. The shield of the cable at the transmitter (for example) should be trimmed and covered with insulating tape or heat shrinkable tubing. The only 'ground' that occurs on the segment is usually at the control room Fieldbus power conditioner.
3. Note that the ground that is connected to the isolated terminator at the far end of the segment does not connect the shields of the Fieldbus. It only allows for a high frequency path for AC currents.
4. There has been no provision made for lightning strikes. However, you should specify a terminator that has some type of spark gap arrestor, which will clamp the shields to about 75 V in such a high voltage surge.
5. A quick way to check that the grounding is correct before powering up is, doing a resistance measurement from the ground bolt on the power conditioner to the earth ground connection point. This measurement should be of the order of megohms. You can then connect the earth protective ground to the power conditioner bolt. Once this has been done, measure the resistance from the cable shields on the isolated terminator at the far end of the segment to a nearby earth ground point. A very low value of resistance should be seen.
6. A standard power supply cannot be used to power a Fieldbus segment. A standard power supply absorbs most of the Fieldbus signals due its low internal impedance. It is possible for a standard power supply to provide

power to a Fieldbus power conditioning device as long as it has sufficient current, is a floating power supply and has very low ripple and noise.

7. Use wiring blocks that hold the wiring securely and will not vibrate loose.

Regular testing of an operational Fieldbus network

A FieldBus tester can be used to get a view on the operation of the network. It is generally connected as follows:

- Red terminal to the (+) wire
- Black terminal to the (–) wire
- Green terminal to the shield.

When the network is operating, the tester will build up a record of operational devices and then builds up a record of their signal characteristics. During later routine network maintenance, the results will be compared. If there is deterioration, this will be indicated and could be wiring problems, additional noise or a device, whose transmitter is starting to fail.

14.9 Troubleshooting

14.9.1 Introduction

Estimates are that 70% of network downtime is caused by physical problems. Foundation Fieldbus is more complicated to troubleshoot than most networks because it can and often does use the communication bus to power devices. The troubleshooter needs to know not only whether the communication is working but also whether there is enough power for the devices. Below is a diagram of a typical system. Notice that the power supply on the left is supplying power to the devices in the system.

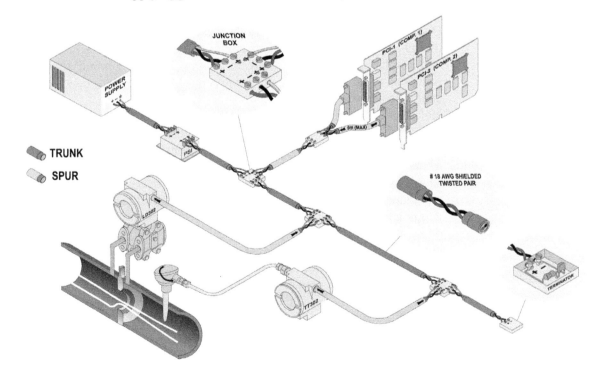

Figure 14.8
A typical foundation Fieldbus system

When troubleshooting a Foundation Fieldbus system, it is necessary to first determine whether the problem is a power problem or a communications problem. In new systems, it may be found that the problem is both. In working systems it is usually one or the other.

14.9.2   Power problems

Power problems in an FF system can be divided into two types. One where the system is new and never worked and the other where the system has been up and running for a while. When new devices are added to an existing system and the communications immediately fails, it is easy to realize that the new device had something to do with the problem. If the system has never worked, then the problem could be anywhere and could be caused by multiple devices. The problem could also be with the design itself. The following need to be known when troubleshooting the power system of an FF system.

- What is the layout of the system? Does each device have at least 9 volts DC?
- What is the supply current?
- What is the supply voltage?
- What is the current draw of each device?
- What is the resistance of each cable leg?

The easiest way of determining a power problem, is to do the following:

- Check each device to see if the power light is on
- Measure the voltage at each device
- Check the connections for opens, corrosion or loose connections
- Measure the current draw of each device to see if it conforms to the manufacturer's specifications.

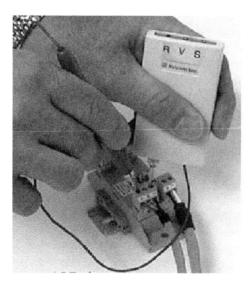

Figure 14.9
Testing the system (courtesy Relcom Inc.)

FIELDBUS SEGMENT

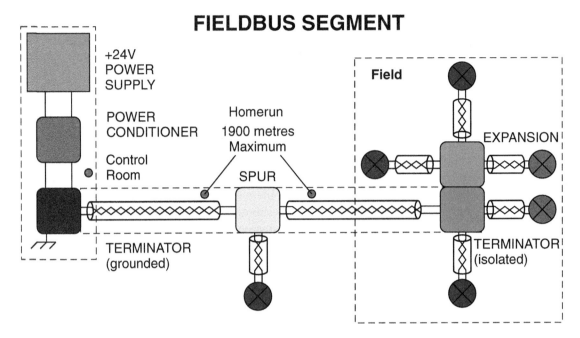

Figure 14.10
Layout of a system

Notice that the home run in the last drawing connects the control room equipment with the devices via common terminal blocks (chickenfoot or crowsfoot). The signal cable also provides the power to the devices. There is a terminator at each end of the cable. Power supplies require power conditioners.

Power example

Here is an example of the power requirements for a system.
- The power supply output is 20 volts
- The two wires are 1 km long with 22 ohms per wire (44 ohms total)
- Each device draws 20 mA
- Minimum voltage at each device 9 volts (20–9 = 11 volts)
- 11 volts/44 ohms = 250 mA

Therefore
- 250 mA/20mA = 12 devices on the system

14.9.3 Communication problems

Once the power is ruled out as a problem, it can be assumed that the communication system is at fault. Initially, it is important to check the following items:
- Are the wires connected correctly?
- Is the shield continuous throughout the system?
- Is the shield grounded at only one place?

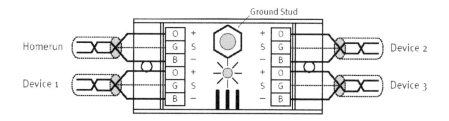

Figure 14.11
Schematic of a terminal block (courtesy Relcom Inc.)

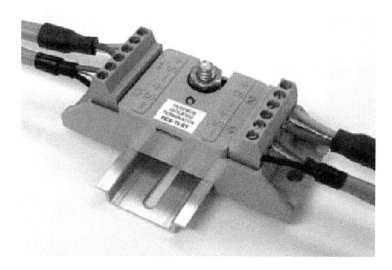

Figure 14.12
Terminal block (courtesy Relcom Inc.)

Once these basics are verified, the next step is to check to make sure that the cables are not too long. To measure the losses through the cable, an FF transmitter device is placed at one end and a receiver test device at the other. The maximum loss is usually around –14 dB. The typical characteristics of a twisted pair cable are:

Impedance:	100 ohms
Wire size:	18 GA (0.8 mm²)
Shield:	90% coverage
Capacitive unbalance:	2 nF/km
Attenuation:	3 dB/km

Using an ungrounded oscilloscope, it is possible to look at the signal. A good transmitter signal might look like this:

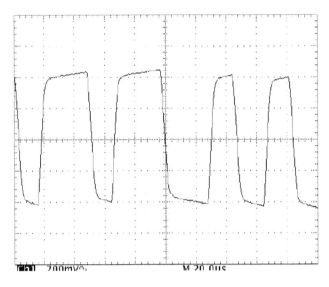

Figure 14.13
A good transmitted signal (courtesy Relcom Inc.)

When it is received it might look like this:

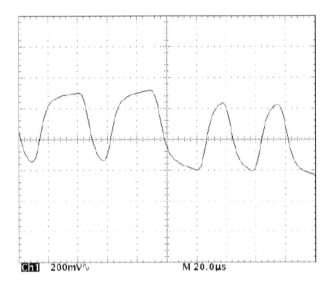

Figure 14.14
Received FF signal (courtesy Relcom Inc.)

Notice that the waveform is a bit distorted and lower in amplitude but still good. The next drawing is what the whole packet might look like.

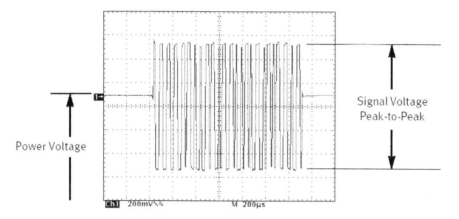

Figure 14.15
Bipolar FF signal (courtesy Relcom Inc.)

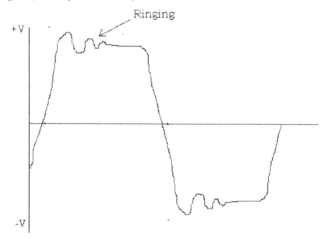

Figure 14.16
Ringing on a line (courtesy Relcom Inc.)

14.9.4   Foundation Fieldbus test equipment

There are a few manufacturers that have brought out test equipment specifically designed for testing FF systems. Some of the test equipment can be used while the system is working and others are used when the system is off-line. Some of the things the test equipment can check for are:

- DC voltage levels
- Link active scheduler probe node frame voltage
- Number of devices on the network
- If devices have been added or removed
- The lowest voltage level transmitted by a device
- Noise level between frames
- Device response noise level

Some of the best troubleshooting tools are the LEDs provided on the devices. These LEDs show many different conditions of the system. If the troubleshooter becomes familiar with them, then LEDs can often indicate what is wrong with the system.

15

Industrial Ethernet overview

Objectives

When you have completed study of this chapter, you will be able to:

- Describe how industrial Ethernet systems operate
- Identify, troubleshoot and fix problems such as:
  - thin and thick coax cable and connectors
  - UTP cabling
  - incorrect media selection
  - jabber
  - too many nodes
  - excessive broadcasting
  - bad frames
  - faulty auto-negotiation
  - 10/100 Mbps mismatch
  - full/half-duplex mismatch
  - faulty hubs
  - switched networks
  - loading

15.1   Introduction

During the mid-seventies, Xerox Corporation (Palo Alto) developed the Ethernet network concept, based on work done by researchers at the University of Hawaii. The University's ALOHA network was setup using radio broadcasts to connect sites on the islands. This was colloquially known as their 'Ethernet' since it used the 'ether' as the transmission medium and created a network between the sites. The philosophy was straightforward: Any station wanting to broadcast would do so immediately. The receiving station then had a responsibility to acknowledge the message, advising the original transmitting station of a successful reception of the original message. This primitive system did not rely on any detection of collisions (two radio stations transmitting at the same time) but depended on an acknowledgment within a predefined time.

The initial Xerox system was so successful that it was soon applied to other sites, typically connecting office equipment to shared resources such as printers and large computers acting as repositories of large databases.

In 1980, the Ethernet Consortium consisting of Xerox, Digital Equipment Corporation and Intel (a.k.a. the DIX consortium) issued a joint specification, based on the Ethernet concepts, known as the Ethernet Blue Book 1 specification. This was later superseded by the Ethernet Blue Book 2 (Ethernet V2) specification, which was offered to the IEEE for ratification as a standard. In 1983, the IEEE issued the 802.3 standard for carrier sense multiple access/collision detect (CSMA/CD) LANs based on the Ethernet standard.

As a result, there are two standards in existence viz. Ethernet V2 (Bluebook) and IEEE 802.3. The differences between these two later standards are minor, yet they are nonetheless significant. Despite the generic term 'Ethernet' being applied to all CSMA/CD networks, it should, technically speaking, be reserved for the original DIX standard. This chapter will continue with popular use and refer to all the LANs of this type as Ethernet, unless it is important to distinguish between them.

Early Ethernet (of the 10 mbps variety) uses the CSMA/CD access method. This gives a system that can operate with little delay, if lightly loaded, but access to the medium can become very slow if the network is heavily loaded. Ethernet network interface cards are relatively cheap and produced in vast quantities. Ethernet has, in fact, become the most widely used networking standard. However, because of its probabilistic access mechanism, there is no guarantee of message transfer and messages cannot be prioritized.

Modern Ethernet systems are a far cry from the original design. From 100BaseT onwards they are capable of full-duplex (sending and receiving at the same time via switches, without collisions) and the ethernet frame has been modified to make provision for prioritization and virtual LANs.

It is assumed that 10 gigabit Ethernet is commercially available since the middle of 2002. Ethernet has also been modified for industrial use and as such, has made vast inroads into the process control environment.

As an introduction to Ethernet, the next section will deal with systems based on the original IEEE 802.3 (CSMA/CD) standard.

15.2   10 Mbps Ethernet

15.2.1   Media systems

The IEEE 802.3 standard (also known as ISO 8802.3) defines a range of media types that can be used for a network based on this standard such as coaxial cable, twisted pair cable and fiber optic cable. It supports various cable media and transmission rates at 10 Mbps, such as:

- 10Base2 – thin wire coaxial cable (6.3 mm/0.25 inch diameter), 10 Mbps baseband operation, bus topology.
- 10Base5 – thick wire coaxial cable (13 mm/0.5 inch diameter), 10 Mbps baseband operation, bus topology.
- 10BaseT – unscreened twisted pair cable (0.4 to 0.6 mm conductor diameter), 10 Mbps baseband operation, hub topology.
- 10BaseF – optical fiber cables, 10 Mbps, 10 Mbps baseband operation, point-to-point topology.

Other variations include 1Base5, 10BaseFB, 10BaseFP and 10Broad36 but these versions would never become commercially viable.

10Base5

This is the original coaxial cable system and is also called 'thicknet'. The coaxial cable (50-ohm characteristic impedance) is yellow or orange in color. The naming convention for 10Base5 means 10 Mbps, baseband signaling on a cable that will support 500-meter (1640 feet) segment lengths. The cable is difficult to work with, and cannot normally be taken to the node directly. Instead, it is laid in a cabling tray and the transceiver electronics (medium attachment unit or MAU) are installed directly on the cable. From there, an intermediate cable, known as an attachment unit interface (AUI) cable is used to connect to the NIC. This cable can be a maximum of 50 meters (164 feet) long, compensating for the lack of flexibility of placement of the coaxial cable. The AUI cable consists of five individually shielded pairs – two each (control and data) for both transmit and receive, plus one for power.

Cutting the cable and inserting N connectors and a coaxial T or more commonly by using a 'bee sting' or 'vampire' tap can make the MAU connection to the cable. The vampire tap is a mechanical connection that clamps directly over the cable. Electrical connection is made via a probe that connects to the center conductor and sharp teeth that physically puncture the cable sheath to connect to the braid. These hardware components are shown in Figure 15.1.

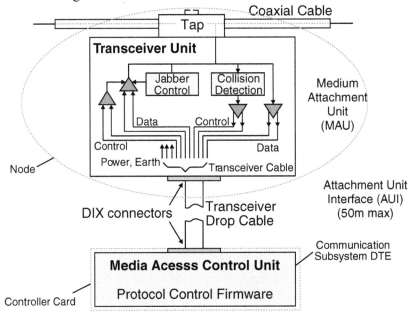

Figure 15.1
10Base5 components

The location of the connection is important to avoid multiple electrical reflections on the cable, and the cable is marked every 2.5 meters (8 feet) with a black or brown ring to indicate where a tap should be placed. Fan-out boxes can be used if there are a number of nodes for connection, allowing a single tap to feed each node as though it was individually connected. The connection at either end of the AUI cable is made through a 25-pin D connector with a slide latch, often called a DIX connector after the original consortium.

There are certain requirements if this cable architecture is used in a network. These include:

- Segments must be less than 500 meters (1640 feet) in length to avoid signal attenuation problems.
- Not more than 100 taps on each segment.
- Taps must be placed at integer multiples of 2.5 meters (8 feet).
- The cable must be terminated with a 50-ohm terminator at each end.
- It must not be bent at a radius exceeding 25.4 cm or 10 inches.
- One end of the cable shield must be earthed.

The physical layout of a 10Base5 Ethernet segment is shown in Figure 15.2.

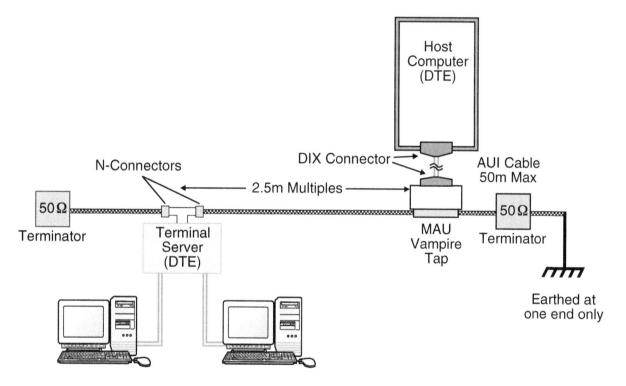

Figure 15.2
10Base5 Ethernet segment

Note that when MAU and AUI cable are used, the on board transceiver on the NIC is not used. Rather, there is a transceiver in the MAU and this is fed with power from the NIC via the AUI cable. Since the transceiver is remote from the NIC, the node needs to be aware that the termination can detect collisions if they occur. A signal quality error (SQE), or heartbeat test function in the MAU performs this confirmation. The SQE signal is sent from the MAU to the node on detecting a collision on the bus. However, on completion of every frame transmission by the MAU, the SQE signal is asserted to ensure that the circuitry remains active, and that collisions can be detected. The SQE pulses occur during the 96-bit inter-frame gap between packets but they are recognized by the NIC and are not confused with a collision. Not all components support SQE test and mixing those that do with those that do not or could not cause problems. If, for example, an MAU is connected to a repeater, the SQE function on that MAU needs to be turned off. If not, the repeater will mistakenly see the SQE signal as a collision, and respond with a jam signal. This is not easily detectable and can radically slow down a network.

10Base2

The other type of coax-based Ethernet network is 10Base2, often referred to as 'thinnet' or 'thinwire Ethernet'. It uses type RG-58 A/U or C/U cable with a 50-ohm characteristic impedance and 5 mm diameter. The cable is normally connected to the NICs in the nodes by means of a BNC T-piece connector. Connectivity requirements include:

- It must be terminated at each end with a 50-ohm terminator
- The maximum length of a cable segment is 185 meters (600 feet) and not 200 meters (650 feet)
- Not more than 30 transceivers can be connected to any one segment
- There must be a minimum spacing of 0.5 meters (1.6 feet) between nodes
- It may not be used as a link segment between two 'thicknet' segments
- The minimum bend radius is 5 cm (2 inches)
- The maximum distance between the medium and the transceiver is 4 inches; this is taken up by the dimensions of the T connector and the pc board tracks, which means that no drop cable may be used – the BNC T-piece has to be located on the front panel of the nic.

The physical layout of a 10Base2 Ethernet segment is shown in Figure 15.3.

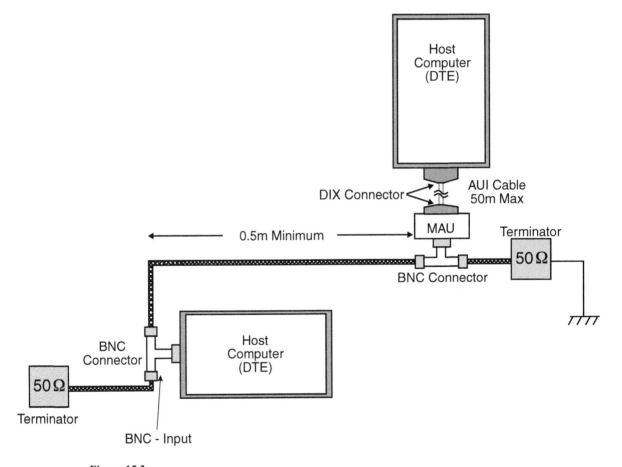

Figure 15.3
10Base2 Ethernet segment

At one stage the use of thinnet cable was very popular as a cheap and relatively easy way to setup a network. However, there are disadvantages with this approach as a cable fault can bring the whole system down.

10BaseT

10BaseT uses AWG24 unshielded twisted pair (UTP) cable for connection to the node. The physical topology of the standard is a star, with nodes connected to a hub. The cable can be category 3 (Cat3) UTP, although that does not support the faster versions of Ethernet.

The node cable (hub to node):

- Has a maximum length of 100 meters (328 feet),
- Consists of four pairs of which only two pairs are used (one for receive and one for transmit) and
- Is connected via RJ-45 plugs.

The hub can be considered as a bus internally (any signal input on a given port are reflected as outputs on all other ports), and so the topology is a logical bus topology although it physically resembles a star (hub) topology. The following figure shows schematically how the hub interconnects the 10BaseT nodes.

Collisions are detected by the NIC and so the hub must retransmit an input signal on all output pairs, so that each NIC can also receive its own transmitted signal. The electronics in the hub must ensure that the stronger retransmitted signal does not interfere with the weaker input signal. The effect is known as far end cross-talk (FEXT), and it is handled by special adaptive cross-talk echo cancellation circuits.

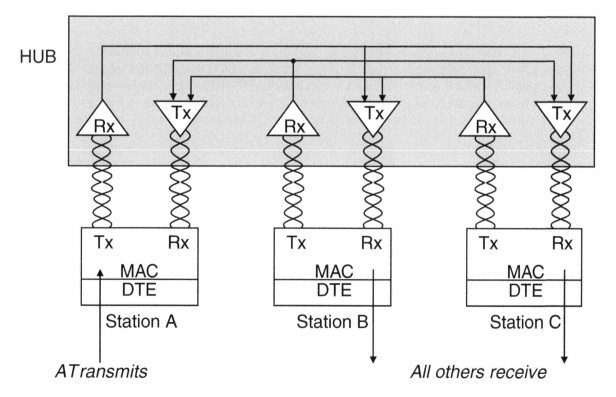

Figure 15.4
10BaseT hub concept

The standard has become increasingly popular for new networks, although there are some disadvantages that should be recognized.

- The unshielded cable is not very resistant to electrostatic electrical noise, and may not be suitable for some industrial environments.
- Whilst the cable is inexpensive, there is the additional cost of the associated wiring hubs to be considered.
- The cable length is limited to 100 m (328 feet).

Advantages of the system include:

- The twisted pair cable provides good electromagnetic noise immunity.
- Intelligent hubs can monitor traffic on each port. This improves on the security of the network – a feature that has often been lacking in a broadcast, common media network such as Ethernet.
- Because of the cheap wire used, flood wiring can be installed in a new building, providing many more wiring points than are initially needed, but giving great flexibility for future expansion. When this is done, patch panels – or punch down blocks – are often installed for even greater flexibility.
- The star-wiring configuration improves physical reliability. A cut cable only affects one node.

10BaseF

This standard makes provision for three architectures viz. 10BaseFL, 10BaseFP and 10BaseFB. The latter two have never gained commercial acceptance and are not currently manufactured by any vendor.

10BaseFL

The fiber link segment standard is basically a 2 km (1.2 miles) upgrade to the existing fiber optic inter repeater link (FOIRL) standard. The original FOIRL as specified in the 802.3 standard was limited to a 1-km (0.6-mile) fiber link between two repeaters, with a maximum length of 2.5-km (1.5 miles) if there are five segments in the link. Note that this is a link between two repeaters in a network, and cannot have any nodes connected to it.

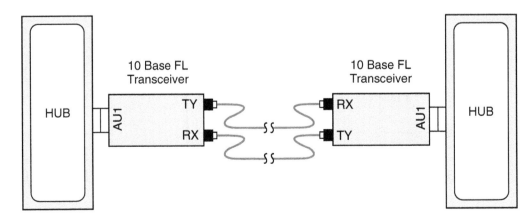

Figure 15.5
10BaseFL segment

15.2.2   Signaling methods

10 Mbps ethernet signals are encoded using the Manchester encoding scheme. This method allows a clock to be extracted at the receiver end to synchronize the transmission/reception process. The encoding is performed by an exclusive or between a 20 MHz clock signal and the data stream. In the resulting signal, a 0 is represented by a high to low change at the center of the bit cell, whilst a 1 is represented by a low to high change at the center of the bit cell. There may or may not be transitions at the beginning of a cell as well, but these are ignored at the receiver. The transitions in every cell allow the clock to be extracted and synchronized with the transmitter. This method is very wasteful of bandwidth (it requires a cable bandwidth of well above twice the bit rate) and is therefore not used on the faster Ethernet versions.

In the 802.3 standard, voltages swing between 0 and –2.05 volts on coax, or between –2.5 V and +2.5 V on twisted pair.

15.2.3   Medium access control

10 Mbps Ethernet operates in half-duplex mode so a station can either receive or transmit, but not both at the same time. The method used to achieve this is called CSMA/CD, or carrier sense multiple access with collision detection.

Essentially, the method used is one of contention. Each node has a connection via a transceiver to the common bus. As a transceiver, it can both transmit and receive at the same time. Each node can be in any one of three states at any time. These states are:

- Idle, or listen
- Transmit
- Contention

In the idle state, the node merely listens to the bus, monitoring all traffic that passes. If a node then wishes to transmit information, it will defer whilst there is any activity on the bus, since this is the 'carrier sense' component of the architecture. At some stage, the bus will become silent, and the node, sensing this, will then commence its transmission. It is now in the transmit mode, and will both transmit and listen at the same time. This is because there is no guarantee that another node at some other point on the bus has not also started transmitting having recognized the absence of traffic.

After a short delay as the two signals propagate towards each other on the cable, there will be a collision of signals. Obviously the two transmissions cannot coexist on the common bus, since there is no mechanism for the mixed analog signals to be 'unscrambled'. The transceiver quickly detects this collision, since it is monitoring both its input and output and recognizes the difference. The node now goes into the third state viz. contention. The node will continue to transmit for a short time – the jam signal – to ensure the other transmitting node detects the contention, and then performs a back-off algorithm to determine when it should again attempt to transmit its waiting frames.

15.2.4   Frame transmission

When a frame is to be transmitted, the medium access control monitors the bus and defers to any passing traffic. After a period of 96-bit times, known as the inter-frame gap (to allow the passing frame to be received and processed by the destination node) the transmission process commences. There is a finite time for this transmission to propagate to the ends of the bus cable. Thus, it ensures that all nodes recognize that the medium is busy. The transceiver turns on a collision detect circuit whilst the transmission takes place. Once a certain number of bits (512) have been transmitted, provided that the network cable segment specifications have been complied with, the collision detection circuitry can be disabled. If a collision should take place after this, it will be the responsibility of higher protocols to request retransmission – a far slower process than the hardware collision detection process.

This is a good reason to comply with cable segment specifications! This initial 'danger' period is known as the collision window and is effectively twice the time interval for the first bit of a transmission to propagate to all parts of the network. The slot time for the network is then defined as the worst case time delay that a node must wait before it can reliably know that a collision has occurred. It is defined as:

Slot time = 2 * (transmission path delay) + safety margin

For a 10 Mbps system the slot time is fixed at 512 bits or 51.2 microseconds.

15.2.5   Frame reception

The transceiver of each node is constantly monitoring the bus for a transmission signal. As soon as one is recognized, the NIC activates a carrier sense signal to indicate that transmissions cannot be made. The first bits of the MAC frame are a preamble and consist of alternating bits of 1010 etc. On recognizing these, the receiver synchronizes its clock, and converts the Manchester encoded signal back into binary form. The eighth octet is a start of frame delimiter, and this is used to indicate to the receiver that it should strip off the first eight octets and commence frame reception into a frame buffer within the NIC.

Further processing then takes place, including the calculation and comparison of the frame CRC with the transmitted CRC. In valid frames, the destination hardware address is checked. If this matches the MAC address in the card firmware or a broadcast address, then the message is passed up to the protocol stack, otherwise the frame is discarded. It also checks that the frame contains an integral number of octets and whether it is either too short or too long. Provided all is correct, the frame is passed to the data link layer for further processing.

15.2.6   MAC frame format

The basic frame format for an IEEE 802.3 network is shown in Figure 15.6. There are eight fields in each frame, and they are described in detail.

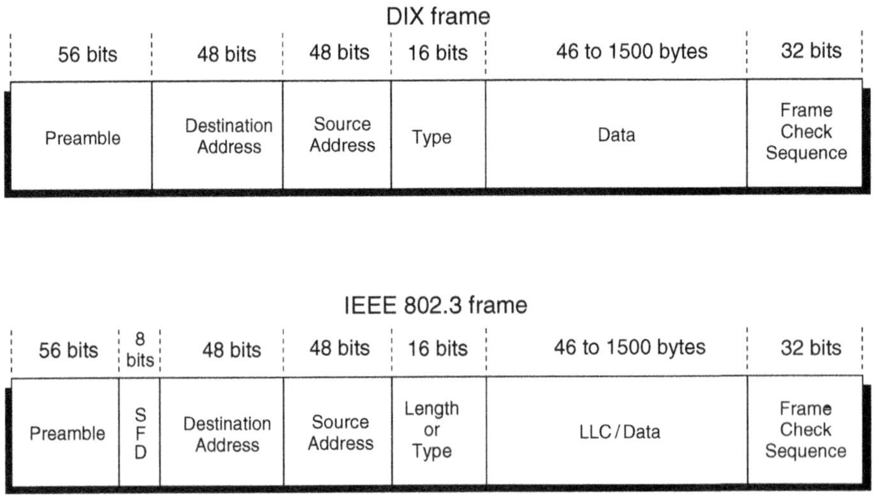

Figure 15.6
Ethernet V2/IEEE 802.3 MAC frame formats

Preamble

This field consists of 7 octets of the data pattern 10101010. The receiver uses this to synchronize its clock to the transmitter.

Start frame delimiter

This single octet field consists of the data 10101011. It enables the receiver to recognize where the address fields commence.

Source and destination address

These are the physical addresses of both the source and destination nodes. The fields are 6 octets long. Assignment of addresses is controlled by the IEEE standards association (IEEE-SA), which administers the allocation of the first three bytes. When assigning blocks of addresses to manufacturers, the IEEE-SA provides a 24-bit organizationally unique identifier (OUI). The OUI (the first three bytes of the hardware address) identifies the manufacturer of the card and is the same for all cards made by it. The second half is the device identifier, and these numbers are assigned sequentially during the manufacturing process so that each card will have a unique address.

The addressing modes according to IEEE 802.3 include:

- Broadcast. The destination address is set to all 1s or FFFFFFFFFFFF, and all nodes on the network will respond to the message.
- Unicast (individual, or point-to-point). If the first bit of the address is 0, the frame is sent to one specific node.

The addresses allocated by a factory are referred to as globally administered addresses. The standard also makes provision for a locally administered address (in which the second bit is set to one) but in practice this is rarely used as cards are invariably bought with preset MAC addresses.

Length

For IEEE 802.3, this two-octet field contains a number representing the length of the data field. For Bluebook Ethernet this field represents 'type' rather than 'length'. 'Type' is a number bigger than 1500 decimal representing the protocol that has submitted the data for delivery.

Data

The information that has been handed down from the layer above between 0 and 1500 bytes.

Pad

Since there is a minimum frame length of 64 octets (512 bits) that must be transmitted to ensure that the collision mechanism works, the pad field will pad out any frame that does not meet this minimum specification by adding up to 46 bytes of 'padding'. This pad, if incorporated, is normally random data. The CRC is calculated over the data in the pad field. Once the CRC checks OK, the receiving node discards the pad data, which it recognizes by the value in the length field.

FCS

A 32-bit CRC value that is computed in hardware at the transmitter and appended to the frame. It is the same algorithm used in the 802.4 and 802.5 standards.

Difference between IEEE 802.3 and Ethernet V2

As mentioned before, there is a difference between 802.3 Blue Book Ethernet. These differences are primarily in the frame structure and are shown below.

802.3 Network	Ethernet network
Star topology supported using UTP, fibre etc, or bus.	Only supports bus topology.
Baseband and broadband signaling.	Baseband only.
Data link layer divided into LLC and MAC.	No subdivision of DLL.
7 octets of preamble plus SFD.	8 bytes of preamble with no separate SFD.
Length field in data frame.	Field used to indicate the higher level protocol using the data link service.

Table 15.1
Differences between IEEE 802.3 and Blue Book Ethernet (V2)

15.2.7   IEEE 802.2 LLC

Ethernet V2 implements both the physical and the data link layer. The frame then contains the detail of the higher-layer protocol being carried in the frame. This information is required for correctly multiplexing and demultiplexing the packets carried by the network, i.e. ensuring that they get delivered to the same type of protocol that sent them.

In the case of IEEE 802.3, however, it was decided to split the data link layer in two, namely, in a lower half that controls access to the medium (medium access control, or MAC) and an upper half that performs the link control between the two NCIs involved (the logical link control or LLC). The MAC part is included together with the physical layer in the IEEE 802.3 specification, but the LLC is implemented separately in the IEEE 802.2 specification. This was done since the LLC functionality is not unique to 802.3, but is used by all other network standards as well.

Since not all LANs necessarily have a 'type' field in their frames, the protocol information required for multiplexing is now carried in the SSAP (source service access point) and DSAP (destination service access point) within the 802.2 LLC header. The entire LLC header is only 3–4 bytes and is carried within the 'data' field of the 802.3 frame, just after the header.

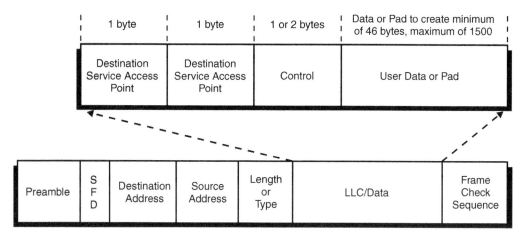

Figure 15.7
LLC PDU carried in an Ethernet frame

15.2.8 Reducing collisions

The main contributors to collisions on an Ethernet network are:

- The number of packets per second
- The signal propagation delay between transmitting nodes
- The number of stations initiating packets
- The bandwidth utilization

A few suggestions on reducing collisions in an Ethernet network are:

- Keep all cables as short as possible.
- Keep all high activity sources and their destinations as close as possible. Possibly, isolate these nodes from the main network backbone with bridges/routers to reduce backbone traffic.
- Separate segments with bridges and switches rather than with repeaters, which do not truncate the collision domain.
- Check for unnecessary broadcast packets.
- Remember that the monitoring equipment to check out network traffic can contribute to the traffic (and the collision rate).

15.2.9 Design rules

The following design rules on length of cable segment, node placement and hardware usage should be strictly observed.

Length of the cable segments

It is important to maintain the overall Ethernet requirements as far as length of the cable is concerned. Each segment has a particular maximum length allowable. For example, 10Base2 allows 185 m maximum length. The recommended maximum length is 80% of this figure. Some manufacturers advise that you can disregard this limit with their equipment. This can be a risky strategy and should be carefully considered.

System	Maximum	Recommended
10Base5	500 m (1640 feet)	400 m (1310 feet)
10Base2	185 m (600 feet)	150 m (500 feet)
10BaseT	100 m (330 feet)	80 m (260 feet)

Table 15.2
Suggested maximum lengths

Cable segments need not be made from a single homogenous length of cable and may comprise multiple lengths joined by coaxial connectors (two male plugs and a connector barrel).

To achieve maximum performance on 10Base5 cable segments, it is preferable that the total segment be made from one length of cable or from sections of the same drum of cable. If multiple sections of cable from different manufacturers are used, then these should be standard lengths of 23.4 m (76.8 feet), 70.2 m (230.3 feet) or 117 m (383.9 feet) (± 0.5 m/1.0 foot), which are odd multiples of 23.4 m/76.8 feet (half wavelength in the cable at 5 MHz). These lengths ensure that reflections from the cable to cable impedance discontinuities are unlikely to add in phase. Using these lengths exclusively a mix of cable sections should be able to be made up to the full 500 m (1640 feet) segment length.

If the cable is from different manufacturers and it is suspected that there are potential mismatch problems, it should be checked for signal reflections, as impedance mismatches do not exceed 7% of the incident wave.

Maximum transceiver cable length

In 10Base5 systems, the maximum length of the transceiver cables is 50 m (164 feet) but it should be noted that this only applies to specified IEEE 802.3 compliant cables. Other AUI cables using ribbon or office grade cables can only be used for short distances (less than 12.5 m/41 feet), so check the manufacturer's specifications for these.

Node placement rules

Connection of the transceiver media access units (MAU) to the cable causes signal reflections due to their bridging impedance. Placement of the MAUs must therefore be controlled to ensure that reflections from them do not significantly add in phase.

In 10Base5 systems, the MAUs are spaced at multiples of 2.5 m (8.2 feet), coinciding with the cable markings.

In 10Base2 systems, the minimum MAU spacing is 0.5 m (1.6 feet).

Maximum transmission path

The maximum transmission path is made of five segments connected by four repeaters. The total number of segments can be made up of a maximum of three coax segments containing station nodes and two link segments. The link segments are defined as [point-to-point full-duplex links that connect 2– and only 2– MAUs (e.g. 10BaseFL)]. This is summarized as the 5–4–3–2 rule.

5 segments 4 repeaters 3 coax segments 2 link segments	**OR**	5 segments 4 repeaters 3 link segments 2 coax segments

Table 15.3
The 5–4–3–2 rule

It is important to verify that the above transmission rules are met by all paths between any two nodes on the network.

Note that the maximum sized network of four repeaters supported by IEEE 802.3 can be susceptible to timing problems. The maximum configuration is limited by propagation delay.

Note that 10Base2 segments should not be used to link 10Base5 segments.

Repeater rules

Repeaters are connected to transceivers that count as one node on the segments.

Special transceivers are used to connect repeaters and these do not implement the signal quality error test (SQE).

Fiber optic repeaters are available giving up to 3000 m links at 10 MB/S. Check the vendor's specifications for adherence with IEEE 802.3 repeater performance and compliance with the fiber optic inter repeater link (FOIRL) standard.

Cable system grounding

Grounding has safety and noise connotations. 802.3 states that the shield conductor of each coaxial cable shall make electrical contact with an effective earth reference at one point only.

The single point earth reference for an Ethernet system is usually located at one of the terminators. Most terminators for Ethernet have a screw terminal to which a ground lug can be attached using a braided cable preferably to ensure good earthing.

Ensure that all other splices, taps or terminators are jacketed so that no contact can be made with any metal objects. Insulating boots or sleeves should be used on all in-line coaxial connectors to avoid unintended earth contacts.

Round trip delay time

The maximum round trip propagation delay is 51.2 µs because of the minimum frame size of 64 bytes (512 bits). This time is determined by adding the propagation delays in all of the electronic components and cables that make up the longest signal path and then doubling this figure to obtain the round trip delay time.

Note that this calculation needs to be done for all components within a collision domain, store-and-forward devices such as bridges, routers or switches terminate collision domains.

Tables 15.4 and 15.5 give typical maximum one-way delay times for various components and cables. Repeater and NIC delays for specific components can be obtained from the manufacturers.

Component	Maximum delay (µs)
Standard transceiver	0.86
Twisted pair transceiver	0.27
Fibre optic transceiver	0.20
Multiport transceiver	0.10
10BaseT concentrator	1.90
Local repeater	0.65
Fibre optic repeater	1.55
Multiport repeater	1.55

Table 15.4
Maximum one-way Ethernet component delays

Cable medium	Maximum delay per meter (µs)
UTP	0.0057
Coax (10Base5)	0.00433
Coax (10Base2)	0.00514
Fiber optic	0.005
AUI	0.00514

Table 15.5
Maximum one-way Ethernet cable delays

15.3 100 Mbps Ethernet

15.3.1 Introduction

100BaseT is the shorthand identifier for all 100 Mbps Ethernet systems, including twisted pair copper and fiber versions. These include 100BaseTX, 100BaseFX, 100BaseT4 and 100BaseT2.

100BaseX is the designator for the 100BaseTX (copper) and 100BaseFX (fiber) systems based on the same 4B/5B block encoding system used for FDDI (ANSI X3T9.5) fiber distributed interface system. 100BaseTX, the most widely used version, is a 100 Mbps baseband system operating over 2 pairs of Cat5 UTP while 100BaseFX operates at 100 Mbps in baseband mode over two multimode optic fibers.

100BaseT4 was designed to operate at 100 Mbps over 4 pairs of Cat3 cable, but this option never gained widespread acceptance. Yet another version, 100BaseT2, was supposed to operate over just 2 pairs of Cat3 cable but was never implemented by any vendor.

One of the limitations of the 100BaseT systems is the size of the collision domain, which is 250 m or 5.12 microseconds. This is the maximum size of a network segment in which collisions can be detected, being one tenth of the size of the maximum size 10 Mbps network. This effectively limits the distance between a workstation and hub to 100 m, the same as for 10BaseT. This means that networks larger than 200 m must be logically connected together by store-and-forward type devices such as bridges, routers or switches. This is not a bad thing, since it segregates the traffic within each collision domain, reducing the number of collisions on the network. The use of bridges and routers for traffic segregation, in this manner, is often done on industrial Ethernet networks.

The format of the frame has been left unchanged. The only difference is that it is transmitted 10 times faster than in 10 Mbps ethernet, hence its length (in time) is 10 times less.

15.3.2   Media access: full-duplex

The original ethernet could only operate in half-duplex mode, using CSMA/CD. Later versions of ethernet (100BaseT and up) can also operate in full-duplex mode, in which case the CSMA/CD mechanism is switched off. Full-duplex means a node can transmit and receive simultaneously and whenever it wants, since there is no possibility of contention. Full-duplex operation is specified in 802.3x.

The 802.3 supplement also describes an optional set of mechanisms used for flow control over full-duplex links, called MAC control and PAUSE. These are unfortunately beyond the scope of this document.

The following requirements must be met for full-duplex operation:

- The media must have separate transmit and receive paths. This is typically true for twisted pair and fiber optic links.
- There must be only two stations on a segment. As a result of this, there can never be any collisions as the two stations can talk and listen simultaneously with their transmit and receive ports cross connected (as in a null modem). Each port of a switching hub exists in its own segment, so if only one station is connected to each switch port, this condition is satisfied.
- Both stations must be capable of and must be setup for, full-duplex operation.

15.3.3   Auto-negotiation

The specifications for auto-negotiation were published in 1995 as part of the 802.3μ fast Ethernet supplement to the IEEE standard based on the NWay system developed by National Semiconductor.

Auto-negotiation can be supported on all Ethernet systems using twistedpair cabling, as well as the gigabit Ethernet fiber optic system. It enables devices to be set to a mutually acceptable speed and operating mode (full-duplex/half-duplex). Auto-negotiation includes the following concepts:

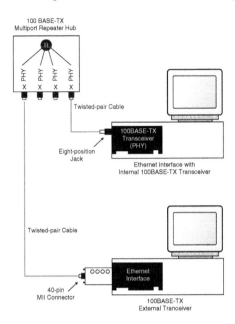

Figure 15.8
100BaseTX connection

Fiber optic cable distances 100BaseFX

The following maximum cable distances apply:

- Node to hub: maximum distance of multimode cable (62.5/125) is 160 meters
  (for connections using a single class II repeater)
- Node to switch: maximum multimode cable distance is 210 meters
- Switch to switch: maximum distance of multimode cable for a backbone connection between two 100BaseFX switch ports is 412 meters
- Switch to switch full-duplex: maximum distance of multimode cable for a full-duplex connection between two 100BaseFX switch ports is 2000 meters

The IEEE has not included the use of single-mode fiber in the 802.3u standard. However numerous vendors have products available enabling switch-to-switch distances of up to 10 to 20 km (6 to 12 miles) using single-mode fiber.

100BaseT repeater rules

The cable distance and the number of repeaters, which can be used in a 100BaseT collision domain, depends on the delay in the cable and the time delay in the repeaters and NIC delays. The maximum round-trip delay for 100BaseT systems is the time to transmit 64 bytes or 512 bits and equals 5.12 µs. A frame has to go from the transmitter to the most remote node then back to the transmitter for collision detection within this round trip time. Therefore the one-way time delay will be half this.

The maximum sized collision domain can then be determined by the following calculation:

Repeater delays + Cable delays + NIC delays + Safety factor (5 bits minimum) <2.56 µs

The following Table 15.6 gives typical maximum one-way delays for various components. Repeater and NIC delays for specific components can be obtained from the manufacturer.

Component	Maximum delay (µs)
Fast Ethernet NIC	0.25
Fast Ethernet switch port	0.25
Class I repeater	0.7 max
Class II repeater	0.46 max
UTP cable (per 100 m/330 ft))	0.55
Multimode fiber (per 100 m/330 ft)	0.50

Table 15.6
Maximum one-way fast Ethernet component delays

Notes

- If the desired distance is too great it is possible to create a new collision domain by using a switch instead of a repeater.
- Most 100BaseT repeaters are stackable, which means multiple units can be placed on top of one another and connected together by means of a fast backplane bus, such connections do not count as a repeater hop and make the ensemble function as a single repeater.

Sample calculation

The following calculation is made to confirm whether it is possible to connect two fast Ethernet nodes together using two class II repeaters connected by 50-m fiber. One node is connected to the first repeater with 50-m UTP while the other has a 100-m fiber connection.

Calculation: Using the time delays in Table 15.6:

The total one-way delay of 2.445 µs is within the required interval (2.56 µs) and allows at least 5 bits safety factor, so this connection is permissible.

15.4   Gigabit Ethernet

15.4.1   Introduction

1000BaseX is the shorthand identifier for the gigabit Ethernet system based on the 8B/10B block encoding scheme adapted from the fiber channel networking standard, developed by ANSI. 1000BaseX includes 1000BaseSX, 1000BaseLX and 1000BaseCX.

- 1000BaseSX is the short wavelength fiber version.
- 1000BaseLX is the long wavelength fiber version.
- 1000BaseCV is a short copper cable version, based on the fiber channel standard.

1000BaseT, on the other hand, is a 1000 Mbps version capable of operating over Cat5 (or better, such as Cat5e) UTP. This system is based on a different encoding scheme.

As with fast Ethernet, gigabit Ethernet supports full-duplex and auto-negotiation. It uses the same 802.3 frame format as 10 Mbps and 100 Mbps Ethernet systems. This operates at ten times the clock speed of fast Ethernet at 1 Gbps. By retaining the same frame format as the earlier versions of Ethernet, backward compatibility is assured. Despite the similar frame format, the system had to undergo a small change to enable it to function effectively at 1 Gbps. The slot time of 64 bytes used with both 10 Mbps and 100 Mbps systems has been increased to 512 bytes. Without this increased slot time the network would have been impracticably small at one tenth of the size of fast Ethernet – only 20 meters. The slot time defines the time during which the transmitting node retains control of the medium, and in particular is responsible for collision detection. With gigabit Ethernet, it was necessary to increase this time by a factor of eight to 6.096 μs to compensate for the tenfold speed increase. This then gives a collision domain of about 200 m (660 feet).

If the transmitted frame is less than 512 bytes the transmitter continues transmitting to fill the 512-byte window. A carrier extension symbol is used to mark frames that are shorter than 512 bytes and to fill the remainder of the frame.

15.4.2 Gigabit Ethernet full-duplex repeaters

Gigabit Ethernet nodes are connected to full-duplex repeaters, also known as non-buffered switches or buffered distributors. These devices have a basic MAC function in each port, which enables them to verify that a complete frame is received and compute its frame check sequence (CRC) to verify the frame validity. Then the frame is buffered in the internal memory of the port before being forwarded to the other ports of the repeater. It is therefore combining the functions of a repeater with some features of a switch.

All ports on the repeater operate at the same speed of 1 Gbps, and operate in full-duplex so it can simultaneously send and receive from any port. The repeater uses 802.3x flow control to ensure the small internal buffers associated with each port do not overflow. When the buffers are filled to a critical level, the repeater tells the transmitting node to stop sending until the buffers have been sufficiently emptied. The repeater does not analyze the packet address fields to determine where to send the packet, like a switch does, but simply sends out all valid packets to all the other ports on the repeater.

The IEEE does allow for half-duplex gigabit repeaters – however none exist at this time.

15.4.3 Gigabit Ethernet design considerations

Fiber optic cable distances

The maximum cable distances that can be used between the node and a full-duplex 1000BaseSX and LX repeater depend mainly on the chosen wavelength, the type of cable, and its bandwidth. The maximum transmission distances on multimode cable are limited by the differential mode delay (DMD). The very narrow beam of laser light injected into the multimode fiber results in a relatively small number of rays going through the fiber core. These rays each have different propagation times because they are going through differing lengths of glass by zigzagging through the core to a greater or lesser extent. These pulses of light can cause jitter and interference at the receiver. This is overcome by using a conditioned launch of the laser into the multimode fiber. This spreads the laser light evenly over the core of the multimedia fiber so the laser source looks more like a light emitting diode (LED) source. This spreads the light in a large number of rays across the fiber resulting in smoother spreading of the pulses, so less interference. This conditioned launch is done in the 1000BaseSX transceivers.

Table 15.7 gives the maximum distances for full-duplex 1000BaseX repeaters.

NIC	0.25 µs
50 m (165 ft) UTP	0.275 µs
Repeater class II	0.46 µs
50 m (165 ft) fiber	0.25 µs
Repeater class II	0.46 µs
100 m (330 ft) fiber	0.50 µs
NIC	0.25 µs
Total delay	**2.445 µs**

Table 15.7
Maximum fiber distances for 1000BaseX (full-duplex)

Gigabit repeater rules

The cable distance and the number of repeaters, which can be used in a half-duplex 1000BaseT collision domain depends on the delay in the cable and the time delay in the repeaters and NIC delays. The maximum round-trip delay for 1000BaseT systems is the time to transmit 512 bytes or 4096 bits and equals 6.096 µs. A frame has to go from the transmitter to the most remote node then back to the transmitter for collision detection within this round trip time. Therefore the one-way time delay will be half this.

The maximum sized collision domain can then be determined by the following calculation:

Repeater delays + Cable delays + NIC delays + Safety factor   (5 bits minimum) <2.048 µs

Table 15.8 gives typical maximum one-way delays for various components. Repeater and NIC delays for your specific components can be obtained from the manufacturer.

Wavelength (nm)	Cable type	Bandwidth (MHz/km)	Attenuation (dB/km)	Maximum distance (m)
850	50/125 multimode	400	3.25	500
850	50/125 multimode	500	3.43	550
850	62.5/125 multimode	160	160	220
850	62.5/125 multimode	200	200	275
1300	50/125 multimode	500	2.32	550
1300	62.5/125 multimode	500	1.0	550
1300	9/125 singlemode	Infinite	0.4	5000

Table 15.8
Maximum one-way gigabit Ethernet component delays

These calculations give the maximum collision diameter for IEEE 802.3z half-duplex gigabit Ethernet systems. The maximum gigabit Ethernet network diameters specified by the IEEE are shown in Table 15.9.

Component	Maximum delay (µs)
Gigabit NIC	0.432
Gigabit repeater	0.488
UTP cable (per 100 m)	0.55
Multimode fiber (per 100 m)	0.50

Table 15.9
Maximum half-duplex gigabit Ethernet network diameters

Note half-duplex gigabit Ethernet repeaters are not available for sale. Use full-duplex repeaters with the point-to-point cable distances between node and repeater or node and switch as listed in Table 15.7.

15.5   Industrial Ethernet

15.5.1   Introduction

Early Ethernet was not entirely suitable for control functions as it was primarily developed for office-type environments. The Ethernet technology has, however, made rapid advances over the past few years. It has gained such widespread acceptance in industry that it is becoming the *de facto* field bus technology.   An indication of this trend is the inclusion of Ethernet as the levels 1 and 2 infrastructure for Modbus/TCP (Schneider), Ethernet/IP (Rockwell Automation and ODVA), ProfiNet (Profibus) and Foundation Fieldbus HSE.

The following sections will deal with problems related to early Ethernet, and how they have been addressed in subsequent upgrades.

15.5.2   Connectors and cabling

Earlier industrial Ethernet systems such as the first-generation Siemens Simatic Net (Sinec-H1) are based on the 10Base5 configuration, and thus the connectors involved include the screw-type N connectors and the D-type connectors, which are both fairly rugged. The heavy-gauge twin-screen (braided) RG-8 coaxial cable is also quite impervious to electrostatic interference.

Most modern industrial Ethernet systems are, however, based on a 10BaseT/100BaseTX configuration and thus have to contend with RJ-45 connectors and (typically) Cat5/Cat5e unshielded twisted pair (UTP) cable. Despite its inherent resistance to electro-magnetically induced interference, the UTP cable is not really suited to industrial applications. The connectors are problematic as well. The RJ-45 connectors are everything but rugged and are suspect when subjected to great temperature extremes, contact with oils and other fluids, dirt, UV radiation, EMI as well as shock, vibration and mechanical loading.

As in interim measure, some manufacturers started using D-type (known also as DB or D-submin) connectors, but currently there is a strong movement towards the M12-based series of connectors. The M12 form factor, governed by IEC 61076, is widely accepted in field bus and automation applications, and is supplied by vendors such as Lumberg, Turck, Escha, InterlinkBT and Hirschmann. Typical M12 connectors for Ethernet are of the 8-pole variety, with threaded connectors. They can accept various types of Cat5/5e twisted pair wiring such as braided or shielded wire (solid or stranded), and offer excellent protection against moisture, dust, corrosion, EMI, RFI, mechanical vibration and shock, UV radiation, and extreme temperatures (–40C to 75C).

As far as the media is concerned, several manufacturers are producing Cat5 and Cat5e wiring systems using braided or shielded twisted pairs. An example of an integrated approach to industrial Ethernet cabling is Lumberg's etherMATETM system, which also includes both the cabling and an M12 connector system.

15.5.3   Deterministic versus stochastic operation

One of the most common complaints with early Ethernet was that it uses CSMA/CD (a probabilistic or stochastic method) as opposed to other automation technologies such as Fieldbus that use deterministic access methods such as token passing or the publisher–subscriber model. CSMA/CD essentially means that it is impossible to guarantee delivery of a possibly critical message within a certain time. This could be due either to congestion on the network (often due to other less critical traffic) or to collisions with other frames. In office applications there is not much difference between 5 seconds and 500 milliseconds, but in industrial applications a millisecond counts. Industrial processes often require scans in a 5 to 20-millisecond range, and some demanding processes could even require 2 to 5 milliseconds. On 10BaseT Ethernet, for example, the access time on a moderately loaded 100-station network could range from 10 to 100 mS, which is acceptable for office applications but not for processes.

With Ethernet, the loading or traffic needs to be carefully analyzed to ensure that the network is not overwhelmed at critical or peak operational times. While a typical utilization factor on a commercial ethernet LAN of 25% to 30% is acceptable, figures of less than 10% utilization on an industrial ethernet LAN are required. Most industrial networks run at 3 or 4% utilization with a fairly large number of I/O points being transferred across the system.

The advent of fast and gigabit Ethernet, switching hubs, VLAN technology, full-duplex operation and deterministic Ethernet has effectively put this concern to rest for most applications.

15.5.4   Size and overhead of Ethernet frame

Data link encoding efficiency is another problem, with the Ethernet frames taking up far more space than for an equivalent Fieldbus frame. If the TCP/IP protocol is used in addition to the Ethernet frame, the overhead increases dramatically. The efficiency of the overall system is, however, more complex than simply the number of bytes on the transmitting cable and issues such as raw speed on the cable and the overall traffic need to be examined carefully. For example, if 2 bytes of data from an instrument had to be packaged in a 60 byte message (because of TCP/IP and Ethernet protocols being used) this would result in an enormous overhead compared to a Fieldbus protocol. However, if the communications link were running at 100 Mbps or 1 Gbps with full-duplex communications, this would put a different light on the problem and make the overhead issue almost irrelevant.

15.5.5   Noise and interference

Due to higher electrical noise near to the industrial LANs some form of electrical shielding and protection is useful to minimize errors in the communication. A good choice of cable is fiber optic (or sometimes coaxial cable). Twisted pair can be used but care should be taken to route the cables far away from any potential sources of noise. If twisted pair cable is selected; a good decision is to use screened twisted pair cable (ScTP) rather than the standard UTP.

It should be noted here that Ethernet-based networks are installed in a wide variety of systems and rarely have problems reported due to reliability issues. The use of fiber ensures that there are minimal problems due to earth loops or electrical noise and interference.

15.5.6   Partitioning of the network

It is very important that the industrial network operates separately from that of the commercial network, as speed of response and real-time operation are often critical attributes of an industrial network. An office type network may not have the same response requirements. In addition, security is another concern where the industrial network is split off from the commercial networks so any problems in the commercial network will not affect the industrial side.

Industrial networks are also often partitioned into individual sub-networks for reasons of security and speed of response, by means of bridges and switches.

In order to reduce network traffic, some PLC manufacturers use exception reporting. This requires only changes in the various digital and analog parameters to be transmitted on the network. For example, if a digital point changes state (from on to off or off to on), this change would be reported. Similarly, an analog value could have associated with it a specified change of span before reporting the new analog value to the master station.

15.5.7   Switching technology

Both the repeating hub and bridge technologies are being superseded by switching technology. This allows traffic between two nodes on the network to be directly connected in a full-duplex fashion. The nodes are connected through a switch with extremely low levels of latency. Furthermore, the switch is capable of handling all the ports communicating simultaneously with one another without any collisions. This means that the overall speed of the switch backplane is considerably greater than the sum of the speeds of the individual Ethernet ports.

Most switches operate at the data link layer and are also referred to as switching hubs and layer 2 switches. Some switches can interpret the network layer addresses (e.g. the IP address) and make routing decisions on that. These are known as layer 3 switches.

Advanced switches can be configured to support virtual LANs. This allows the user to configure a switch so that all the ports on the switch are subdivided into predefined groups. These groups of ports are referred to as virtual LANs (VLANs) – a concept that is very useful for industrial networks. Only switch ports allocated to the same VLAN can communicate with each other.

Switches do have performance limitations that could affect a critical industrial application. If there is traffic on a switch from multiple input ports aimed at one particular output port, the switch may drop some of the packets. Depending on the vendor implementation, it may force a collision back to the transmitting device so that the transmitting node backs off long enough for the congestion to clear. This means that the transmitting node does not have a guarantee on the transmission between two nodes – something that could impact a critical industrial application.

In addition, although switches do not create separate broadcast domains, each virtual LAN effectively forms one (if this is enabled on the switch). An Ethernet broadcast message received on one port is retransmitted onto all ports in the VLAN. Hence a switch will not eliminate the problem of excessive broadcast traffic that can cause severe performance degradation in the operation of the network. TCP/IP uses the Ethernet broadcast frame to obtain MAC addresses and hence broadcasts are fairly prevalent here.

A problem with a switched network is that duplicate paths between two given nodes could cause a frame to be passed around and around the 'ring' caused by the two alternative paths. This possibility is eliminated by the 'spanning tree' algorithm, the IEEE 802.1d standard for layer 2 recovery. However, this method is quite slow and could take from 2 to 5 seconds to detect and bypass a path failure and could leave all networked devices isolated during the process. This is obviously unacceptable for industrial applications.

A solution to the problem is to connect the switches in a dual redundant ring topology, using copper or fiber. This poses a new problem, as an Ethernet broadcast message will be sent around the loop indefinitely. One vendor, Hirschmann, has solved this problem by developing a switch with added redundancy management capabilities. The redundancy manager allows a physical 200 Mbps ring to be created, by terminating both ends of the traditional Ethernet bus in itself. Although the bus is now looped back to itself, the redundancy manager logically breaks the loop, preventing broadcast messages from wreaking havoc. Logically, the redundancy manager behaves like two nodes, sitting back to back, transmitting and receiving messages to the other around the ring using 802.1p/Q frames. This creates a deterministic path through any 802.1p/Q compliant switches (up to 50) in the ring, which results in a real-time 'awareness' of the state of the ring.

When a network failure is detected (i.e. the loop is broken), the redundancy manager interconnects the two segments attached to it, thereby restoring the loop. This takes place in between 20 and 500 milliseconds, depending on the size of the ring.

15.5.8 Active electronics

With the hub-based ethernet topologies (e.g. 10BaseT, 100BaseTX), the hub (be that a repeating hub or a switching hub) is an active electronic device that requires power to operate. If it fails, this will terminate any communications occurring between nodes connected.

To address this problem, vendors such as Hirschmann have introduced the concept of dual-redundant switch rings with the ability to reconfigure itself in a very short time (30 milliseconds for small networks to 600 milliseconds for large networks) after a switch failure.

In addition, the power requirements for an Ethernet card (and other hardware components) are often significantly more than an equivalent Fieldbus card (such as from Foundation Fieldbus). There is no provision within Ethernet for loop powering the instrument through the Ethernet cable.

15.5.9 Fast and gigabit Ethernet

The recent developments in Ethernet technology are making it even more relevant in the industrial market. Fast Ethernet as defined in the IEEE specification 802.3u is essentially Ethernet running at 100 Mbps. The same frame structure, addressing scheme and CSMA/CD access method are used as with the 10 Mbps standard. In addition, fast Ethernet can also operate in full-duplex mode as opposed to CSMA/CD, which means that there are no collisions. Fast Ethernet operates at ten times the speed of that of standard 802.3 Ethernet. Video and audio applications can enjoy substantial improvements in performance using fast Ethernet. Smart instruments that require far smaller frame sizes will not see such an improvement in performance. One area however, where there may be significant throughput improvements, is in the area of collision recovery. The backoff times for 100 Mbps Ethernet will be a tenth that of standard Ethernet. Hence a heavily loaded network with considerable individual messages and nodes would see performance improvements. If loading and collisions are not really an issue on the slower 10 Mbps network, then there will not be many tangible improvements in the higher LAN speed of 100 Mbps.

Note that with the auto-negotiation feature built into standard switches and many Ethernet cards, the device can operate at either the 10 or 100 Mbps speeds. In addition the category 5 wiring installed for 10BaseT Ethernet would be adequate for the 100 Mbps standard as well.

Gigabit Ethernet is another technology that could be used to connect instruments and PLCs. However its speed would probably not be fully exploited by the instruments for the reasons indicated above.

15.5.10   TCP/IP and industrial systems

The TCP/IP suite of protocols provides for a common open protocol. In combination with Ethernet this can be considered to be a truly open standard available to all users and vendors. However, there are some problems at the application layer area. Although TCP/IP implements four layers which are all open (network interface, Internet, transport and application layers), most industrial vendors still implement their own specific application layer. Hence equipment from different vendors can coexist on the factory shop floor but cannot inter-operate. Protocols such as MMS (manufacturing messaging service) have been promoted as truly 'open' automation application layer protocols but with limited acceptance to date.

15.5.11   Industrial Ethernet architectures for high availability

There are several key technology areas involved in the design of Ethernet-based industrial automation architecture. These include available switching technologies, quality of service (QoS) issues, the integration of existing (legacy) field buses, sensor bus integration, high availability and resiliency, security issues, long distance communication and network management – to name but a few.

For high availability systems a single network interface represents a single point of failure (SPOF) that can bring the system down. There are several approaches that can be used on their own or in combination, depending on the amount of resilience required (and hence the cost). The cost of the additional investment in the system has to be weighed against the costs of any downtime.

For a start, the network topology could be changed to represent a switched ring, as implemented by Hirschmann. This necessitates the use of a special controlling switch (redundancy manager) and protects the system against a single failure on the network. It does not, however, guard against a failure of the network interface on one of the network devices.

The next level of resiliency would necessitate two network interfaces on the controller (that is, changing it to a dual-homed system), each one connected to a different switch. This setup would be able to tolerate both a single network failure and a network interface failure.

Ultimately one could protect the system against a total failure by duplicating the switched ring, connecting each port of the dual-homed system to a different ring.

Other factors supporting a high degree of resilience would include hot swappable switches and NICs, dual redundant power supplies and on-line diagnostic software.

15.6   Troubleshooting

15.6.1   Introduction

This section deals with addressing common faults on Ethernet networks. Ethernet encompasses layers 1 and 2, namely the physical and data link layers of the OSI model. This is equivalent to the bottom layer (the network interface layer) in the ARPA model. This section will focus on those layers only, as well as on the actual medium over which the communication takes place.

15.6.2   Common problems and faults

Ethernet hardware is fairly simple and robust, and once a network has been commissioned, providing the cabling has been done professionally and certified, the network should be fairly trouble-free.

Most problems will be experienced at the commissioning phase, and could theoretically be attributed to either the cabling, the LAN devices (such as hubs and switches), the network interface cards (NICs) or the protocol stack configuration on the hosts.

The wiring system should be installed and commissioned by a certified installer (the suppliers of high-speed Ethernet cabling systems such as ITT, will in any case not guarantee their wiring if not installed by an installer certified by them). This effectively rules out wiring problems for new installations, although old installations could be suspect.

If the LAN devices such as hubs and switches are from reputable vendors, it is highly unlikely that they will malfunction in the beginning. Care should nevertheless be taken to ensure that intelligent (managed) hubs and switches are correctly setup.

This applies to NICs also. NICs rarely fail and nine times out of ten the problem lies with a faulty setup or incorrect driver installation or an incorrect configuration of the higher level protocols such as IP.

15.6.3   Tools of the trade

In addition to fundamental understanding of the technologies involved, spending sufficient time, employing a pair of eyes and patience, one can be successful in isolating Ethernet-related problems with the help of the following tools:

15.6.3.1   Multimeters

A simple multimeter can be used to check for continuity and cable resistance, as will be explained in this section.

15.6.3.2   Handheld cable testers

There are many versions available in the market, ranging from simple devices that basically check for wiring continuity to sophisticated devices that comply with all the prerequisites for 1000BaseT wiring infrastructure tests. Testers are available from several vendors such as MicroTest, Fluke, and Scope.

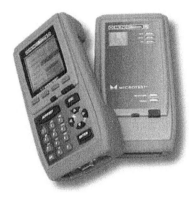

Figure 15.9
OMNIscanner2 cable tester

15.6.3.3 Fiber optic cable testers

Fiber optic testers are simpler than UDP testers, since they basically only have to measure continuity and attenuation loss. Some UDP testers can be turned into fiber optic testers by purchasing an attachment that fits onto the existing tester. For more complex problems such as finding the location of a damaged section on a fiber optic cable, an alternative is to use a proper optical time domain reflectometer (OTDR) but these are expensive instruments and it is often cheaper to employ the services of a professional wire installer (with his own OTDR) if this is required.

15.6.3.4 Traffic generators

A traffic generator is a device that can generate a pre-programmed data pattern on the network. Although they are not used for fault finding strictly speaking, they can be used to predict network behavior due to increased traffic, for example, when planning network changes or upgrades. Traffic generators can be stand-alone devices or they can be integrated into hardware LAN analyzers such as the Hewlett Packard 3217.

15.6.3.5 RMON probes

An RMON (Remote MONitoring) probe is a device that can examine a network at a given point and keep track of captured information at a detailed level. The advantage of a RMON probe is that it can monitor a network at a remote location. The data captured by the RMON probe can then be uploaded and remotely displayed by the appropriate RMON management software. RMON probes and the associated management software are available from several vendors such as 3COM, Bay Networks and NetScout. It is also possible to create an RMON probe by running commercially available RMON software on a normal PC, although the data collection capability will not be as good as that of a dedicated RMON probe.

15.6.3.6 Handheld frame analyzers

Handheld frame analyzers are manufactured by several vendors such as Fluke, Scope, Finisar and PsiberNet, for up to gigabit Ethernet speeds. These little devices can perform link testing, traffic statistics gathering etc and can even break down frames by protocol type. The drawback of these testers is the small display and the lack of memory, which results in a lack of historical or logging functions on these devices.

An interesting feature of the PsiberNet probe is that it is non-intrusive, i.e. it is a clamp-style meter that simply clamps on to the wire and does not have to be attached to a hub port.

Figure 15.10
Psibernet gigabit Ethernet probe

15.6.3.7   Software protocol analyzers

Software protocol analyzers are software packages running on PCs and using either a general purpose or a specialized NIC to capture frames from the network. The NIC is controlled by a so-called promiscuous driver, which enables the NIC to capture all packets on the medium and not only those addressed to it in broadcast or unicast mode.

On the lower end of the scale, simple analyzers are available for download from the Internet as freeware or shareware. Middle of the range packages such as NDG Software's NetBoy Suite offer good value for money but rely heavily on the user for interpreting the captured information. Top of the range software products such as Network Associates' Sniffer or WaveTek Wandel Goltemann's Domino Suite have sophisticated expert systems that can aid in the analysis of the captured software. Unfortunately, this comes at a price.

15.6.3.8   Hardware-based protocol analyzers

Several manufacturers such as Hewlett Packard, Network Associates and WaveTek Wandel Goltemann also supply hardware-based protocol analyzers using their protocol analysis software running on a proprietary hardware infrastructure. This makes them very expensive but dramatically increases the power of the analyzer. For fast and gigabit Ethernet, this is probably the better approach.

15.6.4   Problems and solutions

15.6.4.1   Noise

If excessive noise is suspected on a coax or UTP cable, an oscilloscope can be connected between the signal conductor(s) and ground. This method will show up noise on the conductor, but will not necessarily give a true indication of the amount of power in the noise. A simple and cheap method to pick up noise on the wire is to connect a small loudspeaker between the conductor and ground. A small operational amplifier can be used as an input buffer, so as not to 'load' the wire under observation. The noise will be heard as an audible signal.

The quickest way to get rid of a noise problem, apart from using screened UTP (ScTP), is to change to a fiber-based instead of a wire-based network, for example, by using 100BaseFX instead of 100BaseTX.

Noise can to some extent be counteracted on a coax-based network by earthing the screen AT ONE END ONLY. Earthing it on both sides will create an earth loop. This is normally accomplished by means of an earthing chain or an earthing screw on one of the terminators. Care should also be taken not to allow contact between any of the other connectors on the segment and ground.

15.6.4.2   Thin coax problems

Incorrect cable type

The correct cable for thin Ethernet is RG58A/U or RG58C/U. This is a 5-mm diameter coaxial cable with 50-ohm characteristic impedance and a stranded center conductor. Incorrect cable used in a thin Ethernet system can cause reflections, resulting in CRC errors, and hence many retransmitted frames.

The characteristic impedance of coaxial cable is a function of the ratio between the center conductor diameter and the screen diameter. Hence other types of coax may closely resemble RG58, but may have different characteristic impedance.

Loose connectors

The BNC coaxial connectors used on RG58 should be of the correct diameter and should be properly crimped onto the cable. An incorrect size connector or a poor crimp could lead to intermittent contact problems, which are very hard to locate. Even worse is the 'Radio Shack' hobbyist type screw-on BNC connector that can be used to quickly make up a cable without the use of a crimping tool. These more often than not lead to very poor connections. A good test is to grip the cable in one hand, and the connector in another, and pull very hard. If the connector comes off, the connector mounting procedures need to be seriously reviewed.

Excessive number of connectors

The total length of a thin Ethernet segment is 185 m and the total number of stations on the segment should not exceed 30. However, each station involves a BNC T-piece plus two coax connectors and there could be additional BNC barrel connectors joining the cable. Although the resistance of each BNC connector is small, they are still finite and can add up. The total resistance of the segment (cable plus connectors) should not exceed 10 ohms otherwise problems can surface.

An easy method of checking the loop resistance (the resistance to the other end of the cable and back) is to remove the terminator on one end of the cable and measure the resistance between the connector body and the center contact. The total resistance equals the resistance of the cable plus connectors plus the terminator on the far side. This should be between 50 and 60 ohms. Anything more than this is indicative of a problem.

Overlong cable segments

The maximum length of a thin net segment is 185 m. This constraint is not imposed by collision domain considerations but rather by the attenuation characteristics of the cable. If it is suspected that the cable is too long, its length should be confirmed. Usually, the cable is within a cable trench and hence it cannot be visually measured. In this case, a time domain reflectometer (TDR) can be used to confirm its length.

Stub cables

For thin Ethernet (10Base2), the maximum distance between the bus and the transceiver electronics is 4 cm. In practice, this is taken up by the physical connector plus the PC board tracks leading to the transceiver, which means that there is no scope for a drop cable or 'stub' between the NIC and the bus. The BNC T-piece has to be mounted directly on to the NIC.

Users might occasionally get away with putting a short stub between the T-piece and the NIC but this invariably leads to problems in the long run.

Incorrect terminations

10Base2 is designed around 50-ohm coax and hence requires a 50-ohm terminator at each end. Without the terminators in place, there would be so many reflections from each end that the network would collapse. A slightly incorrect terminator is better than no terminator, yet may still create reflections of such magnitude that it affects the operation of the network.

A 93-ohm terminator looks no different than a 50-ohm terminator; therefore it should not be automatically assumed that a terminator is of the correct value.

If two 10Base2 segments are joined with a repeater, the internal termination on the repeater can be mistakenly left enabled. This leads to three terminators on the segment, creating reflections and hence affecting the network performance.

The easiest way to check for proper termination is by alternatively removing the terminators at each end, and measuring the resistance between connector body and center pin. In each case, the result should be 50 to 60 ohms. Alternatively, one of the T-pieces in the middle of the segment can be removed from its NIC and the resistance between the connector body and the center pin measured. The result should be the value of the two half cable segments (including terminators) in parallel, that is, 25 to 30 ohms.

Invisible insulation damage

If the internal insulation of coax is inadvertently damaged, for example, by placing a heavy point load on the cable, the outer cover could return to its original shape whilst leaving the internal dielectric deformed. This leads to a change of characteristic impedance at the damaged point resulting, in reflections. This, in turn, could lead to standing waves being formed on the cable.

An indication of this problem is when a workstation experiences problems when attached to a specific point on a cable, yet functions normally when moved a few meters to either side. The only solution is to remove the offending section of the cable. It cannot be seen by the naked eye and the position of the damage has to be located with a TDR because of the nature of the damage. Alternatively, the whole cable segment has to be replaced.

Invisible cable break

This problem is similar to the previous one, with the difference that the conductor has been completely severed at a specific point. Despite the terminators at both ends of the cable, the cable break effectively creates two half segments, each with an unterminated end, and hence nothing will work.

The only method to discover the location of the break is by using a TDR.

15.6.4.3   Thick coax problems

Thick coax (RG8), as used for 10Base5 or thick Ethernet, will basically exhibit the same problems as thin coax yet there are a few additional complications.

Loose connectors

10Base5 use N-type male screw-on connectors on the cable. As with BNC connectors, incorrect procedures or a wrong sized crimping tool can cause sloppy joints. This can lead to intermittent problems that are difficult to locate.

Again, a good test is to grab hold of the connector and to try and rip it off the cable with brute force. If the connector comes off, it was not properly installed in the first place.

Dirty taps

The MAU transceiver is often installed on a thick coax by using a vampire tap, which necessitates pre-drilling into the cable in order to allow the center pin of the tap to contact the center conductor of the coax. The hole has to go through two layers of braided screen and two layers of foil. If the hole is not properly cleaned pieces of the foil and braid can remain and cause short circuits between the signal conductor and ground.

Open tap holes

When a transceiver is removed from a location on the cable, the abandoned hole should be sealed. If not, dirt or water could enter the hole and create problems in the long run.

Tight cable bends

The bend radius on a thick coax cable may not exceed 10 inches. If it does, the insulation can deform to such an extent that reflections are created leading to CRC errors. Excessive cable bends can be detected with a TDR.

Excessive loop resistance

The resistance of a cable segment may not exceed 5 ohms. As in the case of thin coax, the easiest way to do this is to remove a terminator at one end and measure the loop resistance. It should be in a range of 50–55 ohms.

15.6.4.4   UTP problems

The most commonly used tool for UTP troubleshooting is a cable meter or pair scanner. At the bottom end of the scale, a cable tester can be an inexpensive tool, only able to check for the presence of wire on the appropriate pins of an RJ-45 connector. High-end cable testers can also test for noise on the cable, cable length, and crosstalk (such as near end signal crosstalk or NEXT) at various frequencies. It can check the cable against CAT5/5e specifications and can download cable test reports to a PC for subsequent evaluation.

The following is a description of some wiring practices that can lead to problems.

Incorrect wire type (solid/stranded)

Patch cords must be made with stranded wire. Solid wire will eventually suffer from metal fatigue and crack right at the RJ-45 connector, leading to permanent or intermittent open connection/s. Some RJ-45 plugs, designed for stranded wire, will actually cut through the solid conductor during installation, leading to an immediate open connection. This can lead to CRC errors resulting in slow network performance, or can even disable a workstation permanently.

The permanently installed cable between hub and workstation, on the other hand, should not exceed 90 m and must be of the solid variety. Not only is stranded wire more expensive for this application, but the capacitance is higher, which may lead to a degradation of performance.

Incorrect wire system components

The performance of the wire link between a hub and a workstation is not only dependent on the grade of wire used, but also on the associated components such as patch panels, surface mount units (SMUs) and RJ-45 type connectors. A single substandard connector on a wire link is sufficient to degrade the performance of the entire link.

High quality fast and gigabit Ethernet wiring systems use high-grade RJ-45 connectors that are visibly different from standard RJ-45 type connectors.

Incorrect cable type

Care must be taken to ensure that the existing UTP wiring is of the correct category for the type of Ethernet being used. For 10BaseT, Cat3 UTP is sufficient, while fast Ethernet (100BaseT) requires Cat5 and gigabit Ethernet requires Cat5e or better. This applies to patch cords as well as the permanently installed ('infrastructure') wiring.

Most industrial Ethernet systems nowadays are 100BaseX based and hence use Cat5 wiring. For such applications, it might be prudent to install screened Cat5 wiring (ScTP) for better noise immunity. ScTP is available with a common foil screen around 4 pairs or with an individual foil screen around each pair.

A common mistake is to use telephone grade patch ('silver satin') cable for the connection between an RJ-45 wall socket (SMU) and the network interface card in a computer. Telephone patch cables use very thin wires that are untwisted, leading to high signal loss and large amounts of crosstalk. This will lead to signal errors causing retransmission of lost packets, which will eventually slow the network down.

'Straight' vs crossover cable

A 10BaseT 100BaseTX patch cable consists of 4 wires (two pairs) with an RJ-45 connector at each end. The pins used for the TX and RX signals are 1, 2 and 3, 6. Although a typical patch cord has 8 wires (4 pairs), the 4 unused wires are nevertheless crimped into the connector for mechanical strength. In order to facilitate communication between computer and hub, the TX and RX ports on the hub are reversed, so that the TX on the computer and the RX on the hub are interconnected whilst the TX on the hub is connected to the RX on the hub. This requires a 'straight' interconnection cable with pin 1 wired to pin 1, pin 2 wired to pin 2 etc.

If the NICs on two computers are to be interconnected without the benefit of a hub, a normal straight cable cannot be used since it will connect TX to TX and RX to RX. For this purpose, a crossover cable has to be used in the same way as a 'null' modem cable.

Crossover cables are normally color coded (for example, green or black) in order to differentiate them from straight cables.

A crossover cable can create problems when it looks like a normal straight cable and the unsuspecting person uses it to connect a NIC to a hub or a wall outlet. A quick way to identify a crossover cable is to hold the two RJ-45 connectors side by side and observe the colors of the 8 wires in the cable through the clear plastic of the connector body. The sequence of the colors should be the same for both connectors.

Hydra cables

Some 10BaseT hubs feature 50-pin connectors to conserve space on the hub. Alternatively, some building wire systems use 50-pin connectors on the wiring panels but the hub equipment has RJ-45 connectors. In both cases, hydra or octopus cable has to be used. This consists of a 50-pin connector connected to a length of 25 pair cable, which is then broken out as a set of 12 small cables, each with an RJ-45 connector. Depending on the vendor the 50-pin connector can be attached through locking clips, velcro strips or screws. It does not always lock down properly, although at a glance it may seem so. This can cause a permanent or intermittent break of contact on some ports.

For 10BaseT systems, near end crosstalk (NEXT), which occurs when a signal is coupled from a transmitting wire pair to a receiving wire pair close to the transmitter, (where the signal is strongest) causes most problems. This is not a serious problem on a single pair cable, as only two pairs are used but on the 25 pair cable, with many signals in close proximity, this can create problems. It can be very difficult to troubleshoot since it will require test equipment that can transmit on all pairs simultaneously.

Excessive untwists

On Cat5 cable, crosstalk is minimized by twisting each cable pair. However, in order to attach a connector at the end the cable has to be untwisted slightly. Great care has to be taken since excessive untwists (more than 1 cm) are enough to create excessive crosstalk, which can lead to signal errors. This problem can be detected with a high quality cable tester.

Stubs

A stub cable is an abandoned telephone cable leading from a punch-down block to some other point. This does not create a problem for telephone systems but if the same Cat3 telephone cabling is used to support 10BaseT, then the stub cables may cause signal reflections that result in bit errors. Again, a high quality cable tester can only detect this problem.

Damaged RJ-45 connectors

On RJ-45 connectors without protective boots, the retaining clip can easily break off especially on cheaper connectors made of brittle plastic. The connector will still mate with the receptacle but will retract with the least amount of pull on the cable, thereby breaking contact. This problem can be checked by alternatively pushing and pulling on the connector and observing the LED on the hub, media coupler or NIC – wherever the suspect connector is inserted. Because of the mechanical deficiencies of RJ-45 connectors, they are not commonly used on industrial Ethernet systems.

T4 on 2 pairs

100BaseTX is a direct replacement for 10BaseT in that it uses the same 2 wire pairs and the same pin allocations. The only prerequisite is that the wiring must be Cat5.

100BaseT4, however, was developed for installations where all the wiring is Cat3, and cannot be replaced. It achieves its high speed over the inferior wire by using all 4 pairs instead of just 2. In the event of deploying 100BaseT4 on a Cat3 wiring infrastructure, a cable tester has to be used to ensure that in fact, all 4 pairs are available for each link and have acceptable crosstalk.

100BaseT4 required the development of a new physical layer technology, as opposed to 100BaseTX/FX that used existing FDDI technology. Therefore, it became commercially available only a year after 100BaseX and never gained real market acceptance. As a result, very few users will actually be faced with this problem.

15.6.4.5   Fiber optic problems

Since fiber does not suffer from noise, interference and crosstalk problems there are basically only two issues to contend with, namely, attenuation and continuity.

The simplest way of checking a link is to plug each end of the cable into a fiber hub, NIC or fiber optic transceiver. If the cable is all right, the LEDs at each end will light up. Another way of checking continuity is by using an inexpensive fiber optic cable tester consisting of a light source and a light meter to test the segment.

More sophisticated tests can be done with an optical time domain reflectometer (OTDR). OTDRs can not only measure losses across a fiber link, but can also determine the nature and location of the losses. Unfortunately, they are very expensive but most professional cable installers will own one.

10BaseFX and 100BaseFX use LED transmitters that are not harmful to the eyes, but gigabit Ethernet use laser devices that can damage the retina of the eye. It is therefore dangerous to try and stare into the fiber (all systems are infrared and therefore invisible anyway).

Incorrect connector installation

Fiber optic connectors can propagate light even if the two connector ends are not touching each other. Eventually, the gap between fiber ends may be so far apart that the link stops working. It is therefore imperative to ensure that the connectors are properly latched.

Dirty cable ends

A speck of dust or some finger oil deposited by touching the connector end is sufficient to affect communication because of the small diameter of the fiber (8–62 microns) and the low light intensity. Dust caps must be left in place when the cable is not in use and a fiber optic cleaning pad must be used to remove dirt and oils from the connector point before installation to avoid this problem.

Component ageing

The amount of power that a fiber optic transmitter can radiate diminishes during the working lifetime of the transmitter. This is taken into account during the design of the link but in the case of a marginal design, the link could start failing intermittently towards the end of the design life of the equipment. A fiber optic power meter can be used to confirm the actual amount of loss across the link but an easy way to troubleshoot the link is to replace the transceivers at both ends of the link with new ones.

15.6.4.6 AUI problems

Excessive cable length

The maximum length of the AUI cable is 50 m assuming that it is a proper IEEE 802.3 cable. Some installations use lightweight office grade cables that are limited to 12 m in length. If these cables are too long, the excessive attenuation can lead to intermittent problems.

DIX latches

The DIX version of the 15-pin D-connector uses a sliding latch. Unfortunately, not all vendors adhere to the IEEE 802 specifications and some use lightweight latch hardware, which results in a connector that can very easily become unstuck. There are basically two solutions to the problem. The first solution is to use a lightweight (office grade) AUI cable, provided the distance would not be a problem. This places less stress on the connector. The second solution is to use a special plastic retainer such as the 'ET Lock' made specifically for this purpose.

SQE test

The signal quality error (SQE) test signal is used on all AUI-based equipment to test the collision circuitry. This method is only used on the old 15-pin AUI-based external transceivers (MAUs) and sends a short signal burst (about 10 bit times in length) to the NIC just after each frame transmission. This tests both the collision detection circuitry and the signal paths. The SQE operation can be observed by means of an LED on the MAU.

The SQE signal is only sent from the transceiver to the NIC and not on to the network itself. It does not delay frame transmissions but occurs during the inter-frame gap and is not interpreted as a collision.

The SQE test signal must, however, be disabled if an external transceiver (MAU) is attached to a repeater hub. If this is not done, the hub will detect the SQE signal as a collision and will issue a jam signal. As this happens after each packet, it can seriously delay transmissions over the network. The problem is that it is not possible to detect this with a protocol analyzer.

15.6.4.7 NIC problems

Basic card diagnostics

The easiest way to check if a particular NIC is faulty is to replace it with another (working) NIC. Modern NICs for desktop PCs usually have auto-diagnostics included and these can be accessed, for example, from the device manager in Windows 95/98. Some cards can even participate in a card-to-card diagnostic. Provided there are two identical cards, one can be setup as an initiator and one as a responder. Since the two cards will communicate at the data link level, the packets exchanged will, to some extent, contribute to the network traffic but will not affect any other devices or protocols present on the network.

The drivers used for card auto-diagnostics will usually conflict with the NDIS and ODI drivers present on the host, and a message is usually generated, advising the user that the Windows drivers will be shut down, or that the user should re-boot in DOS.

With PCMCIA cards, there is an additional complication in that the card diagnostics will only run under DOS, but under DOS the IRQ (interrupt address) of the NIC typically defaults to 5, which happens to be the IRQ for the sound card. Therefore, the diagnostics will usually pass every test, but fail on the IRQ test. This result can then be ignored safely if the card passes the other diagnostics. If the card works, it works!

Incorrect media selection

Some cards support more than one medium, for example, 10Base2/10Base5, or 10Base5/10BaseT, or even all three. It may then happen that the card fails to operate since it fails to 'see' the attached medium.

It is imperative to know how the selection is done. Modern cards usually have an auto-detect function but this only takes place when the machine is booted up. It does NOT re-detect the medium if it is changed afterwards. Therefore, if the connection to a machine is changed from 10BaseT to 10Base2, for example, the machine has to be re-booted.

Some older cards need to have the medium set via a setup program, whilst even older cards have DIPswitches on which the medium has to be selected.

Wire hogging

Older interface cards find it difficult to maintain the minimum 9.6 microsecond inter-frame spacing (IFS) and as a result of this, nodes tend to return to and compete for access to the bus in a random fashion. Modern interface cards are so fast that they can sustain the minimum 9.6 microsecond IFS rate. As a result, it becomes possible for a single card to gain repetitive sequential access to the bus in the face of slower competition and hence 'hogging' the bus.

With a protocol analyzer, this can be detected by displaying a chart of network utilization versus time and looking for broad spikes above 50 per cent. The solution to this problem is to replace shared hubs with switched hubs and increase the bandwidth of the system by migrating from 10–100 megabits per second, for example.

Jabbers

A jabber is a faulty NIC that transmits continuously. NICs have a built-in jabber control that is supposed to detect a situation whereby the card transmits frames longer than the allowed 1518 bytes and shut the card down. However, if this does not happen, the defective card can bring the network down. This situation is indicated by a very high collision rate coupled with a very low or non-existent data transfer rate. A protocol analyzer might not show any packets since the jabbering card is not transmitting any sensible data. The easiest way to detect the offending card is by removing the cables from the NICs or the hub one-by-one until the problem disappears in which case the offending card is located.

Faulty CSMA/CD mechanism

A card with a faulty CSMA/CD mechanism will create a large number of collisions since it transmits legitimate frames but does not wait for the bus to be quiet before transmitting. As in the previous case, the easiest way to detect this problem is to isolate the cards one by one until the culprit is detected.

Too many nodes

A problem with CSMA/CD networks is that the network efficiency decreases as the network traffic increases. Although Ethernet networks can theoretically utilize well over 90% of the available bandwidth, the access time of individual nodes increase dramatically as network loading increases. The problem is similar to that encountered on many urban roads during peak hours. During rush hours, the traffic approaches the design limit of the road. This does not mean that the road stops functioning. In fact, it carries a very large number of vehicles, but to get into the main traffic from a side road becomes problematic.

For office type applications, an average loading of around 30% is deemed acceptable while for industrial applications, 3% is considered maximum. Should the loading of the network be a problem, the network can be segmented using switches instead of shared hubs. In many applications, it will be found that the improvement created by changing from shared to switched hubs, is larger than the improvement to be gained by upgrading from 10 Mbps to fast Ethernet.

Improper packet distribution

Improper packet distribution takes place when one or more nodes dominate most of the bandwidth. This can be monitored by using a protocol analyzer and checking the source address of individual packets. Another way of checking this easily is by using the NDG software Web Boy facility and checking the contribution of the top 10 transmitters.

Nodes like this are typically performing tasks such as video conferencing or database access, which require a large bandwidth. The solution to the problem is to give these nodes separate switch connections or to group them together on a faster 100BaseT or 1000BaseT segment.

Excessive broadcasting

A broadcast packet is intended to reach all the nodes in the network and is sent to a MAC address of ff–ff–ff–ff–ff–ff. Unlike routers, bridges and switches forward broadcast packets throughout the network and therefore cannot contain the broadcast traffic. Too many simultaneous broadcast packets can degrade network performance.

In general, it is considered that if broadcast packets exceed 5% of the total traffic on the network, it would indicate a broadcast overload problem. Broadcasting is a particular problem with Netware servers and networks using NetBIOS/NetBEUI. Again, it is fairly easy to observe the amount of broadcast traffic using the WebBoy utility.

A broadcast overload problem can be addressed by adding routers, layer 3 switches or VLAN switches with broadcast filtering capabilities.

Bad packets

Bad packets can be caused by poor cabling infrastructure, defective NICs, external noise, or faulty devices such as hubs, devices or repeaters. The problem with bad packets is that they cannot be analyzed by software protocol analyzers.

Software protocol analyzers obtain packets that have already been successfully received by the NIC. That means they are one level removed from the actual medium on which the frames exist and hence cannot capture frames that are rejected by the NIC. The only solution to this problem is to use a software protocol analyzer that has a special custom NIC, capable of capturing information regarding packet deformities or by using a more expensive hardware protocol analyzer.

Faulty packets include:

Runts

Runt packets are shorter than the minimum 64 bytes and are typically created by a collision taking place during the slot time.

As a solution, try to determine whether the frames are collisions or under-runs. If they are collisions, the problem can be addressed by segmentation through bridges and switches. If the frames are genuine under-runs, the packet has to be traced back to the generating node that is obviously faulty.

CRC errors

CRC errors occur when the CRC check at the receiving end does not match the CRC checksum calculated by the transmitter.

As a solution, trace the frame back to the transmitting node. The problem is either caused by excessive noise induced into the wire, corrupting some of the bits in the frames, or by a faulty CRC generator in the transmitting node.

Late collisions

Late collisions are typically caused when the network diameter exceeds the maximum permissible size. This problem can be eliminated by ensuring that the collision domains are within specified values, i.e. 2500 meters for 10 Mbps Ethernet, 250 m for fast Ethernet and 200 m for gigabit Ethernet.

Check the network diameter as outlined above by physical inspection or by using a TDR. If that is found to be a problem, segment the network by using bridges or switches.

Misaligned frames

Misaligned frames are frames that get out of sync by a bit or two, due to excessive delays somewhere along the path or frames that have several bits appended after the CRC checksum.

As a solution, try and trace the signal back to its source. The problem could have been introduced anywhere along the path.

Faulty auto-negotiation

Auto-negotiation is specified for

- 10BaseT,
- 100BaseTX,
- 100BaseT2,
- 100BaseT4 and
- 1000BaseT.

It allows two stations on a link segment (a segment with only two devices on it) e.g. an NIC in a computer and a port on a switching hub to negotiate a speed (10/100/1000 Mbps) and an operating mode (full/half-duplex). If auto-negotiation is faulty or switched off on one device, the two devices might be set for different operating modes and as a result, they will not be able to communicate.

On the NIC side the solution might be to run the card diagnostics and to confirm that auto-negotiation is, in fact, enabled.

On the switch side, this depends on the diagnostics available for that particular switch. It might also be an idea to select another port, or to plug the cable into another switch.

10/100 Mbps mismatch

This issue is related to the previous one since auto-negotiation normally takes care of the speed issue.

Some system managers prefer to set the speeds on all NICs manually, for example, to 10 Mbps. If such an NIC is connected to a dual-speed switch port, the switch port will automatically sense the NIC speed and revert to 10 Mbps. If, however, the switch port is only capable of 100 Mbps, then the two devices will not be able to communicate.

This problem can only be resolved by knowing the speed(s) at which the devices are supposed to operate, and then by checking the settings via the setup software.

Full/half-duplex mismatch

This problem is related to the previous two.

A 10BaseT device can only operate in half-duplex (CSMA/CD) whilst a 100BaseTX can operate in full-duplex OR half-duplex.

If, for example, a 100BaseTX device is connected to a 10BaseT hub, its auto-negotiation circuitry will detect the absence of a similar facility on the hub. It will therefore know, by default, that it is 'talking' to 10BaseT and it will set its mode to half-duplex. If, however, the NIC has been set to operate in full-duplex only, communications will be impossible.

15.6.4.8   Host-related problems

Incorrect host setup

Ethernet V2 (or IEEE 802.3 plus IEEE 802.2) only supplies the bottom layer of the DOD model. It is therefore able to convey data from one node to another by placing it in the data field of an Ethernet frame, but nothing more. The additional protocols to implement the protocol stack have to be installed above it, in order to make networked communications possible.

In industrial Ethernet networks, this will typically be the TCP/IP suite, implementing the remaining layers of the ARPA model as follows.

The second layer of the DOD model (the Internet layer) is implemented with IP (as well as its associated protocols such as ARP and ICMP).

The next layer (the host-to-host layer) is implemented with TCP and UDP.

The upper layer (the application layer) is implemented with the various application layer protocols such as FTP, Telnet etc. The host might also require a suitable application layer protocol to support its operating system in communicating with the operating system on other hosts, on Windows, that is NetBIOS by default.

As if this is not enough, each host needs a network 'client' in order to access resources on other hosts, and a network 'service' to allow other hosts to access its own resources in turn. The network client and network service on each host do not form part of the communications stack but reside above it and communicate with each other across the stack.

Finally, the driver software for the specific NIC needs to be installed, in order to create a binding ('link') between the lower layer software (firmware) on the NIC and the next layer software (for example, IP) on the host. The presence of the bindings can be observed, for example, on a Windows 95/98 host by clicking 'settings' –> 'control panel' –> 'networks'–>'configuration', then selecting the appropriate NIC and clicking 'Properties' –> 'Bindings'.

Without these, regardless of the Ethernet NIC installed, networking is not possible.

Failure to log in

When booting a PC, the Windows dialogue will prompt the user to log on to the server, or to log on to his/her own machine. Failure to log in will not prevent Windows from completing its boot-up sequence but the network card will not be enabled. This is clearly visible as the LED on the NIC and hub will not light up.

15.6.4.9 Hub-related problems

Faulty individual port

A port on a hub may simply be 'dead.' Everybody else on the hub can 'see' each other, except the user on the suspect port. Closer inspection will show that the LED for that particular channel does not light up. The quickest way to verify this is to remove the UTP cable from the suspect hub port and plugging it into another port. If the LEDs light up on the alternative port, it means that the original port is not operational.

On managed hubs, the configuration of the hub has to be checked by using the hub's management software to verify that the particular port has not, in fact, been disabled by the network supervisor.

Faulty hub

This will be indicated by the fact that none of the LEDs on the hub are illuminated and that none of the users on that particular hub are able to access the network. The easiest way to check this is by temporarily replacing the hub with a similar one and checking if the problem disappears.

Incorrect hub interconnection

If hubs are interconnected in a daisy-chain fashion by means of interconnecting ports with a UTP cable, care must be taken to ensure that either a crossover cable is used or that the crossover/uplink port on one hub ONLY is used. Failure to comply with this precaution will prevent the interconnected hubs from communicating with each other although it will not damage any electronics.

A symptom of this problem will be that all users on either side of the faulty link will be able to see each other but nobody will be able to see anything across the faulty link. This problem can be rectified by ensuring that a proper crossover cable is being used or, if a straight cable is being used, that it is plugged into the crossover/uplink port on one hub only. On the other hub, it must be plugged into a normal port.

15.6.5 Troubleshooting switched networks

Troubleshooting in a shared network is fairly easy since all packets are visible everywhere in the segment and as a result, the protocol analysis software can run on any host within that segment. In a switched network, the situation changes radically since each switch port effectively resides in its own segment and packets transferred through the switch are not seen by ports for whom they are not intended.

In order to address the problem, many vendors have built traffic monitoring modules into their switches. These modules use either RMON or SNMP to build up statistics on each port and report switch statistics to switched management software.

Capturing the packets on a particular switched port is also a problem, since packets are not forwarded to all ports in a switch hence there is no place to plug in a LAN analyzer and view the packets.

One solution implemented by vendors is port liaising, also known as port mirroring or port spanning. The liaising has to be setup by the user and the switch copies the packets from the port under observation to a designated spare port. This allows the LAN user to plug in a LAN analyzer onto the spare port in order to observe the original port.

Another solution is to insert a shared hub in the segment under observation that is between the host and the switch port to which it was originally connected. The LAN analyzer can then be connected to the hub in order to observe the passing traffic.

15.6.6 Troubleshooting fast Ethernet

The most diagnostic software is PC-based and it uses a NIC with a promiscuous mode driver. This makes it easy to upgrade the system by simply adding a new NIC and driver. However, most PCs are not powerful enough to receive, store and analyze incoming data rates. It might therefore be necessary to rather consider the purchase of a dedicated hardware analyzer.

Most of the typical problems experienced with fast Ethernet, have already been discussed. These include a physical network diameter that is too large, the presence of Cat3 wiring in the system, trying to run 100BaseT4 on 2 pairs, mismatched 10BaseT/100BaseTX ports, and noise.

15.6.7 Troubleshooting gigabit Ethernet

Although gigabit Ethernet is very similar to its predecessors, the packets arrive so fast that they cannot be analyzed by normal means. A gigabit Ethernet link is capable of transporting around 125 MB of data per second and few analyzers have the memory capability to handle this. Gigabit Ethernet analyzers such as those made by Hewlett Packard (LAN Internet Advisor), Network Associates (Gigabit Sniffer Pro) and WaveTech Wandel Goltemann (Domino Gigabit Analyzer) are highly specialized gigabit Ethernet analyzers. They minimize storage requirements by filtering and analyzing capture packets in real-time, looking for a problem. Unfortunately, they come at a price tag of around US$ 50 000.

16

TCP/IP overview

Objectives

When you have completed study of this chapter, you will be able to:

- List the main protocols and operation of TCP/IP
- Correct problems with:
    - Internet layer
    - IP addresses
    - Subnet masks
    - Simple routing problems
    - Transport layer
    - Triple handshake
    - Incorrect ports

16.1 Introduction

TCP/IP is the *de facto* global standard for the Internet (network) and host-to-host (transport) layer implementation of internetwork applications because of the popularity of the Internet. The Internet (known as ARPANet in its early years), was part of a military project commissioned by the Advanced Research Projects Agency (ARPA), later known as the Defense Advanced Research Agency or DARPA. The communications model used to construct the system is known as the ARPA model.

Whereas the OSI model was developed in Europe by the International Standards Organization (ISO), the ARPA model (also known as the DoD model) was developed in the USA by ARPA. Although they were developed by different bodies and at different points in time, both serve as models for a communications infrastructure and hence provide 'abstractions' of the same reality. The remarkable degree of similarity is therefore not surprising.

Whereas the OSI model has 7 layers, the ARPA model has 4 layers. The OSI layers map onto the ARPA model as follows.

- The OSI session, presentation and applications layers are contained in the ARPA process and application layer.

- The OSI transport layer maps onto the ARPA host-to-host layer (sometimes referred to as the service layer).
- The OSI network layer maps onto the ARPA Internet layer.
- The OSI physical and data link layers map onto the ARPA network interface layer.

The relationship between the two models is depicted in Figure 16.1

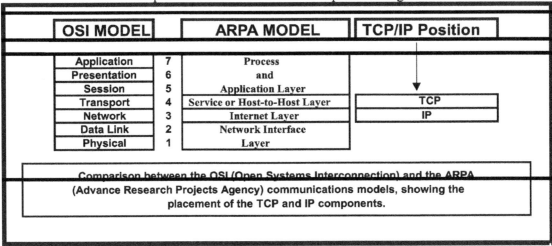

Figure 16.1
OSI vs ARPA models

TCP/IP, or rather – the TCP/IP protocol suite – is not limited to the TCP and IP protocols, but consists of a multitude of interrelated protocols that occupy the upper three layers of the ARPA model. TCP/IP does NOT include the bottom network interface layer, but depends on it for access to the medium.

As depicted in Figure 16.2, an Internet transmission frame originating on a specific host (computer) would contain the local network (for example, Ethernet) header and trailer applicable to that host. As the message proceeds along the Internet, this header and trailer could be replaced depending on the type of network on which the packet finds itself – be that X.25, frame relay or ATM. The IP datagram itself would remain untouched, unless it has to be fragmented and reassembled along the way.

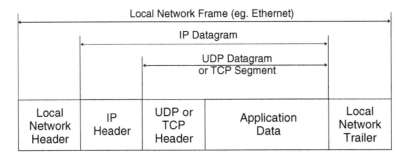

Figure 16.2
Internet frame

Note: RFCs can be obtained from various sources on the Internet such as www.rfc-editor.org if any Internet-related specification is referenced as a request for comments or RFC.

16.1.1   The Internet layer

This layer is primarily responsible for the routing of packets from one host to another. Each packet contains the address information needed for its routing through the internetwork to the destination host. The dominant protocol at this level is the Internet protocol (IP). There are, however, several other additional protocols required at this level such as:

- Address resolution protocol (ARP), RFC 826. This is used for the translation of an IP address to a hardware (MAC) address, such as required by Ethernet.
- Reverse address resolution protocol (RARP), RFC 903. This is the complement of ARP and translates a hardware address to an IP address.
- Internet control message protocol (ICMP), RFC 792. This is a protocol used for exchanging control or error messages between routers or hosts.

16.1.2   The host-to-host layer

This layer is primarily responsible for data integrity between the sender host and receiver host regardless of the path or distance used to convey the message. It has two protocols associated with it, namely:

- User data protocol (UDP), a connectionless (unreliable) protocol used for higher layer port addressing with minimal protocol overhead (RFC 768).
- Transmission control protocol (TCP), a connection-oriented protocol that offers a very reliable method of transferring a stream of data in byte format between applications (RFC 793).

16.1.3   The process/application layer

This layer provides the user or application programs with interfaces to the TCP/IP stack. These include (but are not limited to) file transfer protocol (FTP), trivial file transfer protocol (TFTP), simple mail transfer protocol (SMTP), telecommunications network (TELNET), post office protocol (POP3), remote procedure calls (RPC), remote login (RLOGIN), hypertext transfer protocol (HTTP) and network time protocol (NTP). Users can also develop their own application layer protocols.

16.2   Internet layer protocols (packet transport)

This section will deal with the Internet protocol (IP), the Internet control message protocol (ICMP) and the address resolution protocol (ARP). IPv4 is officially in the process of being replaced by IPv6 but it will still be some time before IPv6 makes its appearance. Therefore, for the sake of simplicity, this section will only deal with IPv4 and its associated protocols.

16.2.1   IP version 4 (IPv4)

IP (RFC 791) is responsible for the delivery of packets ('datarams') between hosts. It is analogous to the postal system, in that it forwards (routes) and delivers datagrams on the basis of IP addresses attached to the datagrams, in the same way the postal service would process a letter based on the postal address. The IP address is a 32-bit entity containing both the network address (the 'zip code') and the host address (the 'street address').

IP also breaks up (fragments) datagrams that are too large. This is often necessary because with LANs and WANs, a datagram may have to traverse on its way to its destination that may have different frame size limitations. For example, Ethernet can handle 1500 bytes but X.25 can handle only 576 bytes. IP on the sending side will fragment a datagram if necessary, attach an IP header to each fragment, and send them off consecutively. On the receiving side, IP will again rebuild the original datagram.

The IPv4 header

The IP header is appended to the information that IP accepts from higher-level protocols, before passing it around the network. This information could, within itself, contain the headers appended by higher level protocols such as TCP. The header consists of at least five 32-bit (4 byte) 'long words' i.e. 20 bytes total and is made up as follows.

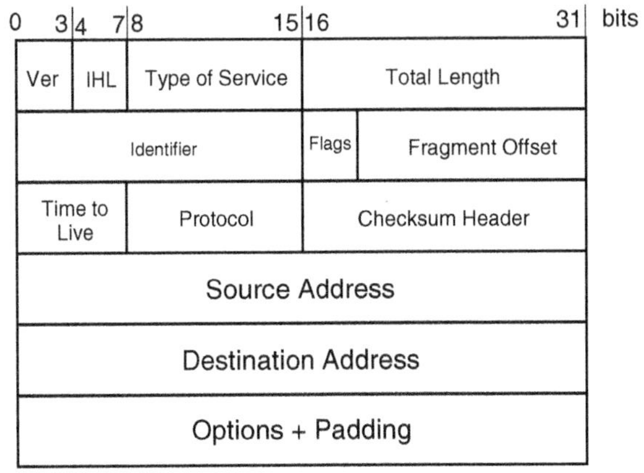

Figure 16.3
IPv4 header

The ver (version) field is 4 bits long and indicates the version of the IP protocol in use. For IPv4, it is 4.

This is followed by the 4-bit IHL (Internet header length) field that indicates the length of the IP header in 32-bit 'long words'. This is necessary since the IP header can contain options and therefore does not have a fixed length.

The 8-bit type of service (ToS) field informs the network about the quality of service required for this datagram. The ToS field is composed of a 3-bit precedence field (which is often ignored) and an unused (LSB) bit that must be 0. The remaining 4 bits may only be turned on (set =1) one at a time, and are allocated as follows:

Bit 3: minimize delay
Bit 4: maximize throughput
Bit 5: maximize reliability
Bit 6: minimize monetary cost

Total length (16 bits) is the length of the entire datagram, measured in bytes. Using this field and the IHL length, it can be determined where the data starts and ends. This field allows the length of a datagram to be up to $2^{16} = 65\ 536$ bytes, although such long datagrams are impractical. All hosts must at least be prepared to accept datagrams of up to 576 octets.

The 16-bit identifier uniquely identifies each datagram sent by a host. It is normally incremented by one for each successive datagram sent. In the case of fragmentation, it is appended to all fragments of the same datagram for the sake of reconstructing the datagram at the receiving end. It can be compared to the 'tracking' number of an item delivered by registered mail or UPS.

The 3-bit flag field contains 2 flags, used in the fragmentation process, viz. DF and MF. The DF (don't fragment) flag is set (=1) by the higher-level protocol (for example, TCP) if IP is NOT allowed to fragment a datagram. If such a situation occurs, IP will not fragment and forward the datagram, but simply return an appropriate ICMP error message to the sending host. If fragmentation does occur, MF=1 will indicate that there are more fragments to follow, whilst MF=0 indicate that it is the last fragment to be sent.

The 13-bit fragment offset field indicates where in the original datagram a particular fragment belongs, for example, how far the beginning of the fragment is removed from the end of the header. The first fragment has offset zero. The fragment offset is measured in units of 8 bytes (64 bits); that is, the transmitted offset is equal to the actual offset divided by eight.

The TTL (time to live) field ensures that undeliverable datagrams are eventually discarded. Every router that processes a datagram must decrease the TTL by one and if this field contains the value zero, then the datagram must be destroyed. Typically, a datagram can be delivered anywhere in the world by traversing fewer than 15 routers.

The 8-bit protocol field indicates the next (higher) level protocol header present in the data portion of the IP datagram, in other words the protocol that resides above IP in the protocol stack and which has passed the datagram down to IP. Typical values are 1 for ICMP, 6 for TCP and 17 for UDP. A more detailed listing is contained in RFC 1700.

The checksum is a 16-bit mathematical checksum on the header only. Since some header fields change all the time (for example, TTL), this checksum is recomputed and verified at each point that the IP header is processed. It is not necessary to cover the data portion of the datagram, as the protocols making use of IP, such as ICMP, IGMP, UDP and TCP, all have a checksum in their headers to cover their own header and data.

Finally, the source and destination addresses are the 32-bit IP addresses of the origin and the destination hosts of the datagram.

IPv4 addressing

The ultimate responsibility for the issuing of IP addresses is vested in the Internet Assigned Numbers Authority (IANA). This responsibility is, in turn, delegated to the three Regional Internet Registries (RIRs) viz. APNIC (Asia-Pacific Network Information Centre), ARIN (American Registry for Internet Numbers) and RIPE NCC (Reseau IP Europeans). RIRs allocate blocks of IP addresses to Internet service providers (ISPs) under their jurisdiction, for subsequent issuing to users or sub-ISPs.

The IPv4 address consists of 32 bits, e.g. 11000000011001000110011001 0000000001. Since this number is fine for computers but a little difficult for human beings, it is divided into four octets w, x, y and z. Each octet is converted to its decimal equivalent. The result of the conversion is written in the format 192.100.100.1. This is known as the 'dotted decimal' or 'dotted quad' notation. As mentioned earlier, one part of the IP address is known as the network ID or 'NetID' while the rest is known as the 'HostID'.

Originally, IP addresses were allocated in so-called address classes. Although the system proved to be problematic, and IP addresses are currently issued 'classless', the legacy of IP address classes remains and has to be understood.

To provide for flexibility in assigning addresses to networks, the interpretation of the address field was coded to specify either a small number of networks with a large number of hosts (class A), or a moderate number of networks with a moderate number of hosts (class B), or a large number of networks with a small number of hosts (class C). There was also provision for extended addressing modes: class D was intended for multicasting whilst E was reserved for future use.

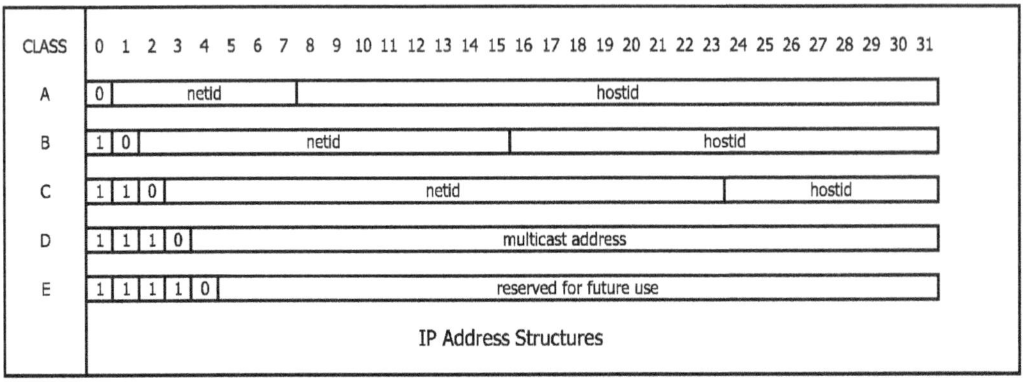

Figure 16.4
Address structure for IPv4

For class A, the first bit is fixed at 0. The values for 'w' can therefore only vary between 0 and 12710. 0 is not allowed and 127 is a reserved number used for testing. This allows for 126 class A NetIDs. The number of HostIDs is determined by octets 'x', 'y' and 'z'. From these 24 bits, $2^{24} = 16\,777\,218$ combinations are available. All zeros and all ones are not permissible, which leaves 16 777 216 usable combinations.

For class B, the first two bits are fixed at 10. The binary values for 'w' can therefore only vary between 128^{10} and 191^{10}. The number of NetIDs is determined by octets 'w' and 'x'. The first 2 bits are used to indicate class B and hence cannot be used. This leaves fourteen usable bits. Fourteen bits allow $2^{14} = 16\,384$ NetIDs. The number of HostIDs is determined by octets 'y' and 'z'. From these 16 bits, $2^{16} = 65\,536$ combinations are available. All zeros and all ones are not permissible, which leaves 65 534 usable combinations.

For class C, the first three bits are fixed at 110. The binary values for 'w' can therefore only vary between 19 210 and 22 310. The number of NetIDs is determined by octets 'w', 'x' and 'y'. The first three bits (110) are used to indicate class C and hence cannot be used. This leaves twenty-two usable bits. Twenty-two bits allow $2^{22} = 2\,097\,152$ combinations for NetIDs. The number of HostIDs is determined by octet 'z'. From these 8 bits, $2^8 = 256$ combinations are available. Once again, all zeros and all ones are not permissible which leaves 254 usable combinations.

In order to determine where the NetID ends and the HostID begins, each IP address is associated with a subnet mask, or, technically more correct, a Netmask. This mask starts with a row of contiguous 1s from the left; one for each bit that forms part of the NetID. This is followed by 0s, one for each bit comprising the HostID.

16.2.2   Address resolution protocol (ARP)

Some network technologies make address resolution difficult. Ethernet interface boards, for example, come with built-in 48-bit hardware addresses. This creates several difficulties:

- No simple correlation, applicable to the whole network, can be created between physical (MAC) addresses and Internet protocol (IP) addresses,
- When the interface board fails and has to be replaced the Internet protocol (IP) address then has to be remapped to a different MAC address,
- The MAC address is too long to be encoded into the 32-bit Internet protocol (IP) address.

To overcome these problems in an efficient manner, and eliminate the need for applications to know about MAC addresses, TCP/IP designers developed the address resolution protocol (ARP), which resolves addresses dynamically.

When a host wishes to communicate with another host on the same physical network, it needs the destination MAC address in order to compose the basic frame. If it does not know what the destination MAC address is, but has its IP address, it broadcasts a special type of datagram in order to resolve the problem. This is called an address resolution protocol (ARP) request. This datagram requests the owner of the unresolved Internet protocol (IP) address to reply with its MAC address. All hosts on the network will receive the broadcast, but only the one that recognizes its own IP address will respond.

While the sender could, of course, just broadcast the original datagram to all hosts on the network, this would impose an unnecessary load on the network, especially if the datagram was large. A small address resolution protocol (ARP) request, followed by a small address resolution protocol (ARP) reply, followed by a direct transmission of the original datagram, is a much more efficient way of resolving the problem.

Address resolution cache

Because communication between two computers usually involves transfer of a succession of datagrams, it is prudent for the sender to 'remember' the MAC information it receives– at least for a while. Thus, when the sender receives an ARP reply, it stores the MAC address it receives as well as the corresponding IP address in its ARP cache. Before sending any message to a specific IP address, it checks first to see if the relevant address binding is in the cache. This saves it from repeatedly broadcasting identical address resolution protocol (ARP) requests.

To further reduce communication overheads, when a host broadcasts an ARP request it includes its own IP address and MAC address, and these are stored in the ARP caches of all other hosts that receive the broadcast. When a new host is added to a network, it can be made to send an ARP broadcast to inform all other hosts on that network of its address.

Some very small networks do not use ARP caches, but the continual traffic of ARP requests and replies on a larger network would have a serious negative impact on the network's performance.

The ARP cache holds 4 fields of information for each device:
- IF index – the physical port
- Physical address – the MAC address of the device
- Internet protocol (IP) address – the corresponding IP address
- Type – the type of entry in the ARP cache; there are 4 possible types:

    4 = static – the entry will not change

    3 = dynamic – the entry can change

    2 = the entry is invalid

    1 = none of the above

ARP datagram

The layout of an ARP request or reply datagram is as follows:

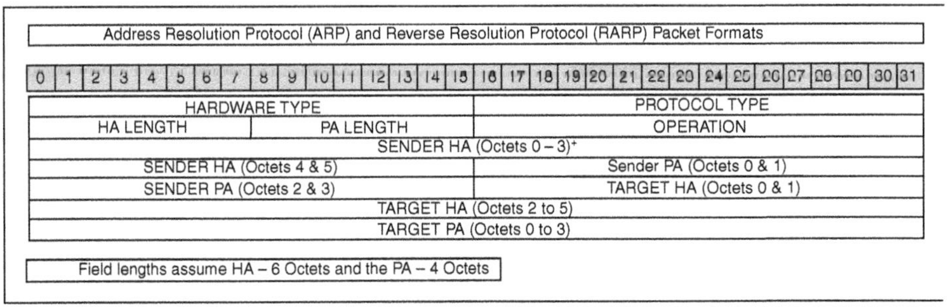

Address Resolution Protocol (ARP) and Reverse Resolution Protocol (RARP) Packet Formats	
0 1 2 3 4 5 6 7 8 9 10 11 12 13 14 15 16 17 18 19 20 21 22 23 24 25 26 27 28 29 30 31	
HARDWARE TYPE	PROTOCOL TYPE
HA LENGTH — PA LENGTH	OPERATION
SENDER HA (Octets 0 – 3)*	
SENDER HA (Octets 4 & 5)	Sender PA (Octets 0 & 1)
SENDER PA (Octets 2 & 3)	TARGET HA (Octets 0 & 1)
TARGET HA (Octets 2 to 5)	
TARGET PA (Octets 0 to 3)	

Field lengths assume HA – 6 Octets and the PA – 4 Octets

Figure 16.5
Address resolution protocol

Hardware type

Specifies the hardware interface type of the target:
1 = Ethernet
2 = Experimental Ethernet
3 = X.25
4 = Proteon ProNET (token ring)
5 = Chaos
6 = IEEE 802.X
7 = ARCnet

Protocol type

Specifies the type of high-level protocol address the sending device is using. For example,

2048 = Internet protocol (IP)
2054 = Address resolution protocol (ARP)
3282 = Reverse ARP (reverse address resolution protocol (RARP))

Hardware address length

The length, in bytes, of the MAC address

Protocol address length

The length, in bytes, of the IP address

Operation code

Indicates the type of ARP datagram:
1 = Address resolution protocol (ARP) request
2 = Address resolution protocol (ARP) reply
3 = Reverse address resolution protocol (RARP) request
4 = Reverse address resolution protocol (RARP) reply

Sender hardware address

The MAC address of the sender

Sender protocol address

The IP address of the sender

Target hardware address

The MAC address of the target host

Target protocol address

The IP address of the target host

Since the fields are used to indicate the lengths of the hardware and protocol addresses, the address fields can be used to carry a variety of address types, making ARP useful for a number of different types of network.

The broadcasting of ARP requests presents some potential problems. Networks such as Ethernet are 'best-effort' delivery systems i.e. the sender does not receive any feedback whether datagrams it has transmitted were received by the target device. If the target is not available, the ARP request destined for it will be lost without trace and no ARP response will be generated. Thus the sender must be programmed to retransmit its ARP request after a certain time period and must be able to store the datagram it is attempting to transmit in the interim. It must also remember what requests it has sent out so that it does not send out multiple ARP requests for the same address. If it does not receive an ARP reply, it will eventually have to discard the outgoing datagrams.

Since it is possible for a machine's hardware address to change, as happens when an Ethernet interface fails and has to be replaced, entries in an ARP cache have a limited life span after which they are deleted. Every time a machine with an ARP cache receives an ARP message, it uses the information to update its own ARP cache. If the incoming address binding already exists, it overwrites the existing entry with the fresh information and resets the timer for that entry.

If a machine is the target of an incoming ARP request, its own ARP software must organize a reply. It must swap the target and sender address pairs in the ARP datagram, insert its physical address into the relevant field, change the operation code to 2 (ARP reply), and send it back to the requester.

16.2.3   ICMP

When nodes fail, or become temporarily unavailable, or when certain routes become overloaded with traffic, a message mechanism called the Internet control message protocol (ICMP) reports errors and other useful information about the performance and operation of the network.

ICMP communicates between the Internet layers on two nodes and is used by routers as well as individual hosts. Although ICMP is viewed as residing within the Internet layer, its messages travel across the network encapsulated in IP datagrams in the same way as higher-layer protocol (such as TCP or UDP) datagrams. The ICMP message, consisting of an ICMP header and ICMP data, is encapsulated as 'data' within an IP datagram that is, in turn, carried as 'payload' by the lower network interface layer (for example, Ethernet).

16.2.4   ICMP datagrams

There are a variety of ICMP messages, each with a different format, yet the first 3 fields as contained in the first 4 bytes or 'long word' are the same for all.

The various ICMP messages are shown in Figure 16.6.

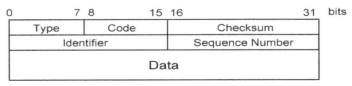

Echo and Echo Reply Messages

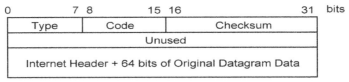

Destination Unreachable, Source Quench and Time Exceeded Messages

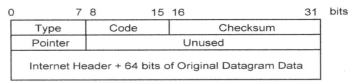

Parameter Problem Message

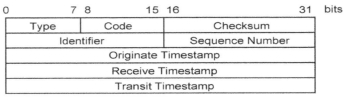

Redirect Message

```
0        7 8        15 16              31  bits
┌────────┬──────────┬───────────────────┐
│  Type  │   Code   │     Checksum      │
├────────┴──────────┼───────────────────┤
│    Identifier     │  Sequence Number  │
├───────────────────┴───────────────────┤
│         Originate Timestamp           │
├───────────────────────────────────────┤
│          Receive Timestamp            │
├───────────────────────────────────────┤
│          Transit Timestamp            │
└───────────────────────────────────────┘
```

Timestamp and Timestamp Reply Messages

```
0        7 8        15 16              31  bits
┌────────┬──────────┬───────────────────┐
│  Type  │   Code   │     Checksum      │
├────────┴──────────┼───────────────────┤
│    Identifier     │  Sequence Number  │
├───────────────────┴───────────────────┤
│             Address Mask              │
└───────────────────────────────────────┘
```

Address mask Request and Address Mask Reply Messages

Figure 16.6
ICMP message formats

The three common fields are:

- An ICMP message type (4 bits), which is a code that identifies the type of ICMP message.
- A code (4 bits) in which interpretation depends on the type of ICMP message.
- A checksum (16 bits) that is calculated on the entire ICMP datagram.

The following table lists the different types of ICMP messages.

Type field	Description
0	Echo, reply
3	Destination unreachable
4	Source quench
5	Redirect (change a route)
8	Echo request
11	Time exceeded (datagram)
12	Parameter problem (datagram)
13	Time-stamp request
14	Time-stamp reply
15	Address mark request
16	Address mark reply
17	
18	

Table 16.1
ICMP message types

ICMP messages can be further subdivided into two broad groups viz. ICMP error messages (destination unreachable, time exceeded, invalid parameters, source quench or redirect) and ICMP query messages (echo request and reply messages, time-stamp request and reply messages, and subnet mask request and reply messages).

16.2.5   Routing

Unlike the host-to-host layer protocols (for example, TCP), which control end-to-end communications, IP is rather 'shortsighted.' Any given IP node (host or router) is only concerned with routing (switching) the datagram to the next node, where the process is repeated. Very few routers have knowledge about the entire internetwork and often, the datagrams are forwarded based on default information without any knowledge of where the destination is actually located.

Direct vs indirect delivery

When the source host prepares to send a message to another host, a fundamental decision has to be made, namely: is the destination host also resident on the local network or not? If the NetID portions of the source and destination IP addresses match, the source host will assume that the destination host is resident on the same network, and will attempt to forward it locally. This is called direct delivery. If not, the message will be forwarded to the default gateway (a router on the local network), which will forward it. This is called indirect delivery. If the router can deliver it directly i.e. the destination host resides on a network directly connected to the router, it will. If not, it will consult its routing tables and forward it to the next appropriate router. This process will repeat itself until the packet is delivered to its final destination.

Static vs dynamic routing

Each router maintains a table with the following format:

Network address	Net mask	Gateway address	Interface	Metric
127.0.0.0	255.0.0.0	127.0.0.1	127.0.0.1	1
207.194.66.0	255.255.255.224	207.194.66.100	207.194.66.100	1
207.194.66.0	255.255.255.255	127.0.0.1	127.0.0.1	1
207.194.66.255	255.255.255.255	207.194.66.100	207.194.66.100	1
224.0.0.0	224.0.0.0	207.194.66.100	207.194.66.100	1
255.255.255.255	255.255.255.255	207.194.66.100	0.0.0.0	1

Table 16.2
Active routes

It basically reads as follows: 'If a packet is destined for network 207.194.66.0, with a Netmask of 255.255.255.224, then forward it to the router port 207.194.66.100,' etc. It is logical that a given router cannot contain the whereabouts of each and every network in the world in its routing tables, hence it will contain default routes as well. If a packet cannot be specifically routed, it will be forwarded on a default route, which should (it is hoped) move it closer to its intended destination.

These routing tables can be maintained in two ways. In most cases, the routing protocols will do this automatically. The routing protocols are implemented in software that runs on the routers, enabling them to communicate on a regular basis and allowing them to share their 'knowledge' about the network with each other. In this way they continuously 'learn' about the topology of the system, and upgrade their routing tables accordingly. This process is called dynamic routing. If, for example, a particular router is removed from the system, the routing tables of all routers containing a reference to that router will change. However, because of the interdependence of the routing tables, a change in any given table will initiate a change in many other routers and it will be a while before the tables stabilize. This process is known as convergence.

Dynamic routing can be further subclassified as distance vector, link state, or hybrid – depending on the method by which the routers calculate the optimum path.

In distance vector dynamic routing, the 'metric' or yardstick used for calculating the optimum routes is simply based on distance, i.e. which route results in the least number of 'hops' to the destination. Each router constructs a table, which indicates the number of hops to each known network. It then periodically passes copies of its tables to its immediate neighbors. Each recipient of the message then simply adjusts its own tables based on the information received from its neighbor.

It is also possible for a network administrator to make static entries into routing tables. These entries will not change, even if a router that they point to is not operational.

Autonomous systems

For the purpose of routing, a TCP/IP-based internetwork can be divided into several autonomous systems (ASs) or domains. An autonomous system consists of hosts, routers and data links that form several physical networks and are administered by a single authority such as a service provider, university, corporation, or government agency.

Routing decisions that are made within an autonomous system (AS) are totally under the control of the administering organization. Any routing protocol, using any type of routing algorithm, can be used within an autonomous system since the routing between two hosts in the system is completely isolated from any routing that occurs in other autonomous systems. Only if a host within one autonomous system communicates with a host outside the system, will another autonomous system (or systems) and possibly the Internet backbone be involved.

An autonomous system can be classified under one of three categories.

Interior, exterior and gateway-to-gateway protocols.

There are three categories of TCP/IP routing protocols namely interior gateway protocols, exterior gateway protocols, and gateway-to-gateway protocols.

Two routers that communicate directly with one another and are both part of the same autonomous system are said to be interior neighbors and are called interior gateways. They communicate with each other using interior gateway protocols such as RIP, HELLO, IS–IS or OSPF.

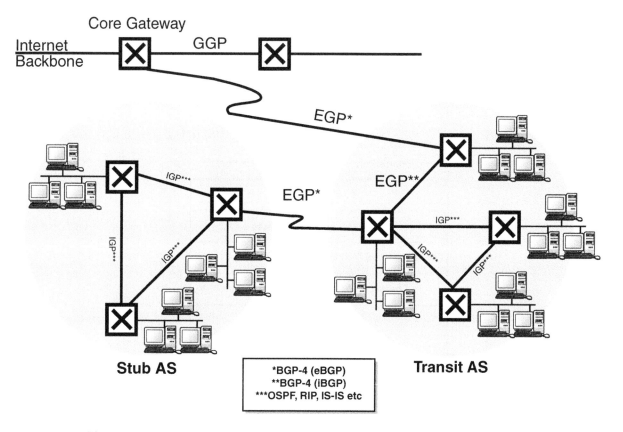

Figure 16.7
Autonomous systems and routing protocols

Routers in different ASs, however, cannot use IGPs for communication for more than one reason. Firstly, IGPs are not optimized for long-distance path determination. Secondly, the owners of ASs (particularly Internet service providers) would find it unacceptable for their routing metrics (which include sensitive information such as error rates and network traffic) to be visible to their competitors. For this reason, routers that communicate with each other and are resident in different ASs communicate with each other using exterior gateway protocols such as BGP-4.

16.3 Host-to-host layer: End to end reliability

16.3.1 TCP

Transmission control protocol (TCP) is a connection-oriented protocol and is said to be 'reliable', although this word is used in a data communications context. TCP establishes a session between two machines before data is transmitted. Because a connection is setup beforehand, it is possible to verify that all packets are received on the other end and to arrange retransmission in case of lost packets. Because of all these built-in functions, TCP involves significant additional overhead in terms of processing time and header size.

TCP fragments large chunks of data into smaller segments if necessary, reconstructs the data stream from packets received, issues acknowledgments of data received, provides socket services for multiple connections to ports on remote hosts, performs packet verification and error control, and performs flow control.

TCP header

The TCP header is structured as follows:

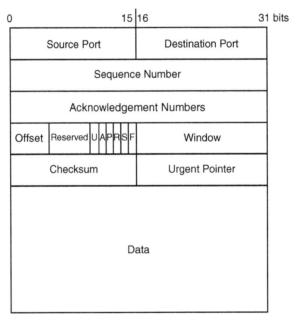

Figure 16.8
TCP header format

The source and destination ports (16 bits each) identify the host processes at each side of the connection. Examples are post office protocol (POP3) at port 110 and simple mail transfer protocol (SMTP) at port 25. Whereas a destination host is identified by its IP address, the process on that host is identified by its port number. A combination of port number and IP address is called socket.

The sequence number (32 bits) ensures the sequentiality of the data stream. TCP by implication associates a 32-bit number with every byte it transmits. The sequence number is the number of the first byte in every segment (or 'chunk') of data sent by TCP. If the SYN flag is set, however, it indicates that the sender wants to establish a connection and the number in the sequence number field becomes the initial sequence number or ISN. The receiver acknowledges this, and the sender then labels the first byte of the transmitted data with a sequence number of ISN+1. The ISN is a pseudo random number with values between 0 and 2^{32}.

The acknowledgments number (32 bits) is used to verify correct receipt of the transmitted data. The receiver checks the incoming data and if the verification is positive, acknowledges it by placing the number of the next byte expected in the acknowledgments number field and setting the ACK flag. The sender, when transmitting, sets a timer and if acknowledgments are not received within a specific time, an error is assumed and the data is retransmitted.

Data offset (4 bits) is the number of 32-bit words in the TCP header. This indicates where the data begins. It is necessary since the header can contain options and thus does not have a fixed length.

Six flags control the connection and data transfer. They are:

- URG: Urgent pointer field significant
- ACK: Acknowledgments field significant
- PSH: Push function
- RST: Reset the connection
- SYN: Synchronize sequence numbers
- FIN: No more data from sender

The window field (16 bits) provides flow control. Whenever a host sends an acknowledgment to the other party in the bi-directional communication, it also sends a window advertisement by placing a number in the window field. The window size indicates the number of bytes, starting with the one in the acknowledgments field, that the host is able accept.

The checksum field (16 bits) is used for error control

The urgent pointer field (16 bits) is used in conjunction with the URG flag and allows for the insertion of a block of 'urgent' data in the beginning of a particular segment. The pointer points to the first byte of the non-urgent data following the urgent data.

16.3.2   UDP

User datagram protocol (UDP) is a 'connectionless' protocol and does not require a connection to be established between two machines prior to data transmission. It is therefore said to be 'unreliable' – the word 'unreliable' used here as opposed to 'reliable' in the case of TCP and should not be interpreted against its everyday context.

Sending a UDP datagram involves very little overhead in that there are no synchronization parameters, no priority options, no sequence numbers, no timers, and no retransmission of packets. The header is small, the protocol is streamlined functionally. The only major drawback is that delivery is not guaranteed. UDP is therefore used for communications that involve broadcasts, for general network announcements, or for real-time data.

The UDP header

The UDP header is significantly smaller than the TCP header and only contains four fields.

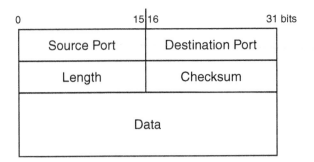

Figure 16.9
UDP header format

UDP source port (16 bits) is an optional field. When meaningful, it indicates the port of the sending process, and may be assumed to be the port to which a reply should be addressed in the absence of any other information. If not used, a value of zero is inserted.

UDP destination port has the same meaning as for the TCP header, and indicates the process on the destination host to which the data is to be sent.

UDP message length is the length in bytes of the datagram including the header and the data.

The 16-bit UDP checksum is used for validation purposes.

16.4 Troubleshooting

16.4.1 Introduction

This section deals with problems related to the TCP/IP protocol suite. The TCP/IP protocols are implemented in software and cover the second (Internet), the third (host-to-host) and the upper (application) layers of the ARPA model. These protocols need a network infrastructure as well as a medium in order to communicate. This infrastructure is typically Ethernet, as dealt with in the previous section.

16.4.2 Common problems

TCP/IP is a complex topic and this section cannot really do justice to it. This section will therefore try to show some common approaches that do not necessarily require an in-depth knowledge of the protocols involved.

16.4.3 Tools of the trade

16.4.3.1 TCP/IP utilities

These utilities are DOS programs that form part of the TCP/IP software. They are not protocols, but simply executable DOS programs that utilize some of the protocols. The utilities that will be shown here are ping, ARP and tracert. This is not the complete list, but it is sufficient for the purposes of this chapter.

16.4.3.2 Third party utilities

Most of the TCP/IP utilities are included in more recent third party software packages. The advantage of these packages such as TJPingPro is that they are Windows-based and hence much easier to use.

16.4.3.3   Software protocol analyzers

These have already been discussed in the section on Ethernet.

16.4.3.4   Hardware-assisted protocol analyzers

These have also been discussed in the section on Ethernet.

16.4.4   Typical network layer problems

16.4.4.1   TCP/IP protocol stack not properly installed on local host (host unable to access the network).

The easiest way to confirm this, apart from checking the network configuration via the control panel and visually confirming that TCP/IP is installed for the particular NIC used on the host, is to perform a loop-back test by pinging the host itself. This is done by executing ping localhost or ping 127.0.0.1. If a response is received, it means that the stack is correctly installed.

16.4.4.2   Remote host (e.g. a web server) not reachable

If it needs to be confirmed that a remote host is available, the particular machine can be checked by pinging it. The format of the command is:

- Ping 193.7.7.3, where 193.7.7.3 is the IP address of the remote machine, or
- Ping www.idc-online.com where www.idconline.com is the domain name of the remote host, or
- Ping john where john (or whatever) has been equated to the IP address of the remote machine in the hosts file of the local machine.

This is an extremely powerful test, since a positive acknowledgment means that the bottom two layers (ARPA) of both the local and the remote hosts as well as all routers and all communication links between the two hosts are operational.

16.4.4.3   A host is unable to obtain an automatically assigned IP address

When TCP/IP is configured and the upper radio button (on Windows 95/98) is highlighted (indicating that an IP address has to be obtained automatically) the host, upon booting up, will broadcast a request for the benefit of the local dynamic host configuration protocol (DHCP) server. Upon hearing the request, the DHCP server will offer an IP address to the requesting host. If the host is unable to obtain such IP address, it can mean one of two hurdles:

- The DHCP server is down. If this is suspected, it can be confirmed by pinging the DHCP server.
- There are no spare IP addresses available. Nothing can be done about this and the user will have to wait until one of the other logged in machines is switched off which will cause it to relinquish its IP address and make it available for reissue.

16.4.4.4   Reserved IP addresses

Reserved IP addresses are IP addresses in the ranges 10.0.0.0–10.255.255.255, 172.16.0.0–172.16.255.255 and 192.168.0.0–192.168.255.255. These IP addresses are only allocated to networks that will never have access to the Internet. All Internet routers are pre-programmed to ignore these addresses. If a user therefore tries to access such an

IP address over the Internet, the message will not be transported across the Internet and hence the desired network cannot be reached.

16.4.4.5   Duplicate IP addresses

Since an IP address is the Internet equivalent of a postal address, it is obvious that duplicate IP addresses cannot be tolerated. If a host is booted up, it tries to establish if any other host with the same IP address is available on the local network. If this is found to be true, the booting up machine will not proceed with logging on to the network and both machines with the duplicate IP address will display error messages in this regard.

16.4.4.6   Incorrect network ID – different netIDs on the same physical network

As explained in the chapter on TCP/IP, an IP address consists of two parts, namely, a network ID (netID), which is the equivalent of a postal zip code and a hostID, which is the equivalent of a street address. If two machines on the same network have different netIDs, their 'zip codes' will differ and hence the system will not recognize them as coexisting on the same network. Even if they are physically connected to the same Ethernet network, they will not be able to communicate directly with each other using TCP/IP.

16.4.4.7   Incorrect subnet mask

As explained in the chapter on TCP/IP, the subnet mask indicates the boundary between the netID and the hostID. A faulty subnet mask, when applied to an IP address, could result in a netID (zip code) that includes bits from the adjacent hostID and hence looks different than the netID of the machine wishing to send a message. The sending host will therefore erroneously believe that the destination host exists on another network and that the packets have to be forwarded to the local router for delivery to the remote network.

If the local router is not present (no default gateway specified) the sender will give up and not even try to deliver the packet. If, on the other hand, a router is present (default gateway specified), the sender will deliver the packet to the router. The router will then realize that there is nothing to forward since the recipient does in fact, live on the local network, and it will try to deliver the packet to the intended recipient. Although the packet eventually gets delivered, it leads to a lot of unnecessary packets transmitted as well as unnecessary time delays.

16.4.4.8   Incorrect or absent default gateway(s)

An incorrect or absent default gateway in the TCP configuration screen means that a host, wishing to send a message to another host on a different network, is not able to do so. The following is an example:

Assuming that a host with IP address 192.168.0.1 wishes to ping a non-existent host on IP address 192.168.0.2. The subnet mask is 255.255.255.0. The pinging host applies the mask to both IP addresses and comes up with a result of 192.168.0.0 in both cases. Realizing that the destination host resides on the same network as the sender, it proceeds to ping 192.168.0.2. Obviously, there will be no response from the missing machine and hence a time-out will occur. The sending host will issue a time-out message in this regard.

Now consider a scenario where the destination host is 193.168.0.2 and there is no valid default gateway entry. After applying the subnet mask, the sending host realizes that the destination resides on another network and that it therefore needs a valid default gateway.

Not being able to find one, it does not even try to ping but simply issues a message to the effect that the destination host is unreachable.

The reason for describing these scenarios is that the user can often figure out the problem by simply observing the error messages returned by the ping utility.

16.4.4.9 MAC address of a device not known to user

The MAC address of a device, such as a PLC, is normally displayed on a sticker attached to the body. If the MAC address of the device is not known, it can be pinged by its IP address (for example, ping 101.3.4.5), whereafter the MAC address can be obtained by displaying the ARP cache on the machine that did the ping. This is done by means of the ARP-a command.

16.4.4.10 IP address of a device not known to user

On a computer, this is not a problem since the IP address can simply be looked up in the TCP/IP configuration screen. Alternatively, the IP address can be displayed by commands such as winipcfg (Windows 95/98) or ipconfig /all (Windows 98/NT).

On a PLC, this might not be so easy unless the user knows how to attach a terminal to the COM (serial) port on the PLC and has the configuration software handy. An easier approach to confirm the IP address setup on the PLC (if any) is to attach it to a network and run a utility such as WebBoy or EtherBoy. The software will pick up the device on the network, regardless of its netID, and display the IP address.

16.4.4.11 Wrong IP address

It is possible that all devices on a network could have valid and correct IP addresses but that a specific host fails to respond to a message sent to it by a client program residing on another host. A typical scenario could be a supervisory computer sending a command to a PLC. Before assuming that the PLC is defective, one has to ascertain that the supervisory computer is in fact using the correct IP address when trying to communicate with the PLC. This can only be ascertained by using a protocol analyzer and capturing the communication or attempt at communication between the computer and the PLC. The packets exchanged between the computer and PLC can be identified by means of the MAC addresses in the appropriate Ethernet headers. It then has to be confirmed that the IP headers carried within these frames do in fact contain the correct IP addresses.

16.4.5   Transport layer problems

A detailed treatment of the TCP protocol is completely beyond the scope of this course but there are a few simple things that even a relatively inexperienced user can check.

16.4.5.1 No connection established

Before any two devices can communicate using TCP/IP, they need to establish a so-called triple handshake. This will be clearly indicated as a SYN, ACK/SYN, ACK sequence to the relevant port. Without a triple handshake the devices cannot communicate at all.

To confirm this, simply try to establish a connection (for example, by using an FTP client to log into an FTP server) and use a protocol analyzer to capture the handshake.

16.4.5.2   Incorrect port number

TCP identifies different programs (processes) on a host by means of so-called port numbers. For example, an FTP server uses ports 21 and 22, a POP3 server (for e-mail) uses port 110 and a web server (HTTP) uses port 80. Any other process wishing to communicate with one of these has to use the correct port number. The port number used is visible right at the beginning of the TCP header that can be captured with a protocol analyzer. Port numbers 1 to 1023 are referred to as 'well known' ports. Port numbers are administered by the Internet Assigned Numbers Authority (IANA – web site http://www.iana.org.). These are detailed in RFC 1700.

17

Radio and wireless communications overview

Objectives

When you have completed study of this chapter, you will be able to:

- List the main features of radio and wireless communications
- Understand and remedy simple problems in:
  - Noise
  - Interference
  - Power
  - Distance

17.1   Introduction

A significant number of industrial protocols are transferred using radio telemetry systems. Radio is often chosen in preference to using landlines for a number of reasons:

- Costs of cable can far exceed that of radio telemetry systems
- Radio systems can be installed faster than landline systems
- Radio equipment is very portable and can be easily moved
- Radio can be used to transmit the data in any format required by the user
- Reasonably high data rates can be achieved compared to some landline applications
- Radio can be used as a backup for landlines

This chapter briefly examines some of the issues that have to be considered if the industrial protocols are used over a radio link. It is broken down into:

- Components of a radio link
- Radio spectrum and frequency allocation
- Summary of radio characteristics for VHF/UHF radio telemetry systems
- Radio modems
- How to prevent intermodulation problems

- Implementing a radio link
- Miscellaneous considerations

17.2 Components of a radio link

A radio link consists of the following components:

- Antennas
- Transmitters
- Receivers
- Antenna support structures
- Cabling
- Interface equipment.

Figure 17.1 illustrates how these elements are connected together to form a complete radio link.

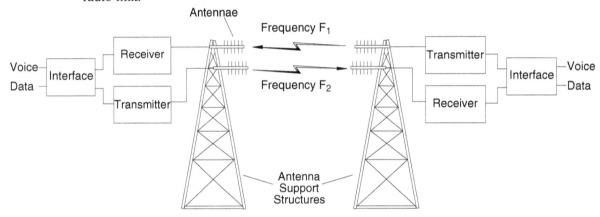

Figure 17.1
Fundamental elements of a radio link

Antenna

It is the device used to radiate or detect the electromagnetic waves. There are many different designs of antennas available. Each one radiates the signal (electromagnetic waves) in a different manner. The type of antenna used depends on the application and on the area of coverage required.

Transmitter

It is the device that converts the voice or data signal into a modified (modulated) higher frequency signal. Then it feeds the signal to the antenna where it is radiated into the free space as an electromagnetic wave at radio frequencies.

Receiver

It is the device that converts the radio frequency signals (fed to it from the antenna detecting the electromagnetic waves from free space) back into voice or data signals.

Antenna support structure

An antenna support structure is used to mount antennas, in order to provide a height advantage, which generally provides increased transmission distance and coverage. It may vary in construction from a three-meter wooden pole to 1000 m steel structure.

A structure, which has guy wires to support it, is generally referred to as a mast. A structure, which is free standing, is generally referred to as a tower.

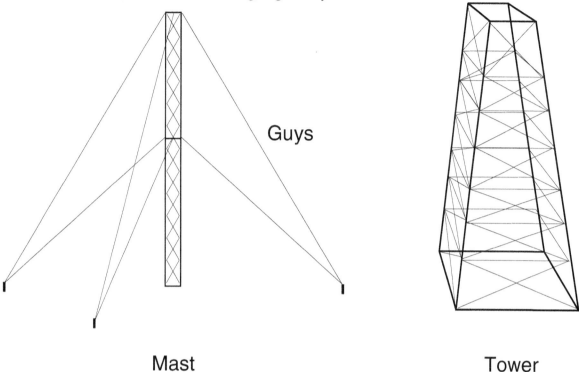

Mast   **Tower**

Figure 17.2
Illustration of mast and tower

Cabling

There are three main types of cabling used in connecting radio systems:

- Coaxial cable for all radio frequency connections
- Twisted pair cables for voice, data and supervisory connections
- Power cables.

Interface equipment

This allows connection of voice and data into the transmitters and receivers from external sources. It also controls the flow of information, timing of operation on the system and control and monitoring of the transmitter and receiver.

17.3   The radio spectrum and frequency allocation

There are very strict regulations that govern the use of various parts of the radio frequency spectrum. Specific sections of the radio frequency spectrum have been allocated for public use. All frequencies are allocated to users by a government regulatory body. Table 17.1 illustrates the typical sections of the radio spectrum allocated for public use around the world. Each section is referred to as a band.

Ultra High Frequency (UHF)	Mid Band UHF	960 Mhz 800 Mhz
	Low Band UHF	520 Mhz 335 Mhz
Very High Frequency (VHF)	High Band VHF	225 Mhz 101 Mhz
	Mid Band VHF	100 Mhz 60 Mhz
	Low Band VHF	59 Mhz 31 Mhz
High Frequency (HF)		30 Mhz 2 Mhz

Table 17.1
The radio spectrum for public use

Certain sections of these bands will have been allocated specifically for telemetry systems.

In some countries, a deregulated telecommunications environment has allowed sections of the spectrum to be sold off to large private organizations to be managed, and then sold to smaller individual users.

Application must be made to the government body, or independent groups that hold larger chunks of the spectrum for on selling, to obtain a frequency and no transmission is allowed on any frequency unless a license is obtained.

17.4 Summary of radio characteristics of VHF/UHF

The following table summarizes the information that was discussed in this section.

	Low band VHF	Mid band VHF	High band VHF
Propagation mode	Mostly L.O.S. some surface wave	L.O.S. Minimal surface wave	L.O.S.
Data rates	600 Baud	1200 Baud	2400 Baud
Diffraction properties	Excellent	Very good	Good
Natural noise environment	High	Medium	Low
Affected by man-made noise	Severe	Bad	Some
Penetration of solids	Excellent	Very good	Good
Fading by ducting	Long term	Medium term	Short term
Absorption by wet vegetation	Negligible	Low	Some
Equipment availability	Minimal	Reasonable	Excellent
Relative equipment cost	High	Medium	Low
Uses	- In Forested areas - Mostly mobile - Very hilly	- Very hilly & Forested areas - Mostly mobile - Over water	- Long Distance / L.O.S. / hilly areas - L.O.S. links - Mobile - Borefields - Over water

Table 17.2
VHF radio characteristics

	L.O.S.	L.O.S.
Propagation mode	L.O.S.	L.O.S.
Data rates	4800 Baud (9600?)	9600 Baud (19.2 K?)
Diffraction properties	Some	Minimal
Natural noise environment	Low	Negligible
Affected by man-made noise	Low	Very low
Penetration of solids	Low	Negligible
Reflection & absorption by solids	Good (enhanced multipathing)	Excellent (excellent multipathing)
Absorption by wet vegetation	High	Very high
Interference by ducting	Some	Some
Equipment availability	Excellent	Reasonable
Relative equipment costs	Low	Medium
Uses	- Telemetry - Mobile	- Telemetry - Mobile - Links

Table 17.3
UHF radio characteristics

17.5 Radio modems

Radio modems are suitable for replacing wire lines to remote sites or as a backup to wire or fiber optic circuits, and are designed to ensure that computers and PLCs, for example, can communicate transparently over a radio link without any specific modifications required.

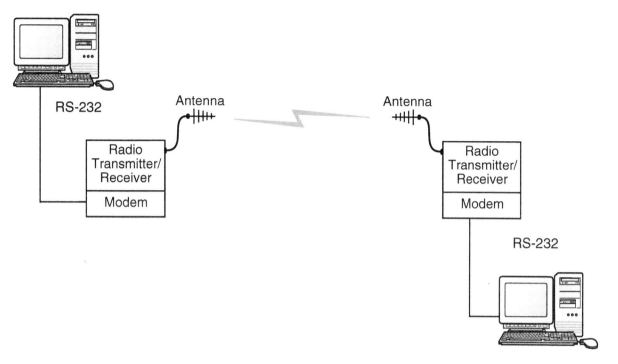

Figure 17.3
Radio modem configuration

Modern radio modems operate in the 400 to 900 MHz band. Propagation in this band requires a free line of sight between transmitting and receiving antennae for reliable communications. Radio modems can be operated in a network, but require a network management software system (protocols) to manage network access and error detection. Often, a master station with hot change-over, communicates with multiple radio field stations. The protocol for these applications can use a simple poll/response technique.

The more sophisticated peer-to-peer network communications applications, require a protocol based on carrier sensing multiple access with collision detection (CSMA/CD). A variation on the standard approach is to use one of the radio modems as a network watchdog to periodically poll all the radio modems on the network and to check their integrity. The radio modem can also be used as a relay station to communicate with other systems, which are out of the range of the master station.

The interface to the radio modem is typically EIA-232 but EIA-422, EIA-485 and fiber optics are also options. Typical speeds of operation are up to 9600 bps. A buffer is required in the modem and is typically a minimum of 32 kilobytes. Hardware and software flow control techniques are normally provided in the radio modem firmware, ensuring that there is no loss of data between the radio modem and the connecting terminal.

Typical modulation techniques are two level direct FM (1200 to 4800 bps) to three level direct FM (9600 bps).

A typical schematic of a radio modem is given in Figure 17.4.

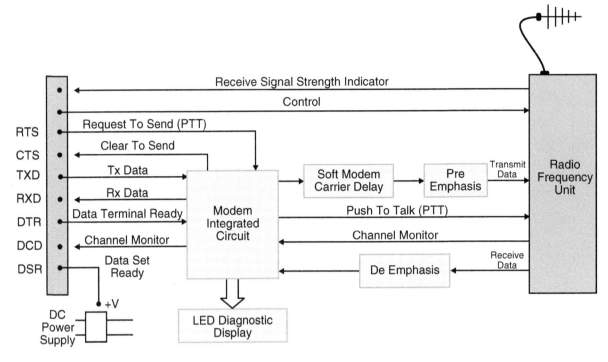

Figure 17.4
Typical block diagram of a radio modem

The following terms are used in relation to radio modems:

PTT	Push to talk signal
RSSI	Receive signal strength indicator – indicates the received signal strength with a proportionally varying DC voltage.
Noise squelch	Attempts to minimize the reception of any noise signal at the discriminator output.
RSSI squelch	Opens the receive audio path when the signal strength of the RF carrier is of a sufficiently high level.
Channel monitor	Indicates if the squelch is open.
Soft carrier delay	Allows the RF transmission to be extended slightly after the actual end of the data message which avoids the end of transmission bursts that occur when the carrier stops and the squelch almost simultaneously disconnects the studio path.
RTS, CTS, DCD, Clock, Transmit data, Receive data	All relate to EIA–232.

The radio modem has a basic timing system for communications between a terminal and the radio modem, indicated in Figure 17.5.

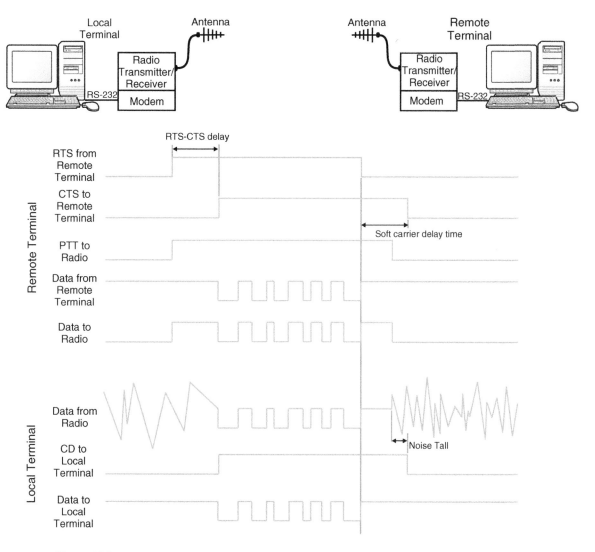

Figure 17.5
Radio modem timing diagram

Data transmission begins with the RTS line becoming active at the remote terminal side. The radio modem then raises the CTS line to indicate that transmission can proceed. At the end of the transmission, the PTT is kept active to ensure that the receiving side detects the remaining useful data before the RF carrier is removed.

17.5.1   Modes of radio modems

Radio modems can be used in two modes:
- Point-to-point
- Point-to-multi-point

A point-to-point system can operate in continuous RF mode, which has a minimal turn on delay in transmission of data, and non continuous mode where there is a considerable energy saving. The RTS to CTS delay for continuous and switched carriers is of the order of 10 ms and 20 ms respectively.

A point-to-multi-point system generally operates with only the master and one radio modem at a time.

In a multi-point system when the data link includes a repeater, data regeneration must be performed to eliminate signal distortion and jitter. Regeneration is not necessary for voice systems where some error is tolerable.

Regeneration is performed by passing the radio signal through the modem which converts the RF analog signal back to a digital signal and then applies this output binary data stream to the other transmitting modem, which repeats the RF analog signal to the next location.

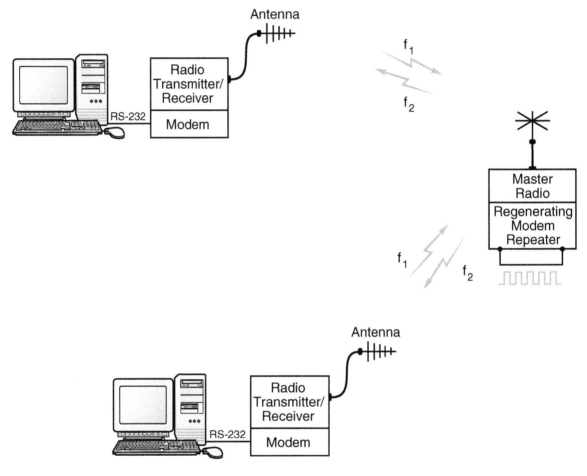

Figure 17.6
Regeneration of a signal with a radio modem

17.5.2 Features of a radio modem

Typical features that have to be configured in the radio modem are:

Transmit/receive radio channel frequency

In a point-to-point configuration running in a dual frequency/split channel assignment, two radios will operate on opposing channel sets.

Host data rate and format

Data rate/character size/parity type and number of stop bits for EIA-232 communications.

Radio channel data rate

Data rate across the radio channel defined by the radio and bandwidth capabilities. Note that these specifications are generally set at the time of manufacture.

Minimum radio frequency signal level

Should not be set too low on the receiver otherwise noise data will also be read.

Supervisory data channel rate

Used for flow control and therefore should not be set too low otherwise the buffer on the receiver will overflow. Typically one flow control bit to 32 bits of serial data is the standard.

Transmitter key up delay

The time for the transmitter to energize and stabilize before useful data is sent over the radio link. Transmitter key up delay should be kept as low as possible to minimize overheads.

17.5.3　Spread spectrum radio modems

Several countries around the world have allocated a section of bandwidth for use with spread spectrum radio modems. In Australia and America, this is in the 900 MHz area.

In brief, a very wide band channel is allocated to the modem, for example, approximately 3.6 MHz wide. The transmitter uses a pseudo random code to place individual bits, or groups of bits, broadly across the bandwidth and the receiver uses the same random code to receive them. Because they are random, a number of transceivers can operate on the same channel and a collision of bits will be received as noise by a receiver in close proximity.

The advantage of 'spread spectrum' radio modems is very high data security and data speeds of up to 19.2 kbps. The disadvantage is the very inefficient use of the radio spectrum.

17.6　Intermodulation and how to prevent it

17.6.1　Introduction

Besides noise and interference that emanates from man-made sources (cars, electrical motors, switches, rectifiers etc) there are three other main causes of RF interference. These are produced from other radio equipment. The first and the most obvious source, is another radio user operating close by on the same frequency as the system suffering from interference. Unfortunately, besides using special coding techniques to minimize the problem there is little that can be done, short of complaining to the regulatory government body that issues licenses, or finding out who it is and asking them to stop transmissions.

The second source of interference comes from noisy transmitters that emit spurious frequencies outside their allocated bandwidth. These spurious emissions will tend to fall on other users' channel bandwidths and cause interference problems. Aging transmitters and those that are not well maintained are normally the culprits.

The third source of interference is known as intermodulation. This is normally the most common source of interference and generally the most difficult to locate and the most costly to eliminate. The following section will examine this phenomenon in more detail.

17.6.2   Intermodulation

Intermodulation occurs where two or more frequencies interact in a non-linear device such as a transmitter, receiver or their environs, or on a rusty bolted joint acting as an RF diode to produce one or more additional frequencies that can potentially cause interference to other users. When two electromagnetic waves meet and intermodulate in a non-linear device, they produce a minimum of two new frequencies – one being the sum of the frequencies and the other being the difference of the frequencies.

A nearby receiver may be on or close to one of the intermodulation frequency products, receive it as noise and interference and then could also retransmit it as further noise and interference. For example, if two frequencies a and b interact, then they will produce two new frequencies c and d where $a + b = c$ and $a - b = d$. c and d are referred to as intermodulation products.

Of course c and d will be of significantly less magnitude than a and b and their exact magnitude depends on the magnitude of a and b, at the point a and b meet, and on the efficiency of the non-linear device at which the intermodulation takes place.

Fortunately, this problem is only really significant when the two transmitters for a and b are within close proximity. But consideration should be given to intermodulation products produced at a distant location as these have been known to cause noticeable background noise.

If there are more than two frequencies at one location then the number of intermodulation products possible increases dramatically.

For example, if there are transmitters on frequencies a, b and c at one location then the intermodulation products become:

$$a + b = f1$$
$$a + b = f2$$
$$b + c = f3$$
$$b - c = f4$$
$$a + c = f5$$
$$a - c = f6$$
$$a + b - c = f7$$
$$a + b - c = f8$$
$$a - b + c = f9$$
$$a - b - c = f10$$

This illustrates that the number of potential intermodulation products becomes prodigious as the number of frequencies increases. Unfortunately, the scenario gets worse. Each frequency from a transmitter will produce a significant harmonic at twice, three times, four times, etc, its carrier frequency (this is particularly true with FM systems). Each sequential harmonic will be of a lesser magnitude than the previous one.

Therefore if the transmitter is operating on frequency a, then harmonics will be produced at 2a, 3a, 4a, etc. The 2a and 3a harmonics can be quite large. These harmonics are produced because of resonant properties of antennas, cables, buildings and tuned circuits in the receivers and transmitters themselves and also due to the harmonic side bands produced in FM.

Taking these harmonics into account, intermodulation products such as:

(i) $2a - b$ and
(ii) $3b - 2c$

These are examples of what can be produced. (i) is referred to as a third order product and (ii) as a fifth order product. This refers to the total of multiples of each frequency.

These intermodulation products can cause severe interference in radio systems that are located at the same site. Taking these into account, there are numerous permutations of a small number of base frequencies that may cause intermodulation interference.

Generally, intermodulation products greater than fifth order are too small to be of consequence and generally radio systems are engineered only to take into account possible intermodulation distortion up to and including fifth order products. At sites where there are many sensitive receivers on different channels in the one building, then calculations to seventh order may be carried out.

Software has been provided with this book for calculation of intermodulation products, from any number of transmitters at a site.

For transmitters that are 1 km or more apart, the intermodulation products are generally so small that it would not be a problem. The exception may be for first or second order products.

If a system is experiencing unexplainable interference problems, a check should be made for distant intermodulation products.

Another source of harmonics sometimes comes from old buildings or masts where rusty bolts or nails act as RF diodes (non-linear devices) to produce intermodulation products. These are then detected by nearby receivers and may also be retransmitted as noise and interference.

A number of devices have been developed to assist in preventing the formation of intermodulation products and to prevent these products, spurious transmissions and harmonics from causing interference to nearby receivers or transmitters. All these devices are connected between the transmitter and the antenna. The following sections provide details on these devices and how they are used to eliminate interference problems.

17.7 Implementing a radio link

There is an important methodology that must be followed when designing and implementing a radio link if it is to work satisfactorily. It is relatively straightforward and will provide successful radio communications if followed closely.

Software has been provided with this manual that carries out the necessary calculations. Reference should be made to the book when using the software. In general, the software will follow a procedure as is described in this section.

The design methodology in a sequential order is as follows:

- Carry out a radio path profile
- Calculate RF losses for the radio path
- Calculate affects of transmitter power
- Decide on required fade margin
- Choose cable and antenna
- Purchase equipment
- Install equipment

17.7.1 Path profile

The first requirement in establishing a successful radio link is to draw up a radio path profile. This is basically a cross-sectional drawing of the earth for the radio propagation path showing all terrain variations, obstructions, terrain type (water, land, trees, buildings, etc) and the masts on which the antenna are mounted. For distances less than a kilometer or two, profiles are not normally required, as the RTU can quite often be clearly seen from the master site (but all other calculations and choices described in the design methodology must be carried out).

The Figure 17.7 illustrates a typical path profile.

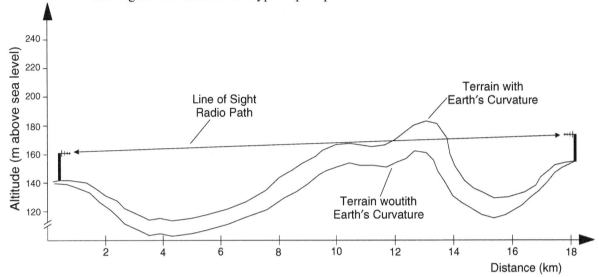

Figure 17.7
Typical path profile

The first step in this process is to obtain a contour map of the location. These survey maps are readily available from government departments that oversee land administration and private companies that carry out surveys and publish their material, for most areas of developed countries and a lot of areas in developing countries. It is recommended that the map have a minimum of 20 m contour lines, with 2 m, 5 m or 10 m being preferred.

Locate the RTU and master site locations on the map and draw a ruled line between the two locations with a pencil. Then assuming that the master site is at distance 0 km, follow the line along noting the kilometer marks and where a contour line occurs and at that point, also note the contour height.

The surface of the earth is of course not flat but curved. Therefore to plot the points you obtained from the map directly would not be a true indication of the path. The formula below provides a height correction factor that can be applied to each point obtained from the map to mark a true earth profile plot.

$$h = \frac{d_1 \times d_2}{12.75K}$$

where:
h = height correction factor that is added to the contour height (in meters)
d_1 = the distance from a contour point to one end of the path (in kilometers)
d_2 = the distance from the same contour point to the other end of the path (in kilometers)
K = the equivalent earth radius factor.

The equivalent earth radius factor K is required to account for the fact that the radio wave is bent towards the earth because of atmospheric refraction. The amount of bending varies with changing atmospheric conditions and therefore the value of K varies to account for this.

For the purposes of radio below 1 GHz it is sufficient to assume that for greater than 90% of the time K will be equal to 4/3. To allow for periods where a changing K will increase signal attenuation, a good fade margin should be allowed for.

The K factor allows the radio path to always be drawn in a straight line and adjusts the earth's contour height to account for the bending radio wave. Once the height has been calculated and added to the contour height, the path profile can be plotted.

From the plot it can now be seen if there are any direct obstructions in the path. As a general rule, the path should have good clearance over all obstructions. There is an area around the radio path that appears as a cone that should be kept as clearance for the radio path. This is referred to as the 'Fresnel zone.'

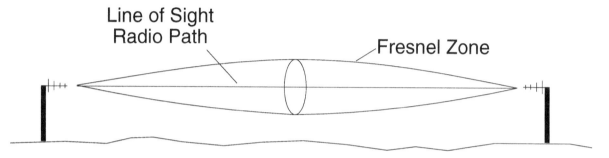

Figure 17.8
Fresnel zone clearance

Fresnel zone clearance is of more relevance to microwave path prediction than to radio path prediction.

The formula for the Fresnel zone clearance required is:

$$F = 0.55\sqrt{\frac{d_1 \times d_2}{F(\text{MHz}) \times D}}$$

Where:
F = Fresnel zone clearance in meters (i.e. radius of cone)
d_1 = distance from contour point to one end of path (in km)
d_2 = distance from contour point to other end of path (in km)
D = total length of path (in km)
F = frequency in MHz

If from the plot it appears that the radio path is going dangerously close to an obstruction, then it is worth doing a Fresnel zone calculation to check for sufficient clearance. Normally, the mast heights are chosen to provide a clearance of $0.6 \times$ the Fresnel zone radius. This figure of 0.6 is chosen because it firstly gives sufficient radio path clearance and secondly assists in preventing cancellation from reflections. At less than $0.6\ F$, attenuation of the line of sight signal occurs. At $0.6\ F$, there is no attenuation of the line of sight signal and therefore there is no gain achieved by the extra cost of providing higher masts.

Another important point to consider is that frequencies below 1 GHz have good diffraction properties. The lower the frequency the more diffraction that occurs. Therefore, for very long paths it is possible to operate the link with a certain amount of obstruction. It is important to calculate the amount of attenuation introduced by the diffraction and determine the affect it has on the availability (i.e. fade margin) of the radio link. The mathematics for doing this calculation is relatively complex and has been included in the software provided.

As an example, Figure 17.10 shows a hill obstructing the radio path. Therefore a calculation is required to be carried out to determine the attenuation due to diffraction at this hill. This would be then added to the total path loss to determine if the link will still operate satisfactorily.

One further point of note is that this discussion and the formulas and software provided are for fixed radio links only. Mobile (moving RTUs) radio uses a completely different set of criteria and formula. The greater majority of telemetry links are fixed and analysis of mobile radio is not provided with this book.

17.7.2 RF path loss calculations

The next step is to calculate the total attenuation of RF signal from the transmitter antenna to the receiver antenna. This includes:

- Free space attenuation
- Diffraction losses
- Rain attenuation
- Reflection losses

Free space attenuation and diffraction losses are calculated using the industry standard formulas. Rain attenuation is negligible at frequencies below 1 GHz.

Reflection losses are difficult to determine. First of all, the strength of the reflected signal depends on the surface it is reflected off (for example, water, rock, sand). Secondly, the reflected signal may arrive in phase, out of phase or at a phase angle in between. So reflected waves can be anything from totally catastrophic to enhancing the signal. Good engineering practice should always assume the worst case, which would be catastrophic failure. Therefore when designing a link, a check is made for reflections and if they exist, measures should be taken to remove the problem.

This can be done by moving antennas or masts to different locations and heights or by placing a barrier in the path of the reflection to absorb it. For example, place the antenna behind a hill, house, billboard, etc.

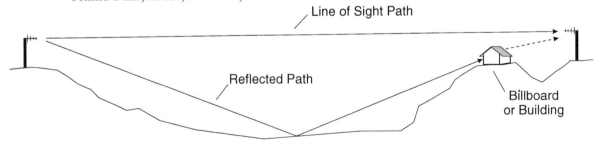

Figure 17.9
Removing potential reflections using barriers

With reference to Figure 17.10, the total loss would be:
Therefore the total RF loss is (a) + (b).

(a) $A = 32.5 + 20\log_{10} F + 20\log_{10} D$ = free space loss

$$= 32.5 + 53.1 + 24.6$$

$$+ 110.2 \text{ dB}$$

(b) *Diffraction loss* = 23 dB

$$(a) + (b) = 133.2 \text{ dB}$$

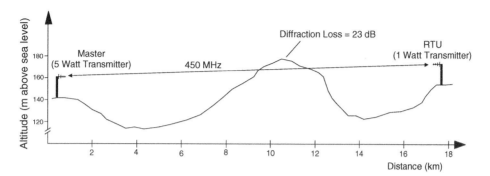

Figure 17.10
Example link

17.7.3 Transmitter power/receiver sensitivity

The next step is to determine the gain provided by the transmitters. If in a link configuration one transmitter operates with less power than the other, the direction with the least power transmitter should be considered. Therefore, as the ACA regulation requires that RTUs be allowed to transmit a maximum of 1 watt into the antenna, while master stations can transmit 5 watts into the antenna (sometimes higher), then the path direction from the RTU to the master should be considered.

The transmit power should be converted to a dBm figure. For an RTU this would be as follows:

$$Power - 10\log\left(\frac{1}{10^1}\right) \text{ dBm}$$
$$Power = +30 \text{ dBm}$$

The next step is to determine the minimum RF level at the receiver input that will open the front end of the receiver (i.e. turn it on). This is referred to as the 'receiver threshold sensitivity level' or sometimes as the 'squelch level'. This figure can be obtained from the manufacturer's specification sheets.

For a radio operating at 450 MHz, this would be approximately −123 dBm. At this level, the signal is only just above noise level and is not very intelligible. Therefore, as a general rule, a figure slightly better than this is used as a receiver sensitivity level.

A *de facto* standard is used where the RF signal is at its lowest but still intelligible. This level is referred to as the 12 dB SINAD level.

Again, this figure is obtained from manufacturer's data sheets. For a typical 450 MHz radio, this level is approximately −117 dBm.

Using these figures, a simple calculation can be performed to determine the link's performance for the example link in Figure 17.10.

Tx Pwr	= Transmit power at RTU
	= + 30 dBm
Loss	= RF path attenuation

	= 133.2 dB
Rx Sen	= Receiver sensitivity for 12 dB SINAD
	= –117 dB
Tx Pwr – Loss	= Available power at receiver
	= + 30 – 133.2 = –103.2 dBm

Since the receiver can accept an RF signal down to –117 dBm, then the RF signal will be accepted by the receiver. In this case, we have 13.8 dBm of spare RF power.

17.7.4   Signal to noise ratio and SINAD

The most common measure of the effect of noise on the performance of a radio system is signal to noise ratio (SNR). This is a measure of the signal power level compared to the noise power level at a chosen point in a circuit. In reality, it is the signal plus noise compared to noise.

$$\therefore SNR = \frac{Psignal + Pnoise}{Pnoise}$$

SNR is often expressed in dBs. Therefore:

$$SNR = 10\log_{10}\left(\frac{Psignal + Pnoise}{Pnoise}\right)$$

The real importance of this measurement is at the radio receiver. As the receive signal at the antenna increases, the noise level at the audio output of the receiver effectively decreases. Therefore, a measure of *SNR* at the receiver audio output for measured low level of RF input to the receiver front end is a good measure of the performance of a receiver.

The highest level of noise at the audio output will be when the RF input signal is at its lowest level. That is when the RF input signal just opens the receiver. At this point, the *SNR* will be between 3 and 6 dB.

A measurement has been developed as a *de facto* standard that measures the receiver input signal level for an *SNR* of 12 dB at the audio output. It is often referred to as the 12 dB SINAD level (signal to noise and distortion). The measurement is made with a device called a SINAD meter.

The SINAD measurement is carried out by feeding the input of the receiver with an RF signal that is modulated by a 1 kHz input audio signal. The 1 kHz signal will produce harmonics and unwanted distortion in the audio output. A SINAD meter is placed at the receiver audio output. The SINAD meter measures the power level of the 1 kHz plus the noise and distortion. It then filters the 1 kHz signal and measures the broadband level of noise power without the 1 kHz signal. It then divides the two measurements and provides a reading of *SNR* in dBs. The level at the receiver RF input is slowly increased until the *SNR* at the audio output is 12 dB. The RF input level is then noted and this is referred to as the 12 dB SINAD level. Some manufacturers refer to it as sensitivity at 12 dB SINAD or sometimes just sensitivity. The latter can be very deceptive and care must be taken.

The equipment configuration for measurement of SINAD is shown in Figure 17.11.

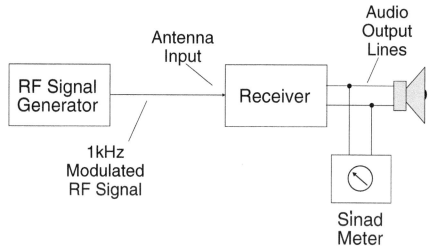

Figure 17.11
Equipment configuration for measuring SINAD

Some typical 12 dB SINAD sensitivity figures for modern radios in the VHF and UHF bands are 0.25 μV to 0.35 μV. The receiver opening sensitivity is normally 0.18 μV to 0.2 μV (i.e. the squelch level).

Another measurement also used to determine receiver performance is the '20 dB quieting' measurement. This is not used as often as the 12 dB SINAD method.

Here, the receiver squelch is set so that it is just open. A level meter is connected to the speaker and the noise level measured. The audio volume control is set for a convenient level (0 dBm). A signal generator is then connected to the receiver input and an unmodulated RF signal is fed into the receiver and increased slowly until the noise level at the audio output has dropped by 20 dB.

The disadvantage of this method is that it only measures the ability of the receiver to receive an unmodulated RF carrier signal. Poor design, component aging or improper alignment can provide a circuit response that admits the carrier signal perfectly but provides poor quality reception of modulated signals.

17.7.5 Fade margin

Radio is statistical by nature and it is therefore impossible to predict with 100% accuracy as to how it is going to perform. For example, due to degrading effects of reflections, multipathing, ducting and RF interference, a link may lose or gain signal by up to 15 dB over short or long periods of time. It is because of this unpredictability, it is important to have a safety margin to allow for intermittent link degradation. This safety margin (or spare RF power) is generally referred to as the fade margin.

IDC recommend that it should be the intention to design most links to have a fade margin of approximately 30 dB. This means that if there was a 30 dB drop in RF signal level then the RF signal at the receiver input would drop below the 12 dB SINAD sensitivity.

Therefore in the example in the previous section there is insufficient fade margin. This is basically overcome by using high gain antennas.

For example, if we use a 13 dB gain Yagi at the RTU and a 6 dB gain omni-directional antenna at the master site, we add an extra 19 dB gain to our signal.

Therefore the total fade margin in our example would become:
$$13.8 + 19 = 32.8 \text{ dB}$$

Finally, we must consider other losses introduced by cables, connectors, multicouplers, etc. In this example, if we have 20 m of 3 dB/100 m loss at each end, total connector losses of 0.5 dB at each end and a multicoupler loss of 3 dB at the master site then:

$$\text{Extra Losses} =$$

$$\text{Cables: } 2 \ (0.2 \times 3 \text{ dB})$$
$$= 1.2 \text{ dB}$$

$$\text{Connectors: } 2 \times 0.5 \text{ dB}$$
$$= 1 \text{ dB}$$

$$\text{Multicoupler} = 3 \text{ dB}$$

$$\text{Total extra losses} = 3 + 1 + 1.2$$
$$= 5.2 \text{ dB}$$

Therefore the fade margin for the link is:

$$32.8 - 5.2 = 27.6 \text{ dB}$$

17.7.6 Summary of calculations

The following equation is a summary of the requirements for calculation of fade margin:
Fade margin $= -$ (free space attenuation) $-$ (diffraction losses) $+$ (transmitter power) $+$ (receiver sensitivity) $+$ (antenna gain at master site) $+$ (antenna gain at RTU) $-$ (cable and connector loss at master) $-$ (cable and connector loss at RTU) $-$ (multicoupler filter or duplexer loss) $+$ (receiver pre-amplifier gain).

17.7.7 Miscellaneous considerations

When implementing a radio link, there are other important considerations. The following is a list of some of the main considerations:
(a) It is important to obtain specification sheets from the radio supplier before purchasing any equipment and ensuring that all parameters meet your requirements.
(b) The audio frequency output and input for a radio is normally a balanced 600-ohm connection. Depending on the equipment provided, it will accept levels of between -30 dBm and $+15$ dB and will output levels from -15 dBm to $+15$ dBm. These are normally adjustable internally.
(c) Most radios operate at $+12$ volt DC (or 13.2 v DC if it is floated across a battery). Depending on the RF output power level the current consumption may vary from 1 amp for 1 watt output to 10 amps for 50 watts output. This should be taken into consideration when sizing power supplies and batteries.
(d) The size and weight of the radio equipment should be noted so that correct mechanical mounting and rack space can be provided. Multicouplers and duplexers can be very large and may be required to be mounted on walls outside the radio equipment racks or in separate racks. In order to minimize intermodulation, multicouplers and filters should be as close as possible to the transmitter and receiver.

(e) It is beneficial to obtain mean time between failure (MTBF) figures from manufacturers because if the radio is to be placed in a remote location that is difficult to regularly access, it is vital that the radio is reliable.

(f) If the telemetry communication protocol requires the radio is to be switched on and off at regular intervals, it is best to avoid having relays in the inline RF circuit. Discrete transistor RF switching is preferred.

17.8   Troubleshooting

When troubleshooting an existing system, it is worth checking on a few issues discussed in the previous section. These are as follows:

Frequency selection

Is the frequency selection good? Look at Table 17.2 to see whether this is correct as far as absorption of wet vegetation, fading issues and diffraction properties are concerned.

Interference from other radio equipment

Besides the issue of intermodulation which can be addressed and corrected, if the interference on the radio equipment is due to another radio user operating close by on the same frequency or from a noisy radio transmitter emitting spurious frequencies outside its allocated bandwidth, the only fix is to terminate the other radio's transmissions in some way.

Intermodulation problems

This is discussed in the previous section. If you find considerable interference from other radio equipment, this could be the problem. Do the analysis using the software to find out whether this is a problem for your particular application.

Incorrect path loss calculation

If you find the received signal is weaker than anticipated, it is worth running a path loss calculation and checking that the fade margin is adequate. A fade margin of at least 30dB should be targeted.

Radio modems

As far as troubleshooting radio modems is concerned, once the issues above have been addressed; often the problems are related to the RS-232 or RS-485 interfaces. The troubleshooting techniques here are indicated in chapters 3 and 4.

Appendix A

Glossary

ABM	asynchronous balanced mode
ACE	association control element
ACE	asynchronous communications element – similar to UART
ACK	acknowledge (ASCII – Control F)
Active filter	are active circuit devices (usually amplifiers) with passive circuit elements (resistors and capacitors) and which have characteristics that more closely match ideal filters than do passive filters
Active passive device	capable of supplying the current for the loop (active) or one that must draw its power from connected equipment (passive)
ADCCP	advanced data communication control procedure
ADDR	this is address field
Address	a normally unique designator for location of data or the identity of a peripheral device that allows each device on a single communications line to respond to its own message
Algorithm	normally used as a basis for writing a computer program – this is a set of rules with a finite number of steps for solving a problem
Alias frequency	a false lower frequency component that appears in data reconstructed from original data acquired at an insufficient sampling rate (which is less than two (2) times the maximum frequency of the original data)
ALU	arithmetic logic unit
Amplitude flatness	a measure of how close to constant the gain of a circuit remains over a range of frequencies
Amplitude modulation	a modulation technique (also referred to as AM or ASK) used to allow data to be transmitted across an analog network, such as a switched telephone network – the amplitude of a single (carrier) frequency is varied or modulated between two levels – one for binary 0 and one for binary 1

Analog   a continuous real time phenomenon, where the information values are represented in a variable and continuous waveform

ANSI   American National Standards Institute – the principal standards development body in the USA

APM   alternating pulse modulation

Appletalk   a proprietary computer networking standard initiated by the Apple computer for use in connecting the Macintosh range of computers and peripherals (including laser writer printers) – this standard operates at 230 kbps

Application layer   the highest layer of the seven layer ISO/OSI reference model structure, which contains all user or application programs

Arithmetic logic unit   the element(s) in a processing system that perform(s) the mathematical functions such as addition, subtraction, multiplication, division, inversion, AND, OR, AND and NOR

ARP   address resolution protocol – a transmission control protocol/ Internet protocol (TCP/IP) process that maps an IP address to Ethernet address, required by TCP/IP for use with Ethernet

ARQ   automatic request for transmission – a request by the receiver for the transmitter to retransmit a block or frame because of errors detected in the originally received message

AS   Australian Standard

ASCII   American Standard Code for Information Interchange – a universal standard for encoding alphanumeric characters into 7 or 8 binary bits – drawn up by ANSI to ensure compatibility between different computer systems

ASI   actuator sensor interface

ASIC   application specific integrated circuit

ASK   amplitude shift keying – *see* amplitude modulation

ASN.1   abstract syntax notation 1– an abstract syntax used to define the structure of the protocol data units associated with a particular protocol entity

Asynchronous   communications where characters can be transmitted at an arbitrary unsynchronized point in time and where the time intervals between transmitted characters may be of varying lengths – communication is controlled by start and stop bits at the beginning and end of each character

Attenuation   the decrease in the magnitude of strength (or power) of a signal – in cables, generally expressed in dB per unit length

AWG   American wire gauge

Balanced circuit   a circuit so arranged that the impressed voltages on each conductor of the pair are equal in magnitude but opposite in polarity with respect to ground

Band pass filter   a filter that allows only a fixed range of frequencies to pass through – all other frequencies outside this range (or band) are sharply reduced in magnitude

Bandwidth   the range of frequencies available expressed as the difference between the highest and lowest frequencies is expressed in hertz (or cycles per second)

Base address	a memory address that serves as the reference point – all other points are located by offsetting in relation to the base address
Baseband	baseband operation is the direct transmission of data over a transmission medium without the prior modulation on a high frequency carrier band
Baud	unit of signaling speed derived from the number of events per second (normally bits per second) – however, if each event has more than one bit associated with it, the baud rate and bits per second are not equal
Baudot	data transmission code in which five bits represent one character – 64 alphanumeric characters can be represented – this code is used in many teleprinter systems with one start bit and 1.42 stop bits added
BCC	block check calculation
BCC	block check character – error checking scheme with one check character; a good example being block sum check
BCD	binary coded decimal – a code used for representing decimal digits in a binary code
BEL	Bell (ASCII for Control-G)
Bell 212	an AT&T specification of full-duplex, asynchronous or synchronous 1200 baud data transmission for use on the public telephone networks
BER	bit error rate
BERT/BLERT	bit error rate/block error rate testing – an error checking technique that compares a received data pattern with a known transmitted data pattern to determine transmission line quality
BIN	binary digits
BIOS	basic input/output system
Bipolar	a signal range that includes both positive and negative values
BISYNC	binary synchronous communications protocol
Bit	derived from 'BInary DigiT,' a one or zero condition in the binary system
Bit stuffing with zero bit insertion	a technique used to allow pure binary data to be transmitted on a synchronous transmission line – each message block (frame) is encapsulated between two flags that are special bit sequences – then if the message data contains a possibly similar sequence, an additional (zero) bit is inserted into the data stream by the sender and is subsequently removed by the receiving device – the transmission method is then said to be data transparent
Bits per second (bps)	unit of data transmission rate
Block sum check	this is used for the detection of errors when data is being transmitted – it comprises a set of binary digits (bits) which are the modulo 2 sum of the individual characters or octets in a frame (block) or message
Bridge	a device to connect similar subnetworks without its own network address – used mostly to reduce the network load

Broadband	a communications channel that has greater bandwidth than a voice grade line and is potentially capable of greater transmission rates – opposite of baseband – in wideband operation the data to be transmitted are first modulated on a high frequency carrier signal – they can then be simultaneously transmitted with other data modulated on a different carrier signal on the same transmission medium
Broadcast	a message on a bus intended for all devices that requires no reply
BS	backspace (ASCII Control-H)
BS	British Standard
BSC	Bisynchronous transmission – a byte or character oriented communication protocol that has become the industry standard (created by IBM) – it uses a defined set of control characters for synchronized transmission of binary coded data between stations in a data communications system
BSP	binary synchronous protocol
Buffer	an intermediate temporary storage device used to compensate for a difference in data rate and data flow between two devices (also called a spooler for interfacing a computer and a printer)
Burst mode	a high-speed data transfer in which the address of the data is sent followed by back-to-back data words while a physical signal is asserted
Bus	a data path shared by many devices with one or more conductors for transmitting signals, data or power
Byte	a term referring to eight associated bits of information – sometimes called a 'character'
CAN	controller area network
Capacitance	storage of electrically separated charges between two plates having different potentials – the value is proportional to the surface area of the plates and inversely proportional to the distance between them
Capacitance (mutual)	the capacitance between two conductors with all other conductors, including shield, short-circuited to the ground
CATV	Community Antenna Television
CCITT	*see* ITU
Cellular polyethylene	expanded or 'foam' polyethylene consisting of individual closed cells suspended in a polyethylene medium
Character	letter, numeral, punctuation, control figure or any other symbol contained in a message
Characteristic impedance	the impedance that, when connected to the output terminals of a transmission line of any length, makes the line appear infinitely long – the ratio of voltage to current at every point along a transmission line on which there are no standing waves
CIC	controller in charge
Clock	the source(s) of timing signals for sequencing electronic events – for example, synchronous data transfer

CMD	command byte
CMR	common mode rejection
CMRR	common mode rejection ratio
CMV	common mode voltage
Common carrier	a private data communications utility company that furnishes communications services to the general public
Composite link	the line or circuit connecting a pair of multiplexes or concentrators – the circuit carrying multiplexed data
Contention	the facility provided by the dial network or a data PABX, which allows multiple terminals to compete on a first come, first served basis for a smaller number of computer posts
CPU	central processing unit
CR	carriage return (ASCII Control-M)
CRC	cyclic redundancy check – an error-checking mechanism using a polynomial algorithm based on the content of a message frame at the transmitter and included in a field appended to the frame – at the receiver, it is then compared with the result of the calculation that is performed by the receiver – also referred to as CRC-16
CRL	communication relationship list
Cross talk	a situation where a signal from a communications channel interferes with an associated channel's signals
Crossed planning	wiring configuration that allows two DTE or DCE devices to communicate – essentially it involves connecting pin 2 to pin 3 of the two devices
Crossover	in communications, a conductor that runs through the cable and connects to a different pin number at each end
CSMA/CD	carrier sense multiple access/collision detection – when two senders transmit at the same time on a local area network, they both cease transmission and signal that a collision has occurred – each then tries again after waiting for a predetermined time period
CTS	clear to send
Current loop	communication method that allows data to be transmitted over a longer distance with a higher noise immunity level than with the standard RS-232-C voltage method – a mark (a binary 1) is represented by current of 20 mA and a space (or binary 0) is represented by the absence of current
DAQ	data acquisition
Data integrity	a performance measure based on the rate of undetected errors
Data link layer	this corresponds to layer 2 of the ISO reference model for open systems interconnection – it is concerned with the reliable transfer of data (no residual transmission errors) across the data link being used
Data reduction	the process of analyzing a large quantities of data in order to extract some statistical summary of the underlying parameters
Datagram	a type of service offered on a packet-switched data network – a datagram is a self-contained packet of information that is sent through the network with minimum protocol overheads

DCD	data carrier detect
DCE	data communications equipment or data circuit-terminating equipment devices that provide the functions required to establish, maintain and terminate a data transmission connection – normally it refers to a modem
DCS	distributed control systems
Decibel (dB)	a logarithmic measure of the ratio of two signal levels where dB = 20log10$V1/V2$ or where dB = 10log10$P1/P2$ and where V refers to voltage or P refers to power – note that it has no units of measurement
Default	a value or setup condition assigned, which is automatically assumed for the system unless otherwise explicitly specified
Delay distortion	distortion of a signal caused by the frequency components making up the signal having different propagation velocities across a transmission medium
DES	data encryption standard
DFM	direct frequency modulation
Dielectric constant (E)	the ratio of the capacitance using the material in question as the dielectric, to the capacitance resulting when the material is replaced by air
Digital	a signal, which has definite states (normally two)
DIN	Deutsches Institut Für Normierung
DIP	dual in line package, referring to integrated circuits and switches
Direct memory access	a technique of transferring data between the computer memory and a device on the computer bus without the intervention of the microprocessor – also, abbreviated to DMA
DISC	disconnect
DLE	data link escape (ASCII character)
DNA	distributed network architecture
DPI	dots per inch
DPLL	digital phase locked loop
DR	dynamic range – the ratio of the full-scale range (FSR) of a data converter to the smallest difference it can resolve – DR = $2n$ where n is the resolution in bits
Driver software	a program that acts as the interface between a higher-level coding structure and the lower level hardware/firmware component of a computer
DSP	digital signal processing
DSR	data set ready or DCE ready in EIA-232D/E – an RS-232 modem interface control signal, which indicates that the terminal is ready for transmission
DTE	data terminal equipment – devices acting as data source or data sink, or both
DTR	data terminal ready or DTE ready in EIA-232D/E
Duplex	the ability to send and receive data simultaneously over the same communications line

EBCDIC	extended binary coded decimal interchange code an eight bit character code used primarily in IBM equipment – the code allows for 256 different bit patterns
EDAC	error detection and correction
EFTPOS	electronic funds transfer at the point of sale
EIA	Electronic Industries Association – a standards organization in the USA specializing in the electrical and functional characteristics of interface equipment
EISA	enhanced industry standard architecture
EMI/RFI	electromagnetic interference/radio frequency interference – 'background noise' that could modify or destroy data transmission
EMS	expanded memory specification
Emulation	the imitation of a computer system performed by a combination of hardware and software that allows programs to run between incompatible systems
ENQ	enquiry (ASCII Control-E)
EOT	end of transmission (ASCII Control-D)
EPA	enhanced performance architecture
EPR	earth potential rise
EPROM	erasable programmable read only memory – non-volatile semiconductor memory that is erasable in ultraviolet light and reprogrammable
Error rate	the ratio of the average number of bits that will be corrupted to the total number of bits that are transmitted for a data link or system
ESC	escape (ASCII character)
ESD	electrostatic discharge
ETB	end of transmission block
Ethernet	name of a widely used LAN, based on the CSMA/CD bus access method (IEEE 802.3) – Ethernet is the basis of the TOP bus topology
ETX	end of text (ASCII Control-C)
Even parity	a data verification method normally implemented in hardware in which each character must have an even number of 'ON' bits
Farad	unit of capacitance whereby a charge of one coulomb produces a one volt potential difference
FAS	fieldbus access sublayer
FCC	Federal Communications Commission
FCS	frame check sequence – a general term given to the additional bits appended to a transmitted frame or message by the source to enable the receiver to detect possible transmission errors
FDM	frequency division multiplexer – a device that divides the available transmission frequency range in narrower bands, each of which is used for a separate channel
FIB	factory information bus
FIFO	first in, first out

Filled cable	a telephone cable construction in which the cable core is filled with a material that will prevent moisture from entering or passing along the cable
FIP	factory instrumentation protocol
Firmware	a computer program or software stored permanently in PROM or ROM or semi-permanently in EPROM
Flame retardancy	the ability of a material not to propagate flame once the flame source is removed.
Flow control	the procedure for regulating the flow of data between two devices preventing the loss of data once a device's buffer has reached its capacity
FMS	fieldbus message specification
FNC	function byte
Frame	the unit of information transferred across a data link – typically, there are control frames for link management and information frames for the transfer of message data
Frequency modulation	a modulation technique (abbreviated to FM) used to allow data to be transmitted across an analog network where the frequency is varied between two levels – one for binary '0' and one for binary '1'– also known as frequency shift keying (or FSK)
Frequency	refers to the number of cycles per second
FRMR	frame reject
FSK	frequency shift keying, see frequency modulation
Full-duplex	simultaneous two-way independent transmission in both directions (4 wire) – see Duplex
G	giga (metric system prefix – 109)
Gateway	a device to connect two different networks that translates the different protocols
GMSK	Gaussian minimum shift keying
GPIB	general purpose interface bus – an interface standard used for parallel data communication, usually used for controlling electronic instruments from a computer – also known as IEEE 488 standard
Ground	an electrically neutral circuit that has the same potential as the earth – a reference point for an electrical system also intended for safety purposes
Half-duplex	transmissions in either direction, but not simultaneously
Hamming distance	a measure of the effectiveness of error checking – the higher the Hamming distance (HD) index, the safer is the data transmission
Handshaking	exchange of predetermined signals between two devices establishing a connection
Hardware	refers to the physical components of a device, such as a computer, sensor, controller or data communications system– these are the physical items that one can see
HART	highway addressable remote transducers

HDLC	high level data link control – the international standard communication protocol defined by ISO to control the exchange of data across either a point-to-point data link or a multidrop data link
Hertz (Hz)	a term replacing cycles per second as a unit of frequency
Hex	hexadecimal
HF	high frequency
Host	this is normally a computer belonging to a user that contains (hosts) the communication hardware and software necessary to connect the computer to a data communications network
HSE	high speed Ethernet
I/O address	a method that allows the CPU to distinguish between different boards in a system – all boards must have different addresses
IA5	international alphabet number 5
IC	integrated circuit
ICS	instrumentation and control system
IDF	intermediate distribution frame
IEC	International Electromechanical Commission
IEE	Institution of Electrical Engineers – an American based international professional society that issues its own standards and is a member of ANSI and ISO
IEEE	Institute of Electrical and Electronic Engineers
IFC	International Fieldbus Consortium
ILD	injection laser diode
Impedance	the total opposition that a circuit offers to the flow of alternating current or any other varying current at a particular frequency – it is a combination of resistance R and reactance X, measured in ohms
Inductance	the property of a circuit or circuit element that opposes a change in current flow, thus causing current changes to lag behind voltage changes – it is measured in henrys
Insulation resistance (IR)	that resistance offered by an insulation to an impressed DC voltage, tending to produce a leakage current though the insulation
Interface	a shared boundary defined by common physical interconnection characteristics, signal characteristics and measurement of interchanged signals
Interrupt handler	the section of the program that performs the necessary operation to service an interrupt when it occurs.
Interrupt	an external event indicating that the CPU should suspend its current task to service a designated activity
IP	Internet protocol
IRQ	interrupt request line
ISA	industry standard architecture (for IBM personal computers)
ISB	intrinsically safe barrier
ISDN	integrated services digital network – the new generation of worldwide telecommunications network that utilizes digital

	techniques for both transmission and switching – it supports both voice and data communications
ISO	International Standards Organization
ISP	interoperable systems project
ISR	interrupt service routine, *see* interrupt handler
ITB	end of intermediate block
ITS	interface terminal strip
ITU	International Telecommunications Union – formerly CCITT (Consultative Committee International Telegraph and Telephone) – an international association that sets worldwide standards (for example, V.21, V.22, V.22bis)
Jumper	a wire connecting one or more pins on the one end of a cable only
k (kilo)	this is 2^{10} or 1024 in computer terminology, for example, 1 kb = 1024 bytes
LAN	local area network – a data communications system confined to a limited geographic area typically about 10 km with moderate to high data rates (100 kbps to 50 Mbps) – same type of switching technology is used but common carrier circuits are not used
LAP-M	link access protocol modem
LAS	link active scheduler
LCD	liquid crystal display – a low-power display system used on many laptops and other digital equipment
LDM	limited distance modem – a signal converter, which conditions and boosts a digital signal so that it may be transmitted further than a standard RS-232 signal
Leased (or private) line	a private telephone line without inter-exchange switching arrangements
LED	light emitting diode – a semiconductor light source that emits visible light or infrared radiation
LF	line feed (ASCII Control-J)
Line driver	a signal converter that conditions a signal to ensure reliable transmission over an extended distance
Line turnaround	the reversing of transmission direction from transmitter to receiver or vice versa when a half-duplex circuit is used
Linearity	a relationship where the output is directly proportional to the input
Link layer	layer 2 of the ISO/OSI reference model – also known as the data link layer
Listener	a device on the GPIB bus that receives information from the bus
LLC	logical link control (IEEE 802)
LLI	lower layer interface
Loaded line	a telephone line equipped with loading coils to add inductance in order to minimize amplitude distortion
Loop resistance	the measured resistance of two conductors forming a circuit

Loopback	type of diagnostic test in which the transmitted signal is returned on the sending device after passing through all, or a portion of, a data communication link or network – a loopback test permits the comparison of a returned signal with the transmitted signal
LRC	longitudinal redundancy check
LSB	least significant bits – the digits on the right-hand side of the written HEX or BIN codes
LSD	least significant digit
M	Mega – metric system prefix for 10^6
m	meter – metric system unit for length
MAC	media access control (IEEE 802)
MAN	metropolitan area network
Manchester encoding	digital technique (specified for the IEEE 802.3 Ethernet baseband network standard) in which each bit period is divided into two complementary halves – a negative to positive voltage transition in the middle of the bit period designates a binary '1', whilst a positive to negative transition represents a '0'– the encoding technique also allows the receiving device to recover the transmitted clock from the incoming data stream (self clocking)
MAP 3.0	standard profile for manufacturing developed by MAP
MAP	manufacturing automation protocol – a suite of network protocols originated by General Motors that follow the seven layers of the OSI mode – a reduced implementation is referred to as a mini-MAP
Mark	this is equivalent to a binary 1
Master/slave	bus access method whereby the right to transmit is assigned to one device only, the master, and all the other devices, the slaves may only transmit when requested
MDF	main distribution frame
MIPS	million instructions per second
MMI	man-machine-interface
MMS	manufacturing message services – a protocol entity forming part of the application layer. It is intended for use specifically in the manufacturing or process control industry – it enables a supervisory computer to control the operation of a distributed community of computer-based devices
MNP	Microcom networking protocol
Modem eliminator	a device used to connect a local terminal and a computer port in lieu of the pair of modems to which they would ordinarily connect, allow DTE to DTE data and control signal connections otherwise not easily achieved by standard cables or connections
Modem	MODulator – DEModulator – a device used to convert serial digital data from a transmitting terminal to a signal suitable for transmission over a telephone channel or to reconvert the transmitted signal to serial digital data for the receiving terminal

MOS	metal oxide semiconductor
MOV	metal oxide varistor
MSB	most significant bits – the digits on the left-hand side of the written HEX or BIN codes
MSD	most significant digit
MTBF	mean time between failures
MTTR	mean time to repair
Multidrop	a single communication line or bus used to connect three or more points
Multiplexer (MUX)	a device used for division of a communication link into two or more channels either by using frequency division or time division
NAK	negative acknowledge (ASCII Control-U)
Network architecture	a set of design principles including the organization of functions and the description of data formats and procedures used as the basis for the design and implementation of a network (ISO)
Network layer	layer 3 in the ISO/OSI reference model, the logical network entity that services the transport layer responsible for ensuring that data passed to it from the transport layer is routed and delivered throughout the network
Network topology	the physical and logical relationship of nodes in a network; the schematic arrangement of the links and nodes of a network typically in the form of a star, ring, tree or bus topology
Network	an interconnected group of nodes or stations
NMRR	normal mode rejection ratio
Node	a point of interconnection to a network
Noise	a name given to the extraneous electrical signals that may be generated or picked up in a transmission line – if the noise signal is large compared with the data carrying signal, the latter may be corrupted resulting in transmission errors
NOS	network operating system
NRM	unbalanced normal response mode
NRZ	non return to zero – pulses in alternating directions for successive 1 bits but no change from existing signal voltage for 0 bits
NRZI	non return to zero inverted
Null modem	a device that connects two DTE devices directly by emulating the physical connections of a DCE device
Nyquist sampling theorem	in order to recover all the information about a specified signal it must be sampled at least at twice the maximum frequency component of the specified signal
OD	object dictionary
Ohm (W)	unit of resistance such that a constant current of one ampere produces a potential difference of one volt across a conductor
Optical isolation	two networks with no electrical continuity in their connection because an optoelectronic transmitter and receiver has been used

OSI	open systems interconnection
Packet	a group of bits (including data and call control signals) transmitted as a whole on a packet switching network – usually smaller than a transmission block
PAD	packet access device – an interface between a terminal or computer and a packet switching network
Parallel transmission	the transmission model where a number of bits are sent simultaneously over separate parallel lines – usually, unidirectional such as the Centronics interface for a printer
Parity bit	a bit that is set to a '0' or '1' to ensure that the total number of 1 bits in the data field is even or odd
Parity check	the addition of non-information bits that make up a transmission block to ensure that the total number of bits is always even (even parity) or odd (odd parity) – used to detect transmission errors but rapidly losing popularity because of its weakness in detecting errors
Passive filter	a circuit using only passive electronic components such as resistors, capacitors and inductors
PBX	private branch exchange
PCIP	personal computer instrument products
PDU	protocol data unit
Peripherals	the input/output and data storage devices attached to a computer e.g. disk drives, printers, keyboards, display, communication boards, etc
Phase modulation	the sine wave or carrier changes phase in accordance with the information to be transmitted
Phase shift keying	a modulation technique (also referred to as PSK) used to convert binary data into an analog form comprising a single sinusoidal frequency signal whose phase varies according to the data being transmitted
Physical layer	layer 1 of the ISO/OSI reference model, concerned with the electrical and mechanical specifications of the network termination equipment
PID	proportional integral derivative – a form of closed loop control
PLC	programmable logic controller
Point-to-point	a connection between only two items of equipment
Polling	a means of controlling devices on a multi-point line – a controller queries devices for a response
Polyethylene	a family of insulators derived from the polymerization of ethylene gas and characterized by outstanding electrical properties, including high IR, low dielectric constant, and low dielectric loss across the frequency spectrum
Polyvinyl chloride (PVC)	a general-purpose family of insulations whose basic constituent is polyvinyl chloride or its copolymer with vinyl acetate – plasticizers, stabilizers, pigments and fillers are added to improve mechanical and/or electrical properties of this material
Port	a place of access to a device or network, used for input/output of digital and analog signals

Presentation layer	layer 6 of the ISO/OSI reference model, concerned with negotiating a suitable transfer syntax for use during an application – if this is different from the local syntax, the translation to/from this syntax
Profibus	process field bus developed by a consortium of mainly German companies with the aim of standardization
Protocol entity	the code that controls the operation of a protocol layer
Protocol	a formal set of conventions governing the formatting, control procedures and relative timing of message exchange between two communicating systems
PSDN	public switched data network – any switching data communications system, such as telex and public telephone networks, which provides circuit switching to many customers
PSK	*see* phase shift keying
PSTN	public switched telephone network – this is the term used to describe the (analog) public telephone network
PTT	Post, Telephone and Telecommunications authority or: push to talk signal
PV	primary variable
QAM	quadrature amplitude modulation
QPSK	quadrature phase shift keying
R/W	read/write
RAM	random access memory – semiconductor read/write volatile memory data is lost if the power is turned off – reactance opposition offered to the flow of alternating current by inductance or capacitance of a component or circuit
REJ	reject
Repeater	an amplifier that regenerates the signal and thus expands the network
Resistance	the ratio of voltage to electrical current for a given circuit measured in ohms
Response time	the elapsed time between the generation of the last character of a message at a terminal and the receipt of the first character of the reply – it includes terminal delay and network delay
RF	radio frequency
RFI	radio frequency interference
Ring	network topology commonly used for interconnection of communities of digital devices distributed over a localized area, for example, a factory or office block – each device is connected to its nearest neighbors until all the devices are connected in a closed loop or ring – data is transmitted in one direction only – as each message circulates around the ring, it is read by each device connected in the ring
RMS	root mean square
RNR	receiver not ready
ROM	read only memory – computer memory in which data can be routinely read but written to only once using special means when the ROM is manufactured – a ROM is used for storing data or programs on a permanent basis

Router	a linking device between network segments which may differ in layers 1, 2a and 2b of the ISO/OSI reference model
RR	receiver ready
RS	recommended standard (e.g. RS-232C) – newer designations use the prefix EIA (e.g. EIA-RS-232C or just EIA-232C)
RS-232-C	interface between DTE and DCE, employing serial binary data exchange – typical maximum specifications are 15 m (50 feet) at 19 200 baud
RS-422	interface between DTE and DCE employing the electrical characteristics of balanced voltage interface circuits
RS-423	interface between DTE and DCE, employing the electrical characteristics of unbalanced voltage digital interface circuits
RS-449	general purpose 37-pin and 9-pin interface for DCE and DTE employing serial binary interchange
RS-485	the recommended standard of the EIA that specifies the electrical characteristics of drivers and receivers for use in balanced digital multi-point systems
RSSI	receiver signal strength indicator
RTS	request to send
RTU	remote terminal unit – terminal unit situated remotely from the main control system
RxRDY	receiver ready
S/N	signal to noise (ratio)
SAA	Standards Association of Australia
SAP	service access point
SDLC	synchronous data link control – IBM standard protocol superseding the bisynchronous standard
SDM	space division multiplexing
SDS	smart distributed system
Serial transmission	the most common transmission mode in which information bits are sent sequentially on a single data channel
Session layer	layer 5 of the ISO/OSI reference model, concerned with the establishment of a logical connection between two application entities and with controlling the dialog (message exchange) between them
SFD	the start of frame delimiter
Short haul modem	a signal converter that conditions a digital signal for transmission over DC continuous private line metallic circuits, without interfering with adjacent pairs of wires in the same telephone cables
Signal to noise ratio	the ratio of signal strength to the level of noise
Simplex transmissions	data transmission in one direction only
Slew rate	this is defined as the rate at which the voltage changes from one value to another
SNA	subnetwork access, or systems network architecture
SNDC	subnetwork dependent convergence
SNIC	subnetwork independent convergence
SNR	signal to noise ratio
Software	refers to the programs that are written by a user to control the actions of a microprocessor or a computer – these may be

	written in one of many different programming languages and may be changed by the user from time to time
SOH	start of header (ASCII Control-A)
Space	absence of signal – this is equivalent to a binary 0
Spark test	a test designed to locate imperfections (usually pinholes) in the insulation of a wire or cable by application of a voltage for a very short period of time while the wire is being drawn through the electrode field
SRC	source node of a message
SREJ	selective reject
Star	a type of network topology in which there is a central node that performs all switching (and hence routing) functions
Statistical multiplexer	a device used to enable a number of lower bit rate devices, normally situated in the same location, to share a single, higher bit rate transmission line – the devices usually have human operators and hence data is transmitted on the shared line on a statistical basis rather than, as is the case with a basic multiplexer, on a pre-allocated basis – it endeavors to exploit the fact that each device operates at a much lower mean rate than its maximum rate
STP	shielded twisted pair
Straight through pinning	RS-232 and RS-422 configuration that match DTE to DCE, pin for pin (pin 1 with pin 1, pin 2 with pin 2, etc)
STX	start of text (ASCII Control–B).
Switched line	a communication link for which the physical path may vary with each usage, such as the public telephone network
SYN	synchronous idle
Synchronization	the coordination of the activities of several circuit elements
Synchronous transmission	transmission in which data bits are sent at a fixed rate, with the transmitter and receiver synchronized – synchronized transmission eliminates the need for start and stop bits
Talker	a device on the GPIB bus that simply sends information on to the bus without actually controlling the bus
TCP	transmission control protocol
TCU	trunk coupling unit
TDM	time division multiplexer – a device that accepts multiple channels on a single transmission line by connecting terminals, one at a time, at regular intervals, interleaving bits (bit TDM) or characters (character TDM) from each terminal
Telegram	in general a data block which is transmitted on the network – usually comprises address, information and check characters
Temperature rating	the maximum, and minimum temperature at which an insulating material may be used in continuous operation without loss of its basic properties
TIA	Telecommunications Industry Association
Time sharing	a method of computer operation that allows several interactive terminals to use one computer
TNS	transaction bytes
Token ring	collision free, deterministic bus access method as per IEEE 802.2 ring topology

TOP	technical office protocol – a user association in USA, which is primarily concerned with open communications in offices
Topology	physical configuration of network nodes, for example, bus, ring, star, tree
Transceiver	transmitter/receiver – network access point for IEEE 803.2 networks
Transient	an abrupt change in voltage of short duration
Transport layer	layer 4 of the ISO/OSI reference model, concerned with providing a network independent reliable message interchange service to the application oriented layers (layers 5 through 7)
Trunk	a single circuit between two points, both of which are switching centers or individual distribution points – a trunk usually handles many channels simultaneously
TTL	transistor–transistor logic
Twisted pair	a data transmission medium, consisting of two insulated copper wires twisted together – this improves its immunity to interference from nearby electrical sources that may corrupt the transmitted signal
UART	universal asynchronous receiver/transmitter – an electronic circuit that translates the data format between a parallel representation, within a computer, and the serial method of transmitting data over a communications line
UHF	ultra high frequency
Unbalanced circuit	a transmission line in which voltages on the two conductors are unequal with respect to ground e.g. a coaxial cable
Unloaded line	a line with no loaded coils that reduce line loss at audio frequencies
UP	unnumbered poll
USB	universal serial bus
USRT	universal synchronous receiver/transmitter – see UART
UTP	unshielded twisted pair
V.35	ITU standard governing the transmission at 48 kbps over 60 to 108 kHz group band circuits
Velocity of propagation	the speed of an electrical signal down a length of cable compared to speed in free space expressed as a percentage
VFD	virtual field device – a software image of a field device describing the objects supplied by it for example, measured data, events, status etc which can be accessed by another network
VHF	very high frequency
VLAN	virtual LAN
Volatile memory	an electronic storage medium that loses all data when power is removed
Voltage rating	the highest voltage that may be continuously applied to a wire in conformance with standards of specifications
VRC	vertical redundancy check
VSD	variable speed drive
VT	virtual terminal

WAN	wide area network
Word	the standard number of bits that a processor or memory manipulates at onc time – typically, a word has 16 bits
X.21	ITU standard governing interface between DTE and DCE devices for synchronous operation on public data networks
X.25	ITU standard governing interface between DTE and DCE device for terminals operating in the packet mode on public data networks
X.25 Pad	a device that permits communication between non X.25 devices and the devices in an X.25 network
X.3/X.28/X.29	a set of internationally agreed standard protocols defined to allow a character oriented device, such as a visual display terminal, to be connected to a packet switched data network
X-ON/X-OFF	transmitter on/transmitter off – control characters used for flow control, instructing a terminal to start transmission (X-ON or Control-S) and end transmission (X-OFF or Control-Q)
XOR	exclusive-OR

Appendix B

Basic terminology

B.1   Concepts

Some of the typical terms that are used in communications and are explained in the following section.

B.1.1   Circuit

In telecommunications, a circuit is a physical electronic path that carries electronic information, be that voice or data, either in digital or analog format, between two points.

B.1.2   Channel

A channel refers to a 'logical' transmission path. For example, a particular radio station is allocated an FM channel centered on a specific frequency of 94.5 MHz. Using multiplexing techniques such as frequency division multiplexing (FDM), several transmission paths i.e. channels (based on different frequencies) can be created across a single medium and allocated to different users.

B.1.3   Line

A line is a telephone connection between a user and an exchange point setup by a telecommunications carrier.

B.1.4   Trunk

In telecommunications, the cable group that forms the primary path between two switching stations is known as a trunk. As such, it handles large volumes of traffic.

B.1.5   Bandwidth

The quantity of information a channel can convey over a given period is determined by its ability to handle the rate of change of the signal, i.e. its frequency. The frequency of an analog signal varies between a minimum and maximum value and the difference between those two frequencies is the bandwidth of that signal.

The bandwidth of an analog channel is the difference between the highest and lowest frequencies that can be reliably transmitted over the channel. Bandwidth is normally specified in terms of those frequencies at which the received signal has fallen to half the power (or 0.707 of the voltage) relative to the mid-band frequencies, referred to as –3 dB points. In this case the bandwidth is known as the –3 dB bandwidth.

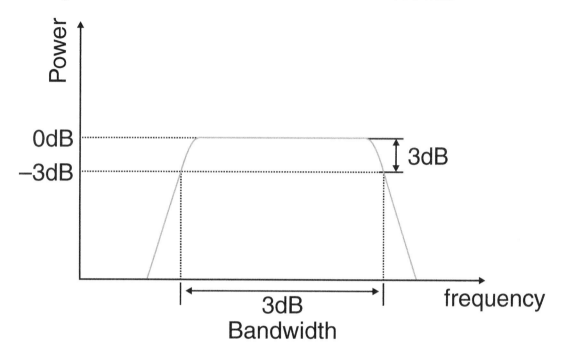

Figure B.1
Channel bandwidth

Digital signals are made up of a large number of frequency components, but only those within the bandwidth of the channel will be able to be received. It follows that the larger the bandwidth of the channel, the higher the data transfer rate can be and more high frequency components of the digital signal can be transported, and so a more accurate reproduction of the transmitted signal can be received.

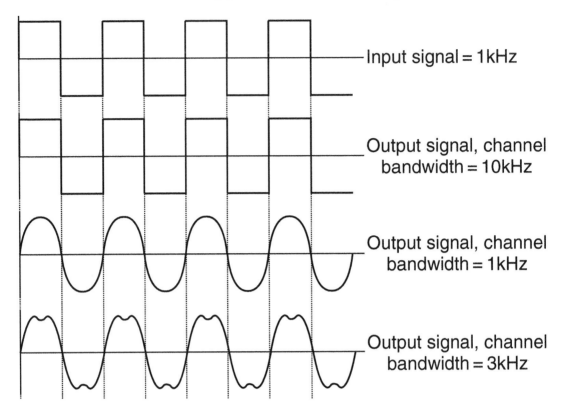

Figure B.2
Effect of channel bandwidth on a digital signal

B.1.6   Channel capacity

The channel capacity i.e. the maximum data transfer rate of the transmission channel can be determined from its bandwidth, by use of the following formula derived by Shannon.
$C = 2B\log_2 M$ bps

where C is the channel capacity, B is the bandwidth of the channel in hertz and M discrete levels are used for each signaling element.

In the special case where only two levels, 'ON' and 'OFF' or 'HIGH' and 'LOW' are used (binary), $M = 2$. Thus $C = 2\ B\log_2 2$ but $\log_2 2 = 1$, therefore $C = 2\ B$. As an example, the maximum data transfer rate for a PSTN channel of bandwidth 3200 hertz carrying a binary signal would theoretically be $2 \times 3200 = 6400$ Bps. In practice this figure is largely reduced by other factors such as the presence of noise on the channel to approximately 4800 Bps.

B.2   Simplex, half-duplex and full-duplex

B.2.1   Simplex

A simplex channel is unidirectional and allows data to flow in one direction only, as shown in Figure B.3. Public radio broadcasting is an example of a simplex transmission. The radio station transmits the broadcast program but does not receive any signals back from the radio receiver.

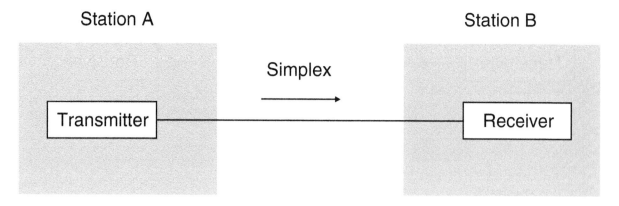

Figure B.3
Simplex transmission

This has limited use for data transfer purposes, as invariably the flow of data is required in both directions in order to control the transfer process, acknowledge data, etc.

B.2.2 Half-duplex

Half-duplex transmission allows simplex communication in both directions over a single channel, as shown in Figure B.4. Here the transmitter at station 'A' sends data to a receiver at station 'B'. A line turnaround procedure takes place whenever transmission is required in the opposite direction. The station 'B' transmitter is then enabled and communicates with the receiver at station 'A'. The delay in the line turnaround procedures reduces the available data throughput of the communications channel. This mode of operation is typical for citizen's band (CB) or marine/aviation VHF radios and necessitates the familiar 'over!' command by both users.

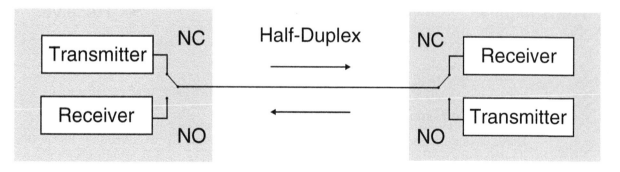

Figure B.4
Half-duplex transmission

B.2.3 Full-duplex

A full-duplex channel gives simultaneous communications in both directions, as shown in Figure B.5.

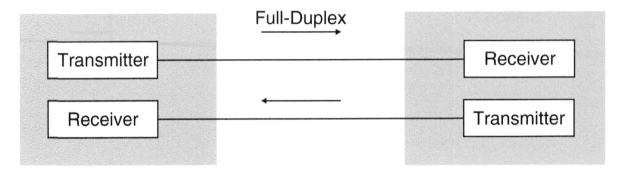

Figure B.5
Full-duplex transmission

B.2.4   Single frequency allocation

If a user is allocated a single frequency, then the system is said to be a simplex system. There are two modes of simplex radio transmissions:

Single direction simplex

Here, information in the form of single frequency radio waves will travel in one direction only, from a transmitter to a receiver.

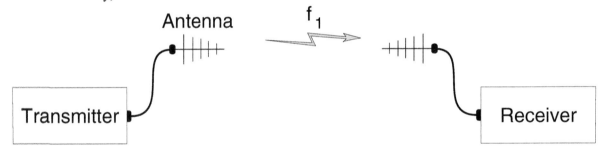

Figure B.6
Single direction simplex

Two direction simplex

Here, the single frequency is used to transmit information in two directions but in one direction at a time only.

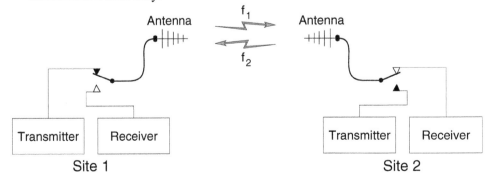

Figure B.7
Two direction simplex

Two frequency allocation

Here two frequencies are allocated, approximately 5 MHz apart, depending on the operating band. This type of system is referred to as a duplex system. There are two modes of duplex radio transmission:

Half-duplex

Here, one frequency is used for transmission in one direction and the other frequency is for transmission in the other direction but transmission is in one direction at a time only.

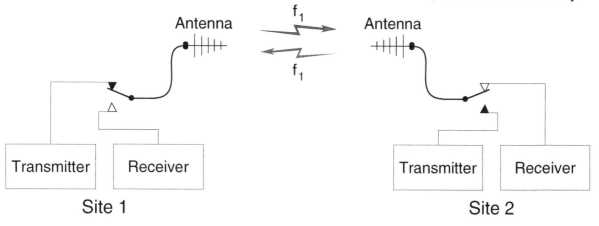

Figure B.8
Half-duplex

This mode is most commonly used for mobile vehicle radio systems where a radio station is used to repeat transmission from one mobile to all other mobiles on the frequencies (i.e. talk through repeaters).

Full-duplex

Here, one frequency is used for transmission in one direction and the other frequency is for transmission in the other direction, with both transmissions occurring simultaneously.

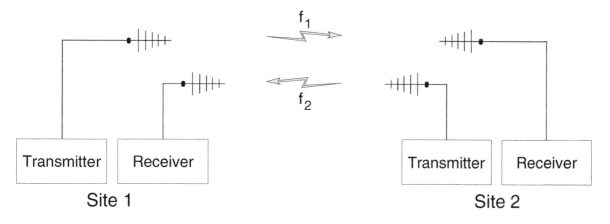

Figure B.9
Full-duplex

B.3 Modulation techniques

Modulation can involve either analog or digital signals, or both. It refers to the process of superimposing the information (modulating) signal on to a carrier signal. In its unmodulated state, the carrier is a constant amplitude, constant frequency signal.

B.3.1 Analog modulation

In this case the analog carrier signal is modulated with an analog information signal. There are basically three methods:

- **Amplitude modulation (AM)**

  The amplitude of the carrier is varied by the amplitude of the information signal.

- **Frequency modulation (FM)**

  The frequency of the carrier is varied by the amplitude of the information signal.

- **Phase modulation (PM)**

  The phase (time displacement) of the carrier signal is varied by the amplitude of the information signal.

B.3.2 RF modulation

RF modulation is similar to analog modulation, with the exception that the input (modulating) signal is digital.

- Amplitude shift keying (ASK): The amplitude of the carrier signal is varied between two values. This is also known as on–off keying (OOK).
- Frequency shift keying (FSK): The frequency of the carrier signal is varied between two values by the modulating signal.

Phase shift keying (PSK) a.k.a. binary phase shift keying (BPSK): The phase of the carrier signal is changed by the modulating signal. Depending on the number of discrete displacements, several bits of data can be transmitted simultaneously. For example, with four shift amounts (0°, 90°, 180° and 270°), two bits e.g. 00, 01,10 and 11 can be sent at a time.

- Quadrature amplitude modulation (QAM): Both the phase AND the amplitude of the carrier are changed, and makes it possible to encode as many as 4 bits at a time.
- Trellis coded modulation (TCM): This is similar to QAM, but includes extra bits for error correction.

The shift keying modulation methods come in plain and differential forms. The differential versions encode values as changes in a parameter, not in a specific value for a parameter. The differential techniques are easier to implement, and more robust than the non-differential ones.

- Differential amplitude shift keying (DASK): similar to ASK, but encoding different digital values as changes in signal amplitude.
- Differential frequency shift keying (DFSK): similar to FSK, but encoding different values as changes in signal frequency.
- Differential phase shift keying: similar to PSK, but encoding different digital values as changes in signal phase.

B.3.3   Digital modulation

Digital modulation converts an analog signal into a serious of binary digits for subsequent transmission. All these methods sample the analog input signal at a pre-defined rate and then generate a binary output based on that sample.

- Delta modulation (DM): The analog signal is represented by a series of bits (1s and 0s) that represent the current amplitude of the input signal relative to the previous amplitude. If the signal is increasing, then 1s are sent. If the signal is decreasing, 0s are sent. If the signal remains constant, alternating 1s and 0s are sent. Variations on the theme include adaptive delta modulation (ADM).
- Pulse code modulation (PCM): Typically used to convey voice signals across a digital channel. The analog signal sample is converted into an n-bit digital number that is subsequently transmitted. A 7-bit code means that the value transmitted will represent the analog voltage sample taken to within $1/(2^7)$ = 1/128th of its original value. This is sufficient for voice applications, but for multimedia applications upto 24 bits may be needed. The transmission rate is typically 64 kbps. Variations on the theme include adaptive differential pulse code modulation (ADPCM).
- Pulse width modulation (PWM) a.k.a. Pulse duration modulation (PDM): An analog value is represented by changing the width (duration) of a discrete pulse.
- Pulse amplitude modulation (PAM): An analog value is represented by the amplitude of the carrier for that interval.
- Pulse position modulation (PPM): An analog signal is represented by varying the position (i.e. the displacement) of a discrete pulse within a bit interval.

B.4   Baseband vs broadband

B.4.1   Baseband

Baseband refers to a communication method where all the traffic shares a single channel, hence when a given user transmits on to the channel, it occupies the entire bandwidth and nobody else can use the channel. If more than one user wishes to use the channel, they need to do it sequentially; i.e. they have to use a time-related multiplexing scheme such as TDM.

The signal is placed on the medium without using a high-frequency carrier signal. In the case of analog signals, such as on an ordinary telephone, the voice signal is simply transmitted 'as is' i.e. no modulation or encoding technique is used. In the case of digital data, such as on an Ethernet network, the consecutive 1s and 0s are simply encoded in the appropriate format (e.g. using Manchester encoding) and sent as a series of high and low voltages. Encoding is not to be confused with modulation. Its only purpose is to enable the receiver to extract the data from the received data stream.

A benefit of baseband is the simplicity of the system. No complex modulation and demodulation equipment is required. On the downside, the signal at the receiving end has to be large enough to be detected by the line receiver. For a digital system operating at, say, 10 Mbps, this may limit the distance to a few hundred meters. Another drawback is that only one signal can be transmitted at a given time.

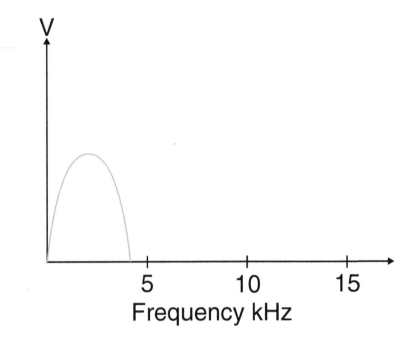

Figure B.10
Baseband signal

B.4.2   Broadband

Broadband refers to a system where multiple signals are transmitted simultaneously over a common physical channel. This is accomplished by using frequency division multiplexing (FDM – for systems transmitting data using copper or radio) or wave division multiplexing (WDM – for systems transmitting data using fiber optics), and transmitting different channels at different frequencies or wavelengths.

The various input signals are used to modulate their allocated carrier signals, using for example, FM or AM, and these modulated carriers are combined and sent across the medium as one composite signal. Even if the input is a digital signal, the result is still a modulated analog signal. At the receiving end, the various signal components are extracted via bandpass filters, and the information is recovered via appropriate demodulators or discriminators.

The advantage of broadband transmission is that several channels of information can be sent concurrently over one physical medium. As analog carrier signals (sine waves) are employed, it is also possible to recover the original signal from a very small received signal, even down to the picovolt range with appropriate technology. Hence broadband can be used across very large distances, in fact across millions of kilometers in the case of discovery satellites.

On the downside, the transmission and reception circuitry is more complex and hence more expensive than in the case of baseband.

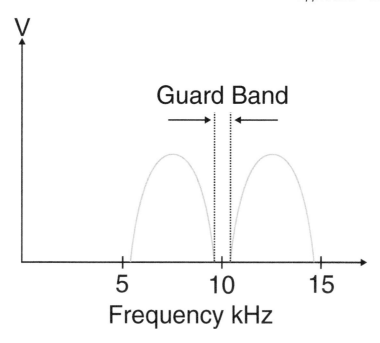

Figure B.11
Broadband signal

B.5 Narrowband vs wideband

B.5.1 Narrowband

Although narrowband and wideband sound very much like baseband and broadband, they have absolutely nothing in common. Narrowband and wideband refer to the relative bandwidth of the channel.

A narrowband system is simply a system with a relatively small bandwidth. The following systems can be classified as narrowband:

- Links designed to connect teletypes with a bandwidth of 300 Hz.
- A plain old telephone system (POTS) with an available bandwidth of about 3 kHz. With the aid of modems they can carry data at speeds up to 56 kbps.
- BRI ISDN (2B+D) with two channels for voice or data, at 64 kbps each.
- A T-1 channel with a bandwidth of between 64 kbps and 1.54 Mbps.

B.5.2 Wideband

The following systems can be classified as wideband.

- T-3 operating at 44.7 Mbps. It is equivalent to 28 T-1 circuits and can transmit 672 conversations simultaneously over fiber optics or microwave.
- ATM, carrying data, voice and video at speeds up to 13.22 Gbps.
- Likewise SONET, an optical multiplexing interface for high-speed data transmission, can operate up to 13.22 Gbps.
- Cable TV (CATV), broadcasting local and satellite TV channels using 500 MHz bandwidth.
- Digital high-definition TV (HDTV), used for broadcasting TV at 6 MHz per channel.

B.6 Analog vs digital transmission

Analog and digital signals can be transmitted using either analog or digital methods. It is therefore possible to transmit:

- Analog signals via analog methods
- Digital signals via analog methods
- Digital signals using digital methods, and
- Analog signals via digital methods

Analog signals change continuously in both frequency and amplitude. These signals are commonly used for audio and video communication as illustrated in Figure B.12.

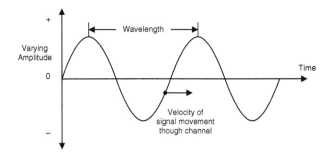

Figure B.12
Analog signal

Digital signals, on the other hand, are characterized by the use of discrete signal amplitudes. A binary digital signal, for example, has only two allowed values representing the binary digits 'ON' and 'OFF'. In fiber optic communications channels the presence or absence of light normally represents these states.

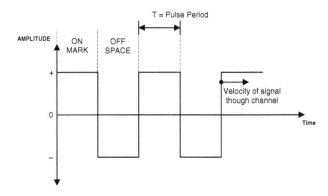

Figure B.13
Digital signal

B.6.1 Analog signals over analog channels

This is typically the case with voice over a POTS telephone line. For this purpose, a normal dial-up or leased analog line is used. No sophisticated conversion or modulation techniques are necessary.

B.6.2 Digital signals via analog methods

In this case, the method used to transmit the signal involves a modulated carrier signal. A sine wave is modulated with the digital data and, if necessary, several channels can be superimposed on the same physical circuit. Commercial solutions to this problem include modems for use on phone lines or one of the broadband alternatives.

B.6.3   Digital signals via digital methods

No special conversion techniques are needed. The transmission channel could be any of the dial-up or leased alternatives such as T1 or ATM.

B.6.4   Analog signals via digital methods

Here the channel expects digital information and for this reason the analog signal first has to be converted to a digital format. A typical solution is pulse code modulation (PCM) as covered elsewhere in this chapter. In the case of voice, the analog signal is typically sampled at 8000 times per second and each sample is then converted to an 8-bit binary number giving a 64 kbps data rate.

B.7   Dial-up vs leased access

B.7.1   Dial-up access

Dial-up access refers to connecting a device to a network via a public switched telephone network and a modem. The procedure is the same as for two people establishing contact via telephone, the difference being that two computer devices are communicating via modems.

The connections are not always good, causing modems to drop out or fail to establish connections, and the data transfer rates are limited to 56 kbps or less, depending on the maximum frequency that can be negotiated by the modems. Provided that multiple connections are supported at both the user and the service provider (e.g. through Multilink PPP), multiple parallel dial-up connections can be established for high volumes of data.

An alternative to a normal telephone connection is an integrated services data network (ISDN) connection. One ISDN BRI (basic rate interface) can handle 128 kbps of data, and if the necessary services are installed, multiple ISDN connections can be established to handle high data volumes for example, in the case of videoconferencing.

B.7.2   Leased access

A leased line (also referred to as a dedicated or a private line) is a permanent line installed between locations, usually to a user's premises, by a telephone company. Unlike a dial-up connection, it is always active. Leased lines are available in 2 and 4 wire versions, which has a bearing on the line quality and obviously also on the line rental costs. Leased lines provide faster throughput of data and better quality connections than dial-up, but at a higher cost.

Examples of leased line services are:

- T-1/E-1 lines. These provide 1.544 Mbps for T-1 (for example, in the United States and Japan or 2.048 Mbps for E-1 in Europe and Mexico).
- Fractional T-1 lines. These build up in units of 64 kbps up to 768 kbps.
- 56/64 kbps lines. In Europe, these lines provide 64 kbps, In the US and other countries 8 kbps is used for control overhead so that only 56 kbps is available to the user.

- Digital data services (DDS). These provide synchronous transmission of digital signals at 2.4, 4.8, 9.6, 19.2 or 56 kbps.

The line is rented on a monthly basis and the rate is affected by factors such as the speed of the circuit and the distance involved. The availability and pricing varies between different service providers.

Ordinary telephone lines were primarily intended for voice traffic and therefore their performance for data transmission is far from optimum. At a cost, specially conditioned analog lines can be leased in order to overcome this problem.

B.8   Multiplexing techniques

Multiplexing is a technique by which the information (voice or data) from more than one source is delivered across one common medium, and delivered to the appropriate recipient at the receiving end. This section describes various methods used to accomplish this.

B.8.1   SDM

Space division switching involves the physical connection of one path to another. This kind of switching was used in all the earliest exchanges, and is still an essential part of modern digital exchanges.

B.8.2   TDM

In time division multiplexing (TDM) the multiplexing takes place in the time domain. Data streams from several sources are combined by sending small slices of data from each input across the common channel sequentially, in time, in a specific sequence. If there are 4 channels to be multiplexed like this, each source will have access to the channel for ¼ (25%) of the time.

Typical examples of TDM are T-1 and T-3 carrier systems, where pulse code modulated (PCM) streams are created for each conversation or data stream, and then combined with TDM on to a common channel.

A variation on the theme is statistical multiplexing (STDM), where the input nodes that have nothing to send are skipped, thereby increasing the available time for the remaining nodes.

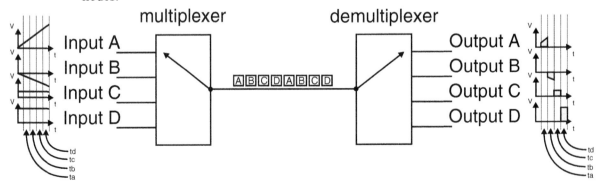

Figure B.14
TDM

B.8.3   FDM

In frequency division multiplexing (FDM), multiplexing takes place in the frequency domain. The various signals are separated not in time, but in frequency. FDM assigns a discrete (fixed) carrier frequency for each channel, and then modulates each carrier with the input for the respective channel. Assuming that there are 4 channels and the total available bandwidth is between 1 and 2 MHz, the maximum available bandwidth per channel is $(2-1)/4 = ¼$ MHz or 250 kHz. In practice this will be slightly less due to a 'guard band' separating the channels on the frequency domain.

Typical examples of FDM applications are commercial TV and radio broadcasts.

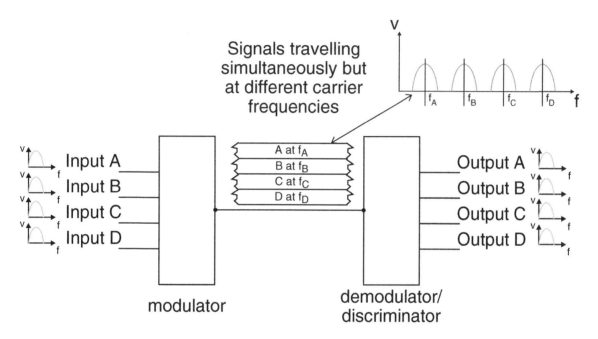

Figure B.15
FDM

B.8.4   WDM

Wavelength division multiplexing (WDM) is used on optical fiber and is the optical equivalent of FDM. Light waves with different light wavelengths (i.e. different 'colors') are used to carry separate streams of information, thereby creating several channels on the same optical fiber. The different modulated light beams are combined and transmitted as one composite light beam down the fiber, and at the receiving end it is optically split into its individual streams, which are subsequently demodulated.

WDM is compatible with many existing technologies such as SDH/SONET and ATM.

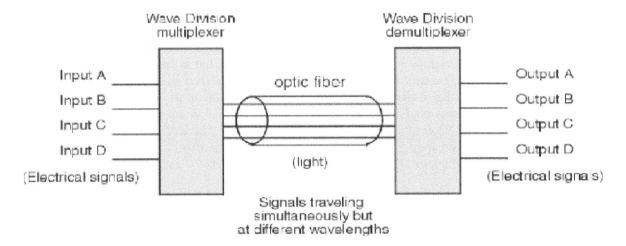

Figure B.16
WDM

B.8.5   DWDM

Dense waveform division multiplexing is related to WDM and offers even higher bandwidths.

It effectively creates several virtual fibers in one physical fiber, each virtual fiber carrying several channels. Thus, for example, eight optical channels of 2.5 Mbps each can be combined to give an effective data rate of 20 Mbps. The technology involved is developing rapidly and by the end of 1999 it was already possible to carry 96 separate data streams.

With a new technology called UDWDM (Ultra-Dense WDM) Bell Labs have prototyped a system that can run 1000 channels over a single fiber, each channel operating at up to 160 Gbps.

B.9   Connection orientated vs connectionless communication

Connectionless communication is a method of communication whereby data is simply sent without confirming whether the recipient is ready and willing to receive it. Because of this situation, the sender cannot get immediate acknowledgement of receipt and has to employ some other method for this purpose. An example of this technique is sending a telegram. In terms of networking protocols, the TCP/IP protocol suite's user datagram protocol (UDP) performs such a function.

Connection-oriented communication is where the sender first establishes contact with the recipient before sending the data. Because of the connection between the two parties, it is easy the obtain acknowledgement of receipt. This method is typical of a normal telephone conversation and in terms of networking protocols, this is performed by the TCP/IP protocol suite's transmission control protocol (TCP).

B.10   Types of transmission

The following transmission types are used for either analog or digital transmissions, or for both.

B.10.1   Point-to-point

Point-to-point transmission takes place between a single pair of stations, normally relatively close to each other. An example of this is a home intercom system with two stations.

B.10.2   Mediated

A message is sent from one station to another but because of the distance, there are intermediate stations. However, there is only one possible path and the signal is forwarded from station to station along the way. Each station along the way handles and could possibly modify the message before it is retransmitted.

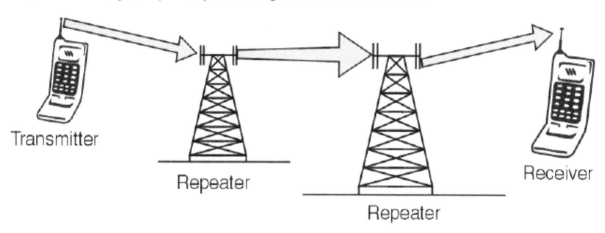

Figure B.17
Mediated transmission

B.10.3   Switched

This is the same as mediated transmission, with the difference that there are multiple paths between source and destination. The nodes in between can therefore switch (i.e. divert) incoming traffic and send it on to an appropriate node. The word 'switch' in this context has the same implication as a 'switch' (a set of 'points') on a railroad track.

There are essentially two basic modes of switching, namely circuit switching and packet switching.

Circuit switching

In a circuit switching system, a continuous connection is made across the network between the two different end points. This is a temporary connection which remains in place as long as both parties wish to communicate; that is until the connection is terminated. All the network resources are available for the exclusive use of these two parties whether they are sending data or not. When the connection is terminated the network resources are released for other users.

The advantage of circuit switching is that the users have an exclusive channel available for the transfer of their data at all times while the connection is made. The obvious disadvantage is the cost of maintaining the connection when there is little or no data being transferred. Such connections can be very inefficient for the bursts of data that are typical of many computer applications.

A telephone call is a good example of a circuit-switching system.

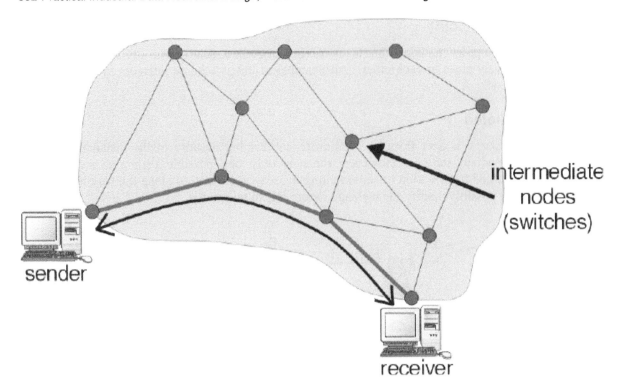

Figure B.18
Circuit switching

Packet switching

Packet switching systems improve the efficiency of the transfer of bursts of data, by sharing communications channels among many similar users. This is analogous to the postal system. The unit of information may vary. It could be an entire message or a block of data with a small part of a long message.

Packet switched messages are broken into a series of packets of certain maximum size, each containing the destination and source addresses and a packet sequence number. The packets are sent over a common communications channel, possibly interleaved with those of other users. All the receivers on the channel check the destination addresses of all packets and accept only those carrying their address. Messages sent in multiple packets are reassembled in the correct order by the destination node.

All packets do not necessarily follow the same path. As they travel through the network they may get separated and handled independently from each other, but eventually arrive at their correct destination. For this reason, packets often arrive at the destination node out of their transmitted sequence. Some packets may even be held up temporarily (stored) at a node, due to unavailable lines or technical problems that might arise on the network. When the time is right, the node then allows the packet to pass or be 'forwarded'.

This method of delivery is obviously more cost efficient than circuit switching since data from any user can travel over any portion of the system at any time, leading to more efficient use of the available infrastructure. On the downside, packets may be delayed, arrive at non-consistent intervals, get lost, or arrive in the wrong sequence. This necessitates the use of additional protocols (sets of rules, implemented in software) to take care of the problem.

The Internet is an example of a packet-switched network.

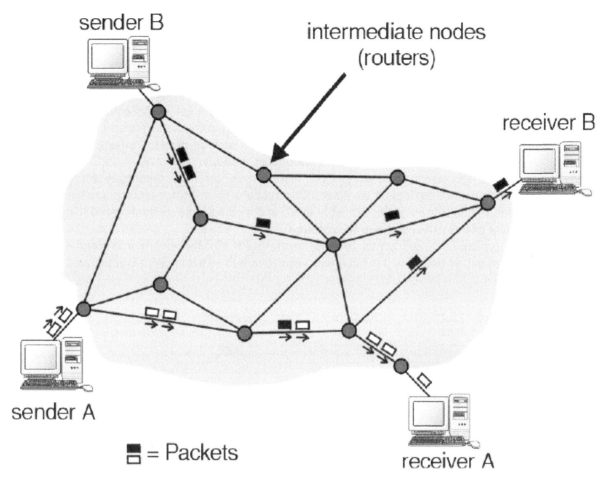

Figure B.19
Packet switching

Broadcast

The information is transmitted to all stations capable of receiving and not to a specific one. This is typical of a radio broadcast.

Multicast

The information is selectively transmitted to a specific sub-group of devices, within a larger group of devices all capable of receiving the same signal. An example is 'pay TV', either cable or satellite based, where all receivers pick up the same signal but only those with a decoder can display it.

Stored and forwarded

Data is sent to a holding location until requested or sent on automatically after a predefined amount of time.

B.11 Local vs wide area networks

B.11.1 Local area networks (LANs)

LANs are characterized by high-speed transmission over a relatively restricted geographical area such as a building or a group of buildings. LANs consist of a common medium (such as a coaxial cable), interconnecting computers, printers, programmable logic controllers, etc. Users can share resources such as printers, transfer data between each other, and communicate via e-mail or chat sessions.

There are many types of LANs, characterized by their topologies (the geometric arrangement of devices on the network), protocols (the rules for sending and receiving the data), media (for example, the cable used) and media access methods (used to control the access of individual nodes to the medium).

A typical LAN is Ethernet 10Base2, operating at 10 Mbps over a maximum distance of 100 m across unshielded twisted pair wire via a hub. This concept is illustrated in Figure B.20.

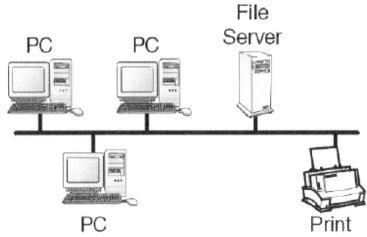

Figure B.20
Local area network

B.11.2 Wide area networks (WANs)

While LANs operate where distances are relatively small, wide area networks (WANs) consist of inter-linked LANs that are separated by large distances ranging from a few tens of kilometers to thousands of kilometers. WANs normally use the public telecommunications carriers to provide cost-effective connection between LANs. Since these links are supplied by independent telecommunications utilities, they are commonly referred to (and illustrated as) a 'communications cloud'. Special equipment called gateways (routers) have been developed for this type of activity, storing the message at LAN speed and transmitting it across the 'communications cloud' at the speed of the interconnecting carrier. When the entire message has been received at the remote LAN, the message is reinserted at the local LAN speed. The speed at which a WAN interconnects is often slower than the LAN speed, but not necessarily so since many of the WAN carrier technologies such as ATM are capable of speeds far in excess of typical LAN speeds. At the end of the day it boils down to cost.

The concept of a WAN is shown in Figure B.21.

Figure B.21
Wide area network

B.11.3   The 'communications cloud'

The various long-distance WAN transport services such as T1, E1, SONET, ATM, etc, are normally drawn as a 'communications cloud' with the end users on the inside and the carriers on the inside. This is done for a reason. The users' systems are normally obscured to the carriers, and the detail of the carriers are normally obscured to the users, hence the 'cloud.'

The carriers are often extended to the premises of the client. In the case of a T-carrier, it is terminated in a channel service unit or CSU on each side, installed on the user's premises. Between the CSUs, the signal is the utility's problem, beyond the CSUs on both sides, it is the user's problem.

B.12   The PSTN vs the Internet

The public switched telephone network (PSTN) refers to the international communications infrastructure for carrying voice and data. The telephone service carried by the PSTN is also referred to as plain old telephone service (POTS). POTS are normally restricted to about 56 kbps.

The high-speed telephone services using high-speed digital lines, such as ISDN, are not classified as POTS.

In contrast, the Internet is a global packet-switching network, originally designed for the transportation of digital data. This digital data is carried on digital PSTN circuits. Almost all Internet users have access to the PSTN as well, and in fact most private users use the PSTN to gain access to their Internet service providers. Because of this co-existence, the idea of using the Internet to carry traditional PSTN information such as voice and fax (in digital format) has emerged.

B.13   The open systems interconnection model

B.13.1   Overview

A communication framework that has had a tremendous impact on the design of communications systems is the open systems interconnection (OSI) model developed by the International Standardization Organization. The objective of the model is to provide a framework for the coordination of standards development and allows both existing and evolving standards activities to be set within that common framework.

The interconnection of two or more devices with digital communication is the first step towards establishing a network. In addition to the hardware requirements, the software problems of communication must also be overcome. Where all the devices on a network are from the same manufacturer, the hardware and software problems are usually easily solved because the system is usually designed within the same guidelines and specifications.

When devices from several manufacturers are used on the same application, the problems seem to multiply. Systems that are specific to one manufacturer and which work with specific hardware connections and protocols are called 'closed systems'. Usually, these systems were developed at a time before standardization or when it was considered unlikely that equipment from other manufacturers would be included in the network.

In contrast, open systems are those that conform to specifications and guidelines which are 'open' to all. This allows equipment from any manufacturer, who complies with that standard, to be used interchangeably on the network. The benefits of open systems include multiple vendors and hence wider availability of equipment, lower prices and easier integration with other components.

In 1978 the ISO, faced with the proliferation of closed systems, defined a 'reference model for communication between open systems' (ISO 7498), which has become known as the open systems interconnection model, or simply as the OSI model. OSI is essentially a data communications management structure, which breaks data communications down into a manageable hierarchy of seven layers. Each layer has a defined purpose and interfaces with the layers above it and below it. By laying down standards for each layer, some flexibility is allowed so that the system designers can develop protocols for each layer independent of each other. By conforming to the OSI standards, a system is able to communicate with any other compliant system, anywhere in the world.

It should be realized at the outset that the OSI reference model is not a protocol or set of rules for how a protocol should be written but rather an overall framework in which to define protocols. The OSI model framework specifically and clearly defines the functions or services that have to be provided at each of the seven layers (or levels).

Since there must be at least two sites to communicate, each layer also appears to converse 'horizontally' with its peer layer at the other end of the communication channel in a virtual (logical) communication. The OSI layering concept is shown in Figure B.22.

OSI Layering

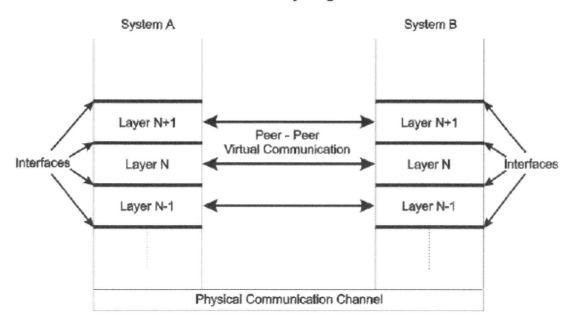

Figure B.22
OSI layering concept

The actual functions within each layer are provided by entities such as programs, functions, or protocols, and implement the services for a particular layer on a single machine. Several entities, for example a protocol entity and a management entity, may exist at a given layer. Entities in adjacent layers interact through the common upper and lower boundaries by passing physical information through service access points (SAPs). A SAP could be compared to a pre-defined 'postbox' where one layer would collect data from the previous layer. The relationship between layers, entities, functions and SAPs is shown in Figure B.22.

In the OSI model, the entity in the next higher layer is referred to as the N+1 entity and the entity in the next lower layer as N–1. The services available to the higher layers are the result of the services provided by all the lower layers.

The functions and capabilities expected at each layer are specified in the model. However, the model does not prescribe how this functionality should be implemented. The focus in the model is on the 'interconnection' and on the information that can be passed over this connection. The OSI model does not concern itself with the internal operations of the systems involved.

The diagram below shows the seven layers of the OSI model.

The OSI Reference Model

Figure B.23
The OSI reference model

Typically, each layer on the transmitting side adds header information, or protocol control information (PCI), to the data before passing it on to the next lower layer. In some cases, especially at the lowest level, a trailer may also be added. At each level, this combined data and header 'packet' is termed a protocol data unit or PDU. The headers are used to establish the peer-to-peer sessions across the sites and some layer implementations use the headers to invoke functions and services at the layers adjacent to the destination layer.

At the receiving site, the opposite occurs with the headers being stripped from the data as it is passed up through the layers. These header and control messages invoke services and a peer-to-peer logical interaction of entities across the sites. Generally, layers in the same site (i.e. within the same host) communicate in software with parameters passed through primitives, whilst peer layers at different sites communicate with the use of the protocol control information, or headers.

At this stage, it should be quite clear that there is NO connection or direct communication between the peer layers of the network. Rather, all physical communication is across the physical layer, or the lowest layer of the stack. Communication is down through the protocol stack on the transmitting stack and up through the stack on the receiving stack. Figure B.24 shows the full architecture of the OSI Model, whilst Figure B.25 shows the effects of the addition of PCI to the respective PDUs at each layer. As will be realized, the net effect of this extra information is to reduce the overall bandwidth of the communications channel, since some of the available bandwidth is used to pass control information.

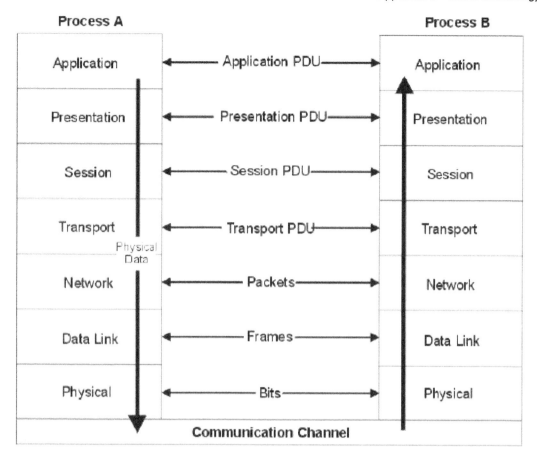

Figure B.24
Communication channel

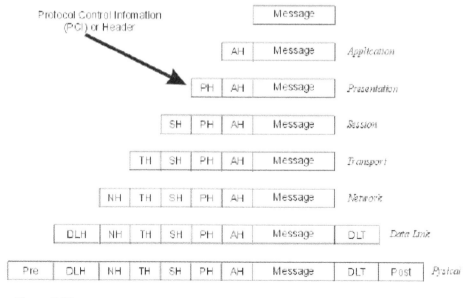

Figure B.25
OSI message passing

The services provided at each layer of the stack are as follows:

B.13.2   Application layer

The application layer is the topmost layer in the OSI reference model. This layer is responsible for giving applications access to the network. Examples of application-layer tasks include file transfer, electronic mail (e-mail) services, and network management. Application-layer services are much more varied than the services in lower layers, because the entire range of application possibilities is available here. Application programs can get access to the application-layer services in software through application service elements (ASEs). There is a variety of such application service elements; each designed for a class of tasks. To accomplish its tasks, the application layer passes program requests and data to the presentation layer, which is responsible for encoding the application layer's data in the appropriate form.

B.13.3   Presentation layer

The presentation layer is responsible for presenting information in a manner suitable for the applications of users dealing with the information. Functions such as data conversion from EBCDIC to ASCII (or vice versa), use of special graphics or character sets, data compression or expansion, and data encryption or decryption are carried out at this layer. The presentation layer provides services for the application layer above it, and uses the session layer below it. In practice, the presentation layer rarely appears in pure form, and it is the least well defined of the OSI layers. Application- or session-layer programs will often encompass some or all of the presentation layer functions.

B.13.4   Session layer

The session layer is responsible for synchronizing and sequencing the dialog and packets in a network connection. This layer is also responsible for making sure that the connection is maintained until the transmission is complete, and ensuring that appropriate security measures are taken during a 'session' (that is, a connection). The session layer is used by the presentation layer above it, and uses the transport layer below it.

B.13.5   Transport layer

In the OSI reference model, the transport layer is responsible for providing data transfer at an agreed-upon level of quality, such as at specified transmission speeds and error rates. To ensure delivery, outgoing packets are sometimes assigned numbers in sequence. These numbers are then included in the packets that are transmitted by lower layers. The transport layer at the receiving end subsequently checks the packet numbers to make sure all have been delivered and to put the packet contents into the proper sequence for the recipient. The transport layer provides services for the session layer above it, and uses the network layer below it to find a route between source and destination. The transport layer is crucial in many ways, because it sits between the upper layers (which are strongly application-dependent) and the lower ones (which are network-based).

The layers below the transport layer are collectively known as the 'subnet' layers. Depending on how well (or not) they perform their function, the transport layer has to interfere less (or more) in order to maintain a reliable connection.

Three types of subnet service (i.e. the service supplied by the underlying physical network between two hosts) are distinguished in the OSI model:

- Type A: very reliable, connection-oriented service.
- Type B: unreliable, connection-oriented service.

- Type C: unreliable, possibly connectionless service.

To provide the capabilities required for the above service types, several classes of transport layer protocols have been defined in the OSI model:

- TP0 (transfer protocol class 0), which is the simplest protocol. It assumes type A service; that is, a subnet that does most of the work for the transport layer. Because the subnet is reliable, TP0 requires neither error detection nor error correction. Because the connection is connection-oriented, packets do not need to be numbered before transmission.
- TP1 (transfer protocol class 1), which assumes a type B subnet; that is, one that may be unreliable. To deal with this, TP1 provides its own error detection, along with facilities for getting the sender to retransmit any erroneous packets.
- TP2 (transfer protocol class 2), which also assumes a type A subnet. However, TP2 can multiplex transmissions, so that multiple transport connections can be sustained over the single network connection.
- TP3 (transfer protocol class 3), which also assumes a type B subnet. TP3 can also multiplex transmissions, so that this protocol has the capabilities of TP1 and TP2.
- TP4 (transfer protocol class 4), which is the most powerful protocol, in that it makes minimal assumptions about the capabilities or reliability of the subnet. TP4 is the only one of the OSI transport-layer protocols that supports connectionless service.

B.13.6 Network layer

The network layer is the third lowest layer, or the uppermost subnet layer. It is responsible for the following tasks:

- Determining addresses or translating from hardware to network addresses. These addresses may be on a local network or they may refer to networks located elsewhere on an intranetwork. One of the functions of the network layer is, in fact, to provide capabilities needed to communicate on an Internet work.
- Finding a route between a source and a destination node or between two intermediate devices.
- Fragmentation of large packets of data into frames which are small enough to be transmitted by the underlying data link layer (fragmentation). The corresponding network layer at the receiving node undertakes re-assembly of the packet.

B.13.7 Data link layer

The data link layer is responsible for creating, transmitting, and receiving data packets. It provides services for the various protocols at the network layer, and uses the physical layer to transmit or receive material. The data link layer creates packets appropriate for the network architecture being used. Requests and data from the network layer are part of the data in these packets (or frames, as they are often called at this layer). These packets are passed down to the physical layer and from there, the data is transmitted to the physical layer on the destination machine. Network architectures (such as ethernet, ARCnet, token ring, and FDDI) encompass the data link and physical layers, which is why these architectures support services at the data-link level. These architectures also represent the most common protocols used at the data-link level.

The IEEE (802.x) networking working groups have refined the data link layer into two sub layers: the logical-link control (LLC) sub layer at the top and the media-access control (MAC) sub layer at the bottom. The LLC sub layer must provide an interface for the network layer protocols, and control the logical communication with its peer at the receiving side. The MAC sub layer must provide access to a particular physical encoding and transport scheme.

B.13.8   Physical layer

The physical layer is the lowest layer in the OSI reference model. This layer gets data packets from the data link layer above it, and converts the contents of these packets into a series of electrical signals that represent 0 and 1 values in a digital transmission. These signals are sent across a transmission medium to the physical layer at the receiving end. At the destination, the physical layer converts the electrical signals into a series of bit values. These values are grouped into packets and passed up to the data link layer.

The mechanical and electrical properties of the transmission medium are defined at this level. These include the following:

- The type of cable and connectors used. Cable may be coaxial, twisted pair, or fiber optic. The types of connectors depend on the type of cable.
- The pin assignments for the cable and connectors. Pin assignments depend on the type of cable and also on the network architecture being used.
- Format for the electrical signals. The encoding scheme used to signal 0 and 1 values in a digital transmission or particular values in an analog transmission depend on the network architecture being used. Most networks use digital signaling, and most use some form of Manchester encoding for the signal.

Note that this layer does NOT include the specifications of the actual medium used, but rather WHICH medium should be used and HOW. The specifications of, for example, unshielded twisted pair as used by Ethernet is contained in specification EIA/TIA 568.

B.14   Summary

As this is important in the following sections, here follows, in conclusion, a brief summary of the seven layers:

- Application – the provision of network services to the user's application programs. Note: the actual application programs do NOT reside here.
- Presentation – primarily takes care of data representation (including encryption).
- Session – control of the communications (sessions) between the users.
- Transport – the management of the communications between the two end systems.
- Network – primarily responsible for the routing of messages.
- Data link – responsible for assembling and sending a frame of data from one system to another.
- Physical – defines the electrical signals and mechanical connections at the physical level.

Appendix C

Practicals

Practical and exercise sessions

There are over 20 practical sessions listed below. Each session should not take longer than 20 minutes (10 minutes if you know the material).

The practical sheets below are by necessity briefly written; so some initiative is required in connecting and reconfiguring equipment. The common theme running through the practical sessions and exercises is troubleshooting and remedying communication problems. They will show you how to use some common troubleshooting equipment and software.

Notes

Please be careful with the PCMCIA cards. They are not very robust and tend to fail very easily. There are a number of different ones used; and the connectors are not interchangeable.

Please bear in mind that memory conflicts/interrupt settings/faulty cables are a real part of setting up any communications system.

List of practicals and exercises

The troubleshooting sessions are listed below:

1. RS-232.
2. Use of a breakout box.
3. Monitoring communication with a breakout box.
4. RS-485.
5. Using RS-232 to 485 converters.
6. RS-485 testing.
7. Fiber optics.
8. Continuity test.
9. Insertion loss testing.
10. Modbus.
11. Logging Modbus data.

12. HART.
13. Noise problem.
14. ASI bus.
15. Incorrect configuration of network.
16. DeviceNet, installation and configuration.
17. DeviceNet voltage drop problems.
18. Profibus.
19. Wiring problems.
20. Foundation fieldbus.
21. Foundation fieldbus voltage drop problems.
22. Industrial Ethernet.
23. Wiring problems/hub problems.
24. TCP/IP.
25. TCP/IP network setup and problems.
26. Using ICMP echo messages (PING).
27. Address resolution protocol.
28. Duplicate IP addresses.
29. Incorrect subnet mask.
30. Radio and wireless communications.
31. Path loss calculation.

Practical 1: Use of a breakout box

Objective

The objective of this practical is to give you an understanding of how to use a breakout box for troubleshooting a 232E communication system.

Overview

You will learn how to read and use a breakout box in a real 232 system.

Hardware/software required

- One computer with PAT software.
- One breakout box.
- One 9-pin to 25-pin adapter cable.
- One voltmeter.
- One jumper wire for breakout box.

Equipment setup

Connect the breakout box to the adapter cable.
Connect the adapter cable to COM 1 on the computer.
Make sure all switches on the breakout box are in the ON position.
When the system is connected turn on the computer.
Run the PAT software on the computer.
Put the computer into interactive mode under the display menu.

Interpretation

1. Verify that there are lights illuminated on the breakout box.

2. Which LEDs are on? PIN _____ PIN _____ PIN _____
3. Which color indicates which voltage?_____
4. Change the baud rate in the PAT software to a baud rate of 2.
5. Press one key on the keyboard.
6. View the breakout box. Which LED lights? _____
7. Measure the voltage as the LED flashes.
8. What is the idle voltage? _____
9. What is the voltage when the light is red? _____
10. Short pins 2 and 3. What do you see on the screen when you type a letter?_____
11. What do you see on the breakout box?

Practical 2: Monitoring communication with a breakout box

Objective

The objective of this practical is to show you how to monitor data communications using a breakout box on an EIA-232E communication system.

Overview

You will learn how to monitor data communications using a breakout box on a real 232 system. You will also learn how to alter the wiring at the breakout box and the effect of an incorrectly wired connection.

Hardware/software required

- Two computers with PAT software.
- One breakout box.
- One 9-pin to 25-pin adapter cable.
- One 9-pin to 25-pin null modem cable.
- One voltmeter.
- Two jumper wires for breakout box.

Implementation

1. Connect the breakout box to the adapter cable.
2. Connect the adapter cable to COM 1 on one computer.
3. Connect the null modem cable between the other side of the breakout box and COM 1 on the other computer.
4. Make sure all switches on the breakout box are in the ON position.
5. When the system is connected turn on both computers.
6. Run the PAT software on both computers.
7. Configure COM 1 on both computers to 4800 baud, even parity, 7 data bits, 1 stop bit using the configuration/ports menu.
8. Put both computers into interactive mode under the display menu.
9. Verify that there are lights illuminated on the breakout box.
10. Type a character on one computer and verify that it is correctly received on the other computer. Observe the lights on the breakout box.
11. Which lights flash?
12. Repeat by typing on the other machine.

13. Open the switch on pin 2 and note the effect on transmission.
14. Repeat, closing pin 2 and opening pin 3.
15. Open both switches on pins 2 and 3 and then use two jumper wires to cross connect pin 2 to 3 and similarly 3 to 2.
16. Observe the lights on the breakout box and the effect on transmission.

Practical 3: Using RS-232 to 485 converters

Objective

The objective of this practical is to demonstrate how an RS-422 and RS-485 communications system operates.

Overview

The practical will show the delegate how to install and run a two wire half-duplex and four wire full-duplex system. The delegate should be familiar with the PAT and Modbus software used in this practical. The software will be used to show how a packet of information is sent over both systems.

Installation

Locate the following parts and tools:

- Six laptop computers.
- Six DB9 to DB25 adapters.
- Six EIA-232 breakout boxes.
- Six EIA-232 to 485 adapters.
- Various figure 8 cables.
- Two power supplies.
- One small flat blade screwdriver.
- Wire strippers.

RS-422 PAT communications test (point-to-point full-duplex)

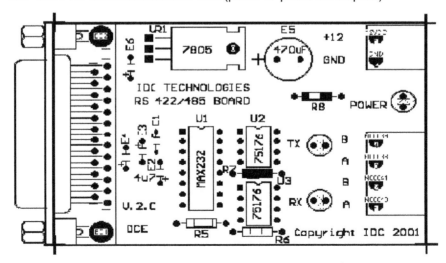

Figure C.1
RS-422 PAT communications test (point-to-point full-duplex)

Note: In the following practical, all B connections are connected to 'Bs' on the other end. And all A connections are connected to 'As' on the other end.

32. Plug in the 9 to 25 adapters into the com1 ports on the computers.
33. Plug the breakout box into the adapters.
34. Plug in the 232 to 422/485 adapters into the breakout box.
35. Plug in the power cable into the ground and +12 V DC connectors on the adapters (wire strippers + voltage).
36. Connect a wire between the TX B on one adapter to the RX B on the other adapter.
37. Connect a wire between the TX A on one adapter to the RX A on the other adapter.
38. Connect a wire between the RX B on one adapter to the TX B on the other adapter.
39. Connect a wire between the RX A on one adapter to the TX A on the other adapter.
40. Plug in the power pack to both adapters.
41. Run PAT on both computers and then assert RTS under config/ports.
42. Get into the interactive mode.
43. Send characters to and from the each computer.
44. Are the characters shown on both computers?

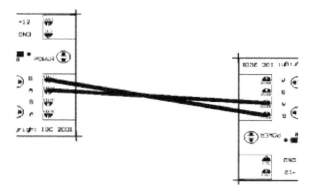

Figure C.2
RS-485 wiring connections

RS-485 PAT test (point-to-point half-duplex)

1. On one computer assert the RTS line using the PAT software.
2. Disconnect the power from the interface boards.
3. Remove the bottom 75176 chip from both interface boards.
4. Remove both sets of wires between the interfaces.
5. Connect a pair of wires between TX B and A on one interface and the
6. TX B and A on the other interface.
7. Apply power again and send data from the computer that has the RTS asserted.

Can you send or receive data from the other computer?
1. Un-assert RTS on the one computer and assert it on the other.
2. Try to send and receive data from both computers.
3. Can you now send data from the other computer?

RS-422 Modbus test (point-to-point)

1. Replace the 75176 chip on each interface.
2. Exit completely from the PAT software.
3. Run Modbus up both computers and set one for master and the other for slave (F10).
4. Press F11 on each computer and setup for 9600 8N1 slave address 1.
5. On the master press F3 and type 000A 0002.
6. On the slave press the page down keys and set the A and B under coils as 1s.
7. On the slave press the page up key and wait for data to come in.
8. On the master press F9 and follow the instructions on the screen.
9. Is the data transferred correctly?

The following should show up on the other computer:

```
01 01 000A 0002 C189.
```

Figure C.3
RS-422 point-to-point wiring connections

RS-422 Modbus test (point-to-multipoint)

1. The instructor will designate one computer as the master. All other computers will be slaves.
2. Rewire the interfaces in the following manner:
3. TX B and A on all the slaves to the TX B and A on all the other slaves and to the RX B and A on the master.
4. RX B and A on all the slaves to the RX B and A on all the other slaves and to the TX B and A on the master.
5. On the master press F3 and type 000A 0002.
6. On the slaves press F11 and set each slave for a different address.
7. On all the slaves press the page down key and set the A and B under coils as 1s.
8. On the slave press the page up key and wait for data to come in.

9. Press F11 on the master and set the slave address for 1. Press F9 and follow the instructions on the screen.
10. After each successful transmission, change the slave address on the master and try again.
11. Is the communication correct?

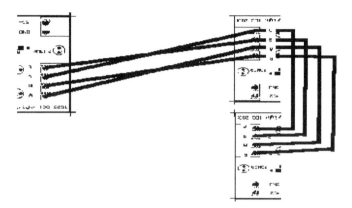

Figure C.4
RS-422 multidrop wiring connections

When finished, shut down the software and disassemble the hardware.

Practical 4: Logging data on RS-485 system

Objective

The objective of this practical is to give the delegate practice at logging data on an RS-485 communication system and using the PAT software to analyze the transmission.

Overview

The delegates learn how to monitor, log, time and analyze a 485 transmission utilizing the PAT software. This will be accomplished by using the working simulation of a computer to equipment communication system in Practical 3. Two computers will be connected together to simulate the computer to equipment system. The delegate will use the third computer to monitor and log the communication and analyze the transmission.

Hardware/software required

- Three computers with PAT software.
- Three 232 to 485 adapters.
- Wire to connect the adapters.
- Small blade screwdriver.
- One voltmeter.

Implementation

1. Setup the equipment as for Practical 3.
2. Setup the monitoring computer using the PAT software for 4800 8N1. Enter the monitor mode in the PAT software under the display menu.

3. Type on each of the other computers and verify on the monitoring computer that the correct information is displayed on the screen.
4. Set up the monitoring computer to log the packets by entering the log data under the display menu in the PAT software (single log/COM 1).
5. Send data from one of the computers by holding down a key to send a continuous string of characters.
6. View the data on the monitor computer by ending out of log data and moving down one line to view log file under the display menu.

Timing analysis

1. View the data logged in 'logging data'.
2. Notice the clock in the left–hand corner. Also notice that as the cursor is moved across the data the time changes.
3. Work out the transmission time for the message.
4. Calculate the baud rate for the transmission, noting that each character is sent as 10 bits (1 start, 8 data and 1 stop).
5. Repeat practical for other baud rates if time allows.

Practical 5: Fiber optic continuity test

Objective

The objective of this practical is to give the delegate an opportunity to undertake a simple continuity test on a fiber optic cable.

Overview

The delegates learn how to use a visual fault locator to check some fiber optic cables and subjectively determine whether they are functioning. This test can also be used to identify one fiber from several in a cable.

Hardware required

- A fiber optic visual fault locator.
- Several fibers to test.

Implementation

1. Remove the plastic connector covers from both ends of the test fiber cable.
2. Connect the fiber optic visual fault locator one end of the fiber. Press the tester button and observe that light emanates from the other end of the fiber. This gives a simple indication of the continuity of the fiber link.
3. Repeat with several other fibers. Check for light that can be seen leaking from a faulty splice. This may illustrate an easy way of carrying out visual fault finding on bad splices or joints.
4. Disconnect all equipment, put the plastic covers back on the connector ends and return everything to the state it was in before you started the practical so that the next group can carry out the practical in full.

Practical 6: Fiber optic insertion loss testing

Objective

The objective of this practical is to illustrate the use of an optical source and an optical power meter to measure insertion loss.

Hardware required

- One optical light source.
- One optical power meter.
- Test adapter cable (SC to ST).
- SC through adapter.
- Reference fiber.
- Test fibers.

Implementation

1. Remove the plastic connector covers from the optical connectors.
2. Connect the optical source (850 nm) to the power meter using the reference fiber, through adapter and test adapter fiber.
3. Set the wavelength of the power meter to match the source wavelength in use. Turn on both devices and note the output reading on the power meter.
4. Replace the reference fiber with a test fiber and repeat the measurement. Calculate the relative loss of the test fiber, as the difference between the two readings.
5. Turn off the light source and power meter, disconnect all equipment, put the plastic covers back on the connector ends, and return everything to the state it was in before you started the practical so that the next group can carry out the practical in full.

Practical 7: Logging Modbus data

Objective

The objective of this practical is to give the delegate an understanding how to monitor, log and analyze data communications between two devices using EIA-232 communication.

Overview

The delegates learn how to monitor, log, time and analyze the communications path utilizing the PAT software. This will be accomplished by building a working simulation of a computer to equipment communication system using the Modbus protocol. Two computers will be connected together to simulate the computer to equipment system. The delegate will insert a third computer using the PAT adapter. This third computer will be used to monitor and log the communication and analyze the Modbus protocol.

Hardware required

- Two computers with PAT software.
- One dual computer with two com ports and PAT software.

- Two laplink cables.
- One PAT interface unit.
- One 25-pin male to male gender changer.

Implementation

1. Connect the laplink cables to the com port of the two computers.
2. Connect the PAT interface unit to the dual computer using the two com ports.
3. Connect the other end of the laplink cables to the PAT interface cable with one gender changer.
4. When the system is connected turn on the computers.
5. Run the PAT software on all computers.
6. On the two simulation computers get into the interactive mode. In the dual computer, get into the dual monitor mode.
7. Type something on each computer and verify that the data is correct on all computers. If it is not correct, why is it so?
8. The instructor will designate one computer as the slave and another as the master.
9. The third computer will be used to log the data as it is transferred from the master to the slave and back.
10. The master and slave computers will run the Modbus software.
11. The monitoring computer will run the PAT software.
12. Toggle between master and slave by pressing F10.
13. On the master and slave press F11 and setup the communications for 9600 8N1 address 1.
14. Setup the monitoring computer using the PAT software for 9600 8N1 using hex as a code. Enter the dual monitor mode in the PAT software under the menu display.
15. On the master computer configure the Modbus software by pressing F3. Enter 000A0002.
16. On the slave computer configure the Modbus software by pressing the page down key. Change coil data for A and B to 1s. Press the page up key to return to the main screen when the change has been made.
17. On the master computer press F9. This will send the packet 0101000A00029DC9.
18. On the slave computer, follow the instructions on the screen and press any key. This will send the packet 010101031189 back to the master computer.
19. Verify on the monitoring computer that the correct information is displayed on the screen.
20. Setup the monitoring computer to log the packets by entering the log data under the display menu in the PAT software (dual log).
21. The master will send the packet again by hitting F9.
22. How many bytes of data did the monitor see?
23. View the data on the monitor computer by ending out of log data and moving down one line to view log file under the display menu.
24. How would you setup the monitor to log based on a string?
25. Setup the monitor computer to do a pre trigger log based on a CRC. What problems can be encountered when using hex that would not be encountered using ASCII when logging based on a string?

Timing analysis

1. View the data logged in 'logging data'.
2. Notice the clock in the left-hand corner. Also notice that as the cursor is moved across the data the time changes.
3. On a piece of paper calculate the following times:

Transmitter packet transmission time _____

Receiver packet transmission time _____

Delay time between TX and RX _____

Total turn around time _____

What would be the turn around time be if the BPS rate of the data were doubled?_____

What if it was doubled again? _____

Practical 8: HART protocol noise calculation

Objective

The objective of this practical is to give the delegate an understanding of signal strength calculation on a HART system.

Overview

The delegate will use the HSIM software package to the signal strength in a HART scenario, to determine the likelihood of noise interference.

Hardware/software required

- One computer with HSIM software.

Implementation

A HART system uses 800 m of Belden 8441 cable and has one device connected with a load resistor of 250 ohms. You have measured peak-to-peak noise on this cable of 30 mV. Use the HSIM program to calculate the HART signal level and assess whether this is likely to be a problem.

Practical 9: AS-i bus configuration

Objectives

- To demonstrate a true binary sensor level interface
- To illustrate the polarized wiring
- To introduce the programming techniques

Hardware/software required

- One computer with the program ASI Tools loaded up.
- One AS–i power supply (Pepperl & Fuchs)
- One VAM module (Interface to RS-232)

- One RS-232 cable between VAM module and notebook computer
- One digital input/output module (part no. 37398)
- One proximity detector module (part no. 34016)
- One digital output module (4 digital outputs) (part no. 34170)

Implementation

1. Connect up the modules together either with the flat AS-i yellow cable.
2. Identify each module clearly.
3. Connect computer to the VAM gateway module using an RS-232 cable.
4. Be careful about installation of power cable to the gateway.
5. Boot up PC and select AS-i tools icon.
6. Confirm that lights are illuminated on the AS-i gateway.
7. Go to master new and check configuration under communications and ensure that it is set up for COM 1 standard communications.
8. Select file from open file called blank. AS-i (C:\act\demo\)
9. Load in file.
10. Confirm that the mode switch light is off on the gateway module.
11. Download the program to the gateway module by using the downward arrow on the menu.
12. Remove the RS-232 cable from the gateway module.
13. Click on the mode switch for approximately 15 seconds until the project light goes out and observe it scanning the stations on the network.
14. Hold down again until the project light goes again.
15. The system has now been reset.
16. Go to master new and set up comms link appropriately.
17. Upload the current status of the network.
18. Read the parameter settings and comment on the symbols used under slaves.
19. Alternatively, delete everything after assembler and type in the following sequence:

    > Program
    > A I 2.0
    > = Q 1.0
    > ***
    >
    > be
    > end.

Ensure that the program input and output declarations match up with the original declarations at the beginning of the program.

1. Save and download the program.
2. Activate the program by setting it to run.
3. Confirm program works.
4. Modify program and note the effect of an incorrect configuration.

Practical 10: DeviceNet installation and configuration

Objectives

- To show the ease of cabling a DeviceNet system
- To configure a simple DeviceNet system

- To demonstrate the use of DeviceNet in a simple network

There are three parts to this practical exercise.

Hardware/software required

- One computer with DeviceNet software installed
- Three trunk cables (gray)
- Four spur cables (yellow)
- One small gray interface cable
- One interface box (1770-KFD)
- One power supply
- One power tap connection box
- Photoelectric cell
- One RS-232 cable
- One male terminator
- One female terminator

Implementation

1. Connect up cables with photoelectric cell installed. Remember to put terminators at each end.
2. What do you notice about the cabling and connections? Identify the colors used for power and communications. Is there much similarity with these cables and RS-485?
3. Plug in power supply leads to power tap box. Be very careful about power leads. In case of doubt, get instructor to assist.
4. Connect up the RS-232 cable to the RS-232 interface box and notebook computer.
5. Click on the DeviceNet icon. Note that the DeviceManager is being installed.
6. Open a new project. Give a new name (for example, 'test') and description (whatever you need to identify it).
7. Go to utilities and setup on line connection. Click on OK and then OK again under driver configuration. Note the baud rate settings.
8. Then click on 'on line' build. Observe it building. Be patient as it identifies all the devices on the network.
9. Confirm that all the devices on the network are identified. Remove a terminator and note the results. Why is this the case? If you have time, reconfigure all the devices later for a higher baud rate and note the results.
10. To demonstrate a simple configuration of a device, select the photoelectric cell (9000 series – diffuse).
11. Double click on the description and wait for it to upload. Click then on start monitoring. Be patient as it takes a few seconds.
12. Move your hand in front of the photoelectric cell and observe the results.
13. Stop the device by typing stop monitoring.
14. Refer to the specification sheet and modify the parameters; specifically the operate mode. Change this to the other setting and save to device and click OK. Commence monitoring and note what happens. Adjust operating margin as well. Observe the results.
15. Review the specification sheets and identify the differences between this approach and using standard 'hard wiring'.

16. Click on switch and note the results.
17. If you have time connect up the other devices and identify the results.
18. What is still needed to complete this system? Why? What is one design flaw with the system? How can the design of configurator and controller be improved?

Practical 11: DeviceNet

Overview

The objective of this practical is to give the delegate an opportunity to calculate the correct power and cable configurations for DeviceNet.

Objective

The delegate will learn how to calculate the current and cable requirement for a sample DeviceNet network.

Implementation

A system is connected up as in the diagram below. Using the relevant table for thick DeviceNet cable length and power capacity calculate whether this will work.
What happens when the cable of section 2 is increased in length by another 100 m?

Thick Cable Network Current Distribution and Allowable Current Loading												
Network length in meters	0	25	50	100	150	200	250	300	350	400	450	500
Network length in feet	0	82	164	328	492	656	820	984	1148	1312	1476	1640
Maximum current in Amps	8	6	5.42	2.93	2.01	1.53	1.23	1.03	0.89	0.78	0.69	0.63

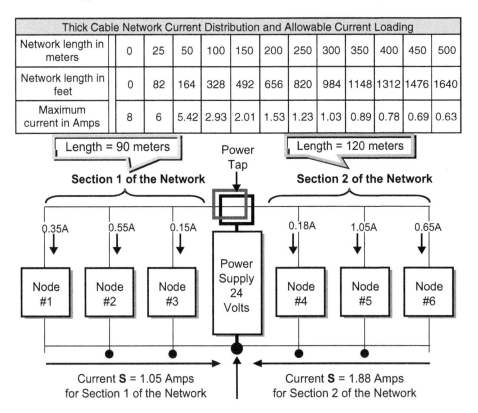

Figure C.5
Singles supply – center connected

Conclusion

- Current in section 1 = 1.35 amps over a length of 90 meters.
- Current in section 2 = 2.33 amps over a length of 120 meters.
- Using interpolation from the thick cable data:

  Current limits for a distance of 90 meters = 3.43 amps and for 120 meters is 2.56 amps.

- Power for both sections is correct, and a 3.5 amp (minimum) power supply is required.

Practical 12: Profibus wiring problem

The problem

You are provided with the schematic of a network indicated below connected up in a star type configuration. It is a Profibus-DP type network with a master–slave type operation. You notice that there is a problem with intermittent communications. Investigate further and identify the following situation:

When the baud rate of operation is below 30 kbaud, the communications appears to be OK i.e. the messages tend to get through between the various devices on the network. However when the speed is increased over 100 kbaud, there is again significant communications failure.

What do you think the problem is likely to be and how would you correct it?

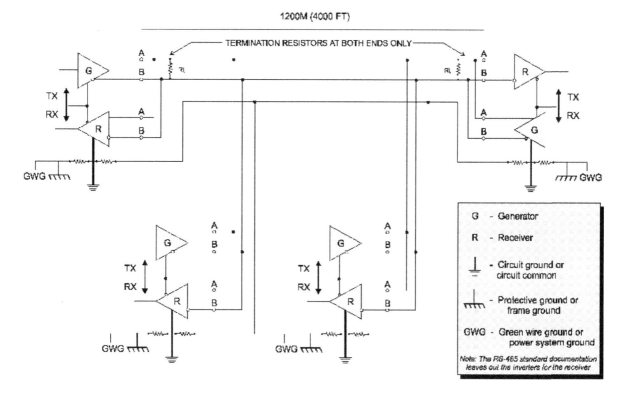

Figure C.6
Schematic of full-duplex RS-485 network

Conclusion

There are three main issues that could be a problem.

- Firstly, the stub length of each of the bottom two stations is probably longer than the maximum length (preferably 0 m or as close to this as possible). This stub length must be minimized because the impedance mismatch will cause reflections down the line as well as reducing the effective power transfer. When the baud rate is low, reflection is not a major problem but as the cable length is increased and the baud rate is increased over 30 kBaud, this becomes a major problem (due to reflections at the stub connection points).
- Secondly, the terminating resistors may not be connected properly. They are often connected in using DIP switches (on the last PLC at the end of a bus, for example) or simply resistors, which have come adrift. Check that these are correctly installed. If you disconnect the power you can do a simple resistance check with a meter.
- Finally, check that there are no common mode voltage problems (often intermittent). This requires excellent grounding or earthing practice or the use of the third wire (GWG – green wire ground) to connect all the devices on the bus together. Failure to do this could mean that the receiver's are exceeding their +12 V to –7 V common mode voltage limit.

Practical 13: Foundation Fieldbus topology

Objectives

The objective of this practical is to give the delegate an understanding of the design requirements for Foundation Fieldbus networks.

Implementation

The following example network is given below. Calculate whether the network sketched out below is satisfactory.

What happens if the trunk is extended by another 700 m? Is this allowable?

Assumptions

Each device must have a minimum of 9 V DC in order to operate.

You need to know:

- The current consumption of each device.
- Location on the network.
- Location of power supply on network.
- Resistance of each cable section.
- Power supply voltage.

Assume that type B cable is used throughout. The resistance per meter is 0.1 ohm. Assume that the power supply is 12 V. The bridge is separately powered and does not draw any power.

The total length is 500 m for the trunk. The other lengths are as indicated in the diagram below.

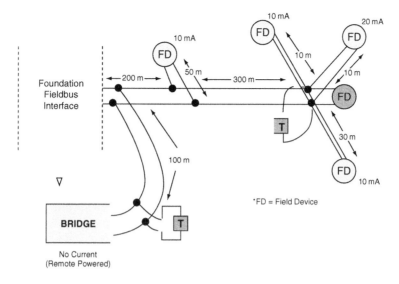

Figure C.7
Foundation Fieldbus network

Conclusion

First of all neglect temperature effects. The resistance per meter is 0.1 ohm. The following table shows the electrical structure.

Section length	Resistance	Current (Amps)	Voltage drop
200 m	20 ohms	0.05 A	1 V
50 m	5 ohms	0.01 A	0.05 V
300 m	30 ohms	0.04 A	1.2 V
10 m	1 ohm	0.02 A	0.02 V
30 m	3 ohms	0.01 A	0.03 V

The voltages at each of the network-powered devices are:
50 m spur   10.95 V
Trunk end   9.8 V
10 m spur   9.78 V
30 m spur   9.77 V

These are all over the 9 V minimum; so the setup is acceptable.

From a troubleshooting perspective the results of these calculations can be easily checked on a working system by measuring the voltage at the devices.

(Courtesy: Fieldbus Foundation Wiring and Installation 31.25 kbps, Voltage Mode Wire Medium Application Guide AG-140, Revision 1.0)

Practical 14: Industrial Ethernet wiring/hub problems

Overview

This practical illustrates some problems that can be encountered with Ethernet wiring and 10BaseT hubs. Simple troubleshooting techniques are demonstrated.

Objectives

The delegates will learn how to setup a network and identify faulty UTP cable segments and/or faulty ports on the 10BaseT hubs.

Hardware/software required

The hardware required for this practical is as follows:

- Two or more personal computers or laptops.
- One network interface card (NIC) or Ethernet card per computer.
- One unshielded twisted pair (UTP) patch cable per computer.
- UTP hub(s) with requisite number of ports.
- Power supplies for the various computers and hubs as required.
- One crossover UTP patch cable.

The software required is as follows:

- All the computers loaded with Windows 98 second edition operating system (Windows 95, NT, 2000 and ME will also do, this book shall refer to Windows 98 SE).
- Driver software for the NIC as provided by the manufacturer.

Implementation

To setup a basic network insert the NIC into each computer as recommended by the manufacturer. Then connect the UTP cables to the NICs and hub(s) as shown in the diagram below.

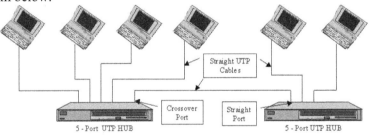

Figure C.8
Basic network

In addition, there are four software network components that are required to be setup within each host computer. They are:

- **CLIENT:** enables the host computer to connect to other computers
- **ADAPTER:** is a hardware device that physically connects the host computer to the network
- **PROTOCOL:** is the 'language' the host computer uses. All hosts must use the same protocol in order to communicate
- **SERVICE:** enables the host computer to share its resources such as files, printers and other services with other computers on the network

The network dialog box may be opened by clicking start, pointing to settings, clicking control panel, and then double-clicking network within.

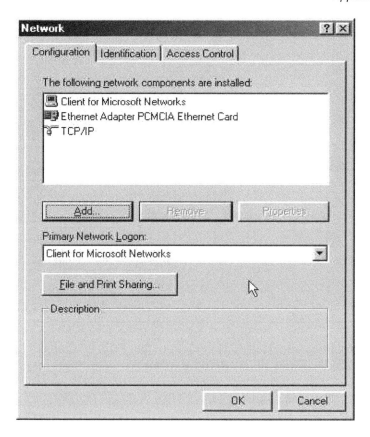

Note that as soon as the NIC is inserted into the host computer, the operating system prompts the user to provide the necessary device drivers and the operating system CDROM in order to install and configure the device. Thus the client, adapter and protocol network components are automatically installed. For the purposes of this practical, we shall install protocol that requires minimal configuration, NetBEUI being a good example. This can be installed by clicking on the Add button in the network window.

Select protocol as the network component type to install and click on the Add button, as shown.

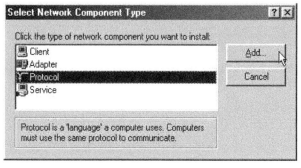

Select Microsoft as the network protocol manufacturer on the left-hand side and scroll down to NetBEUI and select it on the right-hand side, and click on the OK button to install.

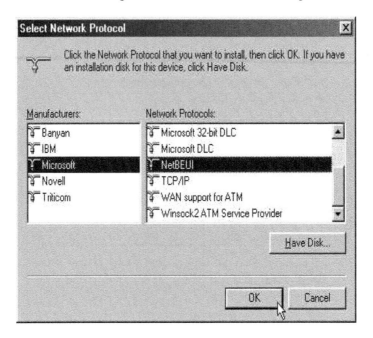

After installing the NetBEUI protocol, the TCP/IP protocol may be removed, by selecting it and clicking on the Remove button.

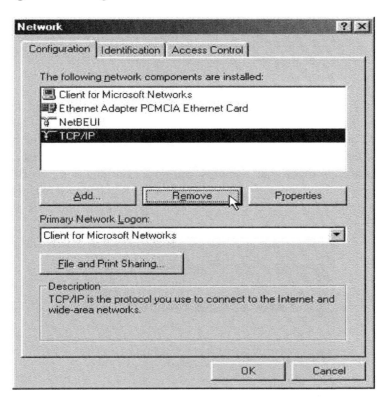

To add the final network component, service, click on the File and Print Sharing button.

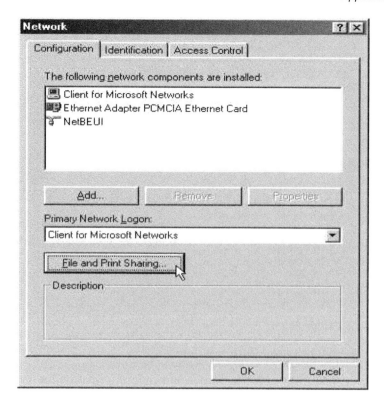

Check the box alongside 'I want to be able to give others access to my files' and click on OK.

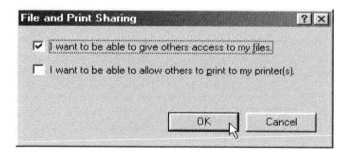

The four basic network components are now installed.

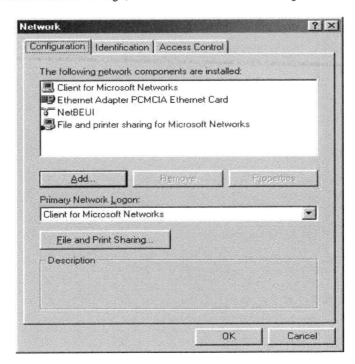

Click on the identification tab as shown and enter information for the following:
Computer name: a unique identity for your host computer on the network
Workgroup: the name that the host computer will appear in, and
Computer description: a short description of the host computer.

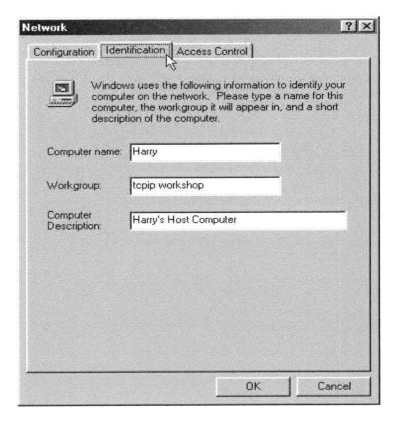

Finally, click on the Access Control tab and select Share-level access control that enables the user to supply a password for each shared resource, and click on OK.

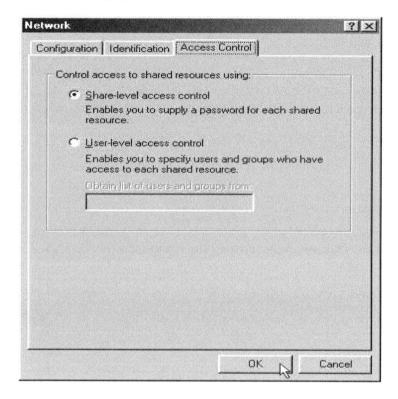

The system will then require a restart for the new network settings to take effect. Click on yes when prompted to restart the computer.

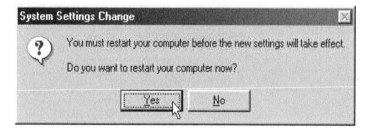

The user will be prompted for a user name and a password after restart.

The computer should now be suitably configured for networking, and the user can explore some troubleshooting of the wiring and hubs.

The user will notice that the NIC and associated port on the hub both have their link integrity lights illuminated. Unplug the UTP cable and observe both lights extinguish. The link integrity lights are illuminated only when both devices are operational and correctly wired.

Unplug the UTP cable at both the NIC and hub and replace it with the crossover cable. Note that both sets of lights are extinguished because the wiring is now incorrect. If the hub has a crossover (uplink) port then plug the crossover cable into this and observe the link integrity lights illuminate as the wiring is once again correct.

Plug the crossover cable directly into another NIC and observe both sets of lights illuminate. This is a useful way of connecting two computers without a hub, for file transfers etc.

If the link integrity light at the hub is not illuminated simply swap that cable to another working port on the hub to check that the port itself is not faulty.

Whenever data is being sent on the network the network activity lights on ALL ports and NICs will flicker since all messages on an Ethernet network are sent to all users. As other computers are being booted up onto the network activity should be noted.

Now replace the UTP connection at the hub and continue to use the network to share data.

Double click on My Computer icon on the desktop.

Select the C drive and then using the right mouse button right click on the same and select Sharing.

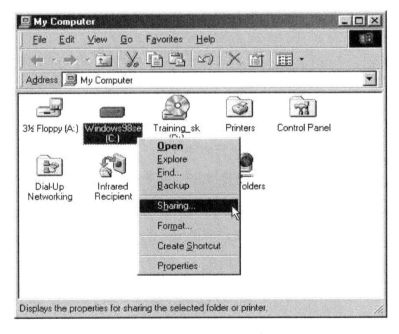

Select shared as and provide a shared name for this resource and a short comment. Choose the access type desired and supply the relevant passwords. Click on Apply followed by OK.

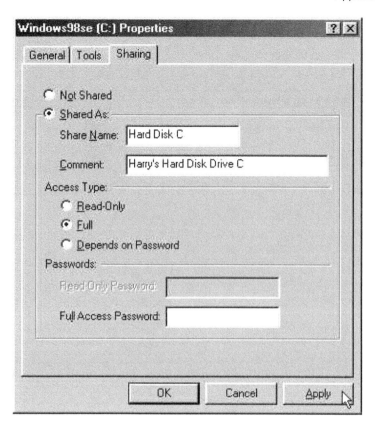

Notice the small hand symbol under the C drive icon indicating that this drive is now shared.

Other resources may similarly be shared.

This procedure thus far concludes the basic network setup of host computers using simple configuration information. To verify whether the network setup is successful, double click on the Network Neighborhood icon to view the other host computers on the network.

This folder contains links to all the computers in the workgroup and on the entire network.

To see the shared resources available on a specific computer, just click the computer icon.

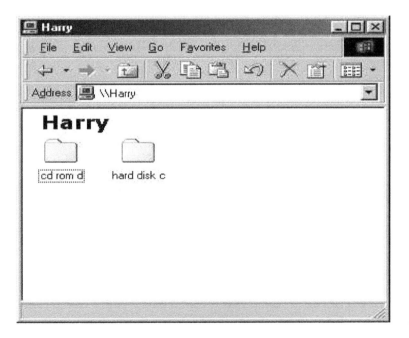

You may now browse the Network Neighborhood and the other computers' shared resources. You may also transfer files back and forth across the network, view a remote file, execute a remote program etc, all from the comfort of the local host computer.

Note

1. During boot-up, there may be errors displayed stating that there is already another computer on the network with the same name, and all networking

services would be disabled. Follow the procedure described above and choose another computer name and restart.

2. Some users may not be able to browse the Network Neighborhood if the user has not logged in with an appropriate username and password at the time of boot-up. Log-off and log-on as a valid user.

3. Others may not be able to either see their own host computer or other's host computers. Not sharing any resource leads to this problem. Follow the procedure described above and share some resource.

Conclusion

As you can see, within a few steps you can easily setup and configure a simple network and exchange files & other resources. The link integrity check makes trouble shooting at the physical connection level very easy.

Practical 15: TCP/IP network setup

Overview

This practical session will review the basic steps in configuring the IP addressing details and introduces some important troubleshooting utilities.

Objectives

This practical demonstrates to you the procedure of setting up a simple TCP/IP network.

Hardware/software required

The hardware required for this practical is as follows:
- Two or more personal computers or laptops.
- One network interface card (NIC) or Ethernet card per computer.
- One unshielded twisted pair (UTP) patch cable per computer.
- UTP hub(s) with requisite number of ports.
- Power supplies for the various computers and hubs as required.
- One additional Ethernet card with a UTP patch cable.

The software required is as follows:
- All the computers loaded with Windows 98 second edition operating system (Windows 95, NT, 2000 and ME will also do, this book shall refer to Windows 98 SE).
- Driver software for the NIC as provided by the manufacturer.

Implementation

Call up the network dialog box and add a new Microsoft protocol – TCP/IP, if not already present.

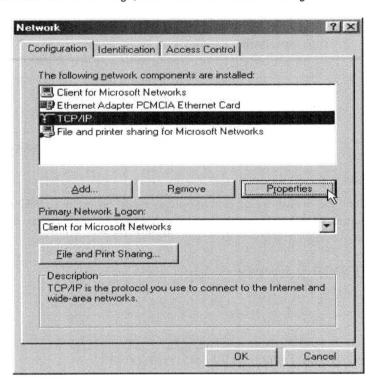

Select TCP/IP and click on the Properties button.

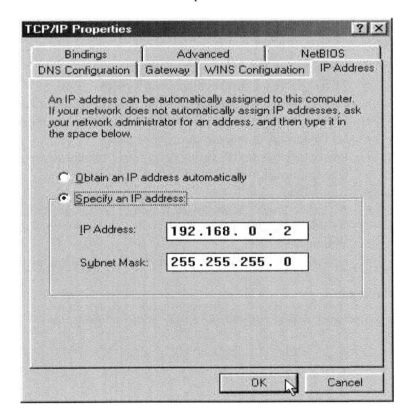

Choose Specify an IP address and then supply an IP address of the form: W.X.Y.Z (for example, 192.168.0.1) where the first three dotted decimals (W.X.Y) represent the network identity and the last dotted decimal (Z) represents the unique host computer identity for this network. (Note that W, X, Y and Z can each have values ranging from 1 through 254.)

Thus we could have IP addresses in the range:

```
192.168.0.1

192.168.0.2

192.168.0.3

192.168.0.4

192.168.0.5
```

and so on … for each host computer

Enter default Subnet mask of 255.255.255.0 to signify a class 'C' type of a network, which permits over 2 million network identities with each network containing 254 hosts. (Refer to the Practical TCP/IP and Networking manual for other classes of networks.) Click on OK to return to the network dialog box.

All other configuration information remains unchanged from the previous practical. Click on OK again to save all information and restart the computer.

To verify whether all configuration information has been entered correctly, one of the utilities that is bundled with the Microsoft's TCP/IP is WINIPCFG.exe. This can be called-up by clicking on start, and selecting run. Then typing winipcfg in the run dialog box and clicking on OK.

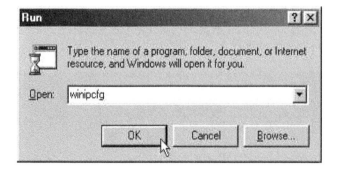

The IP configuration will be displayed along with the Ethernet adapter information. Note the adapter address.

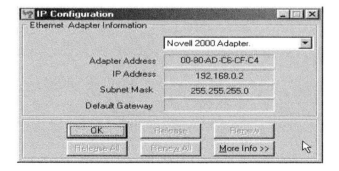

(Windows NT users need to use the DOS-based IPCONFIG.exe)

```
MS-DOS Prompt

Auto

C:\>ipconfig

Windows 98 IP Configuration

0 Ethernet adapter :

        IP Address. . . . . . . . . : 192.168.0.2
        Subnet Mask . . . . . . . . : 255.255.255.0
        Default Gateway . . . . . . :

C:\>
```

To verify whether this entire configuration is working correctly within the local host computer, the ping localhost command shall be employed within the DOS environment.

```
C:\>ping localhost

Pinging harry [127.0.0.1] with 32 bytes of data:

Reply from 127.0.0.1: bytes=32 time=1ms TTL=128
Reply from 127.0.0.1: bytes=32 time=1ms TTL=128
Reply from 127.0.0.1: bytes=32 time<10ms TTL=128
Reply from 127.0.0.1: bytes=32 time=1ms TTL=128

Ping statistics for 127.0.0.1:
    Packets: Sent = 4, Received = 4, Lost = 0 (0% loss),
Approximate round trip times in milli-seconds:
    Minimum = 0ms, Maximum =  1ms, Average =  0ms

C:\>_
```

If there is a response received from 127.0.0.1(a universal local loopback IP address, common to all computers), then it may be assumed that TCP/IP is now correctly configured. Note that the host need not be connected to the network for this command to function and to receive a reply.

The user is left to ping the local host computer's name and the IP address individually.

To verify communication between the local host computer and remote host computers, ping the remote host computer's IP address and/or name. If a reply is received, then it is an indication that the remote host computer is powered on, connected to the network and the TCP/IP configuration on that remote host computer is correct.

```
C:\>ping 192.168.0.3

Pinging 192.168.0.3 with 32 bytes of data:

Reply from 192.168.0.3: bytes=32 time=6ms TTL=128
Reply from 192.168.0.3: bytes=32 time=2ms TTL=128
Reply from 192.168.0.3: bytes=32 time=1ms TTL=128
Reply from 192.168.0.3: bytes=32 time=2ms TTL=128

Ping statistics for 192.168.0.3:
    Packets: Sent = 4, Received = 4, Lost = 0 (0% loss),
Approximate round trip times in milli-seconds:
    Minimum = 1ms, Maximum =  6ms, Average =  2ms

C:\>_
```

```
C:\>ping tom

Pinging tom [192.168.0.1] with 32 bytes of data:

Reply from 192.168.0.1: bytes=32 time=1ms TTL=255
Reply from 192.168.0.1: bytes=32 time=1ms TTL=255
Reply from 192.168.0.1: bytes=32 time=1ms TTL=255
Reply from 192.168.0.1: bytes=32 time=1ms TTL=255

Ping statistics for 192.168.0.1:
    Packets: Sent = 4, Received = 4, Lost = 0 (0% loss),
Approximate round trip times in milli-seconds:
    Minimum = 1ms, Maximum =  1ms, Average =  1ms

C:\>_
```

Note

1. During boot-up, there may be errors displayed stating that there is already another computer on the network with the same IP address, and all networking services would be disabled. Follow the procedure described above and choose another IP address and restart.
2. Some users may not be able to browse the Network Neighborhood if the user has not logged in with an appropriate username and password at the time of boot-up. Log-off and log-on as a valid user.
3. Others may not be able to either see their own host computer or other's host computers. Not sharing any resource leads to this problem. Follow the procedure described in Practical-1 and share some resource.

Conclusion

Notice that even though we enter only one ping command, we receive four replies. This is due to the default programmed within the ping.exe executable that instructs to send out 4 requests one after another. The ping command sends out 32 bytes of data to the remote host and therefore we observe bytes=32. The time=1 ms denotes the total round trip time taken for the ping message to be transmitted and a reply to be received, in milliseconds. TTL=255, indicates that the replying host is permitting this reply to traverse through a total of 255 networks. Why do we need to limit the number of networks traversed?

At the DOS prompt type ping without any options and it will display a list of the various switches that are available to customize the ping message.

Practical 16: Using ICMP echo messages (PING)

Overview

The Internet control message protocol (ICMP) is the protocol used by the TCP/IP network layers to communicate with each other, monitor one another and exchange information. This practical reviews the workings of the PING command.

Objectives

This practical shows you how to use ICMP echo messages using the PING command over a TCP/IP network. This is a very important troubleshooting tool as it allows us to check whether a particular computer is accessible on the network.

Hardware/software required

The hardware required for this practical is as follows:
- Two or more personal computers or laptops.
- One network interface card (NIC) or Ethernet card per computer.
- One unshielded twisted pair (UTP) patch cable per computer.
- UTP hub(s) with requisite number of ports.
- Power supplies for the various computers and hubs as required.
- One additional Ethernet card with a UTP patch cable.

The software required is as follows:
- All the computers loaded with Windows 98 second edition operating system (Windows 95, NT, 2000 and ME will also do, this book shall refer to Windows 98 SE).

- Driver software for the NIC as provided by the manufacturer.
- Additional third-party Ethernet packet capturing utility called PacketBoy a part of the NetBoy Suite of applications from NDG software.

Implementation

Startup the PacketBoy utility by clicking on Start, selecting Programs, NDG Software, PacketBoy, PacketBoy x.x.

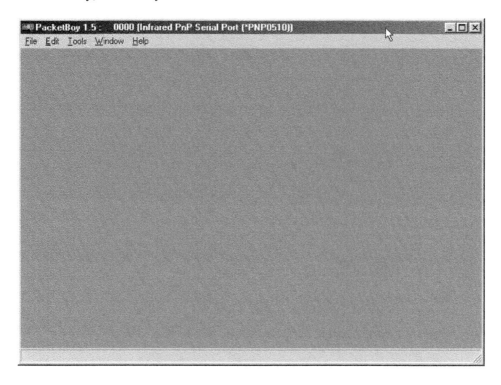

Maximize the window by clicking on the title-bar and then ensure that the correct adapter is selected, by clicking on file on the pull-down menu and selecting choose adapter.

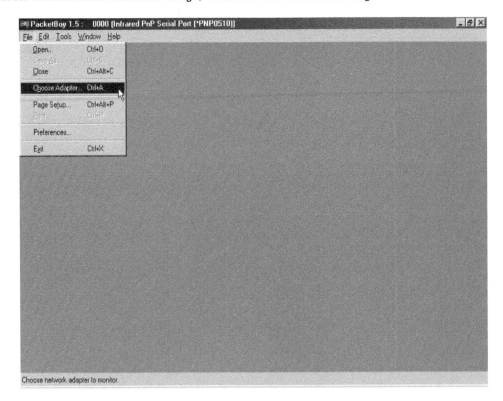

Select the correct adapter from the display.

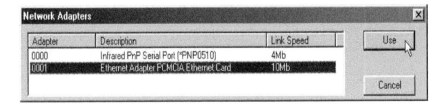

Then click on use to select the desired adapter. The correct adapter will now be displayed in the title-bar of the PacketBoy utility.

Select the capture console from the tools menu.

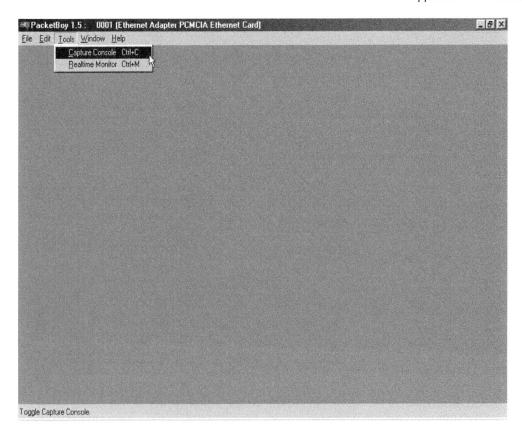

Click on the button with the red circle to start capture.

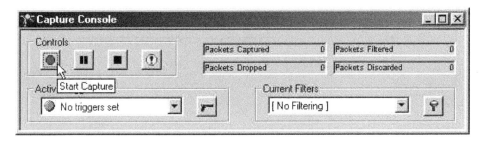

Call-up the DOS prompt and ping a remote host computer.

```
C:\>ping 192.168.0.1

Pinging 192.168.0.1 with 32 bytes of data:

Reply from 192.168.0.1: bytes=32 time=2ms TTL=255
Reply from 192.168.0.1: bytes=32 time=2ms TTL=255
Reply from 192.168.0.1: bytes=32 time=2ms TTL=255
Reply from 192.168.0.1: bytes=32 time=2ms TTL=255

Ping statistics for 192.168.0.1:
    Packets: Sent = 4, Received = 4, Lost = 0 (0% loss),
Approximate round trip times in milli-seconds:
    Minimum = 2ms, Maximum =  2ms, Average =  2ms

C:\>_
```

Switch back to PacketBoy and click on the button with the black square to stop capture and view the packets captured.

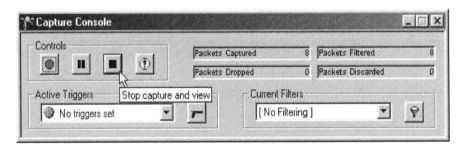

The captured packets would be displayed.

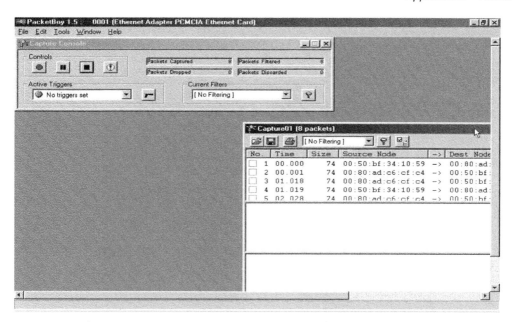

To view the entire captured packets' window, double click on the title-bar as shown.

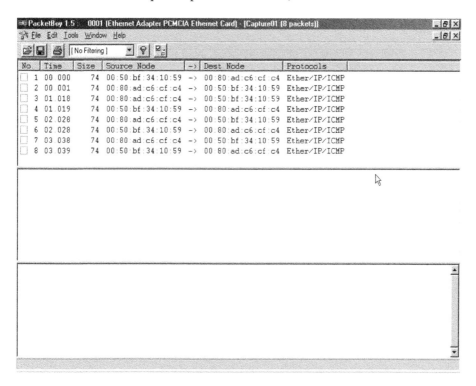

Notice that the screen is divided into three sections. The top section displays the various captured packets, the middle section displays the various protocol headers and their details, and the bottom section displays the actual data that is contained within the entire packet. Select the first packet to display the various headers within.

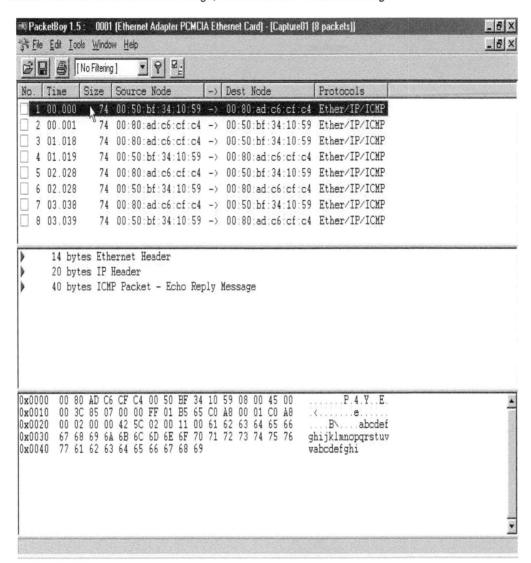

Resize the sections such that it is convenient to browse through the middle section comfortably, and then double click on the ICMP packet header in order to expand it and display the details of the header, as illustrated below.

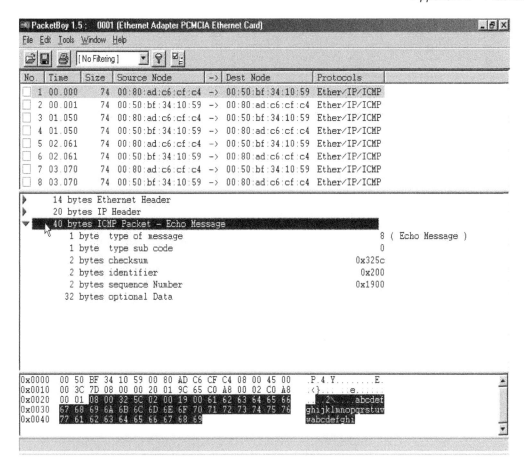

Note the identifier and the sequence number as displayed. Select the second packet to display the echo reply, expand the ICMP header and note the identifier and the sequence number. This is the same as that of the first packet.

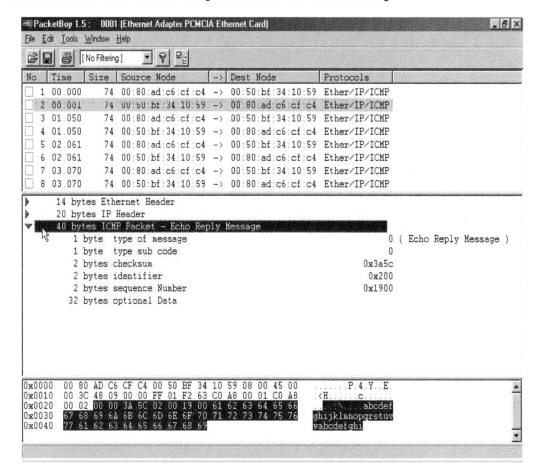

Conclusion

In order to originate the host to identify that it is receiving a reply to a particular echo message that it has transmitted, it sets an echo message identifier and sequence number, that is displayed under the ICMP header of the first packet. When the destination host replies to the echo message, it uses the same identifier and sequence number, indicating to the originating host that it is receiving a reply to this particular ping (ICMP) message.

Thus for every ICMP echo message there would be one ICMP echo reply message. The ping.exe transmits four ICMP messages, therefore 8 messages are captured and displayed.

You should also browse the other headers that are displayed.

Practical 17: Address resolution protocol (ARP)

Overview

Thus far, it has been observed that most applications communicate with each other using IP addresses. So how do the physical and the data link layers know what the media address is for a particular IP address?

Objectives

This practical is designed to introduce to the user the operation of address resolution protocol over a TCP/IP network.

Hardware/software required

The hardware required for this practical is as follows:

- Two or more personal computers or laptops.
- One network interface card (NIC) or Ethernet card per computer.
- One unshielded twisted pair (UTP) patch cable per computer.
- UTP hub(s) with requisite number of ports.
- Power supplies for the various computers and hubs as required.
- One additional Ethernet card with a UTP patch cable.

The software required is as follows:

- All the computers loaded with Windows 98 second edition operating system (Windows 95, NT, 2000 and ME will also do, this book shall refer to Windows 98 SE).
- Driver software for the NIC as provided by the manufacturer.
- Additional third-party Ethernet packet capturing utility called PacketBoy a part of the NetBoy Suite of applications from NDG software.

Implementation

Startup the PacketBoy utility by clicking on Start, selecting Programs, NDG Software, PacketBoy, PacketBoy x.x. Maximize the window by clicking on the title-bar and then ensure that the correct adapter is selected. If not, follow the procedure as described in Practical 3. Select the Capture Console from the Tools menu. Click on the button with the red circle to Start Capture.

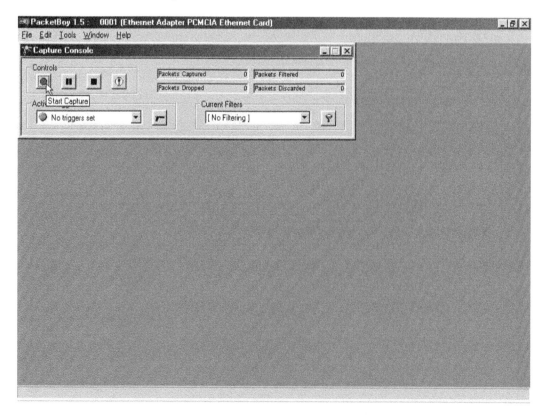

Call-up the DOS prompt and ping a remote host computer.

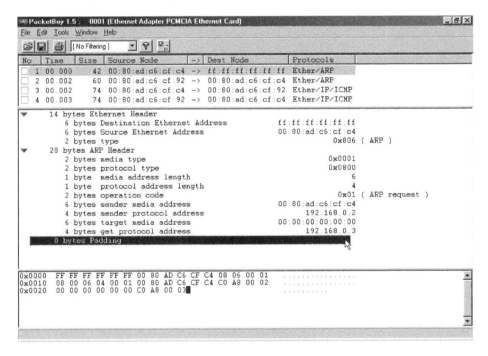

```
C:\>ping 192.168.0.3 -n 1

Pinging 192.168.0.3 with 32 bytes of data:

Reply from 192.168.0.3: bytes=32 time=4ms TTL=128

Ping statistics for 192.168.0.3:
    Packets: Sent = 1, Received = 1, Lost = 0 (0% loss),
Approximate round trip times in milli-seconds:
    Minimum = 4ms, Maximum = 4ms, Average = 4ms

C:\>_
```

Switch back to PacketBoy and click on the button with the black square to Stop capture and view the packets captured. Select the first packet to display the various headers within. Resize the sections such that it is convenient to browse through the middle section comfortably, and then expand both the Ethernet Header and the ARP Header.

Note the destination Ethernet address and the source Ethernet address within the Ethernet Header, and the sender media address, sender protocol address, target media address and target protocol address, within the ARP Header.

Notice that the destination Ethernet address is all 1s (FF:FF:FF:FF:FF:FF), indicating that this message is a broadcast to all hosts on the network from the originating host as indicated by the source Ethernet address.

Select the second packet to display the ARP response, expand both headers and note the destination Ethernet address and the source Ethernet address within the Ethernet Header, and the sender media address, sender protocol address, target media address and target protocol address, within the ARP Header.

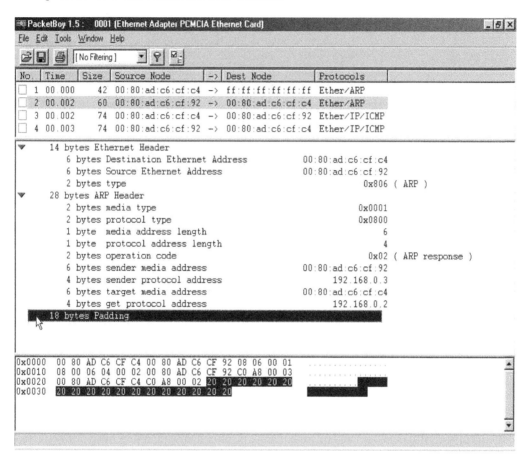

Conclusion

As covered in the Practical TCP/IP and Networking workshop, all Ethernet messages require the physical address of the destination adapter. Therefore, if the destination host adapter address is unknown, then ARP is first invoked to resolve the IP address of the destination host to its Ethernet address.

Within the ARP Header, notice that the source host is introducing itself by supplying its media address and its IP address. In turn, it is also indicating to the host that answers to the IP address as indicated by the target protocol address, that the target's media address is unknown, as indicated by all 0s (00:00:00:00:00:00). This is known as a gratuitous ARP.

The fact that the first message is a broadcast, all active hosts on the network would pick up the message via the physical layer and the data link layer and pass the information onto the network layer for further processing. As the ARP resides at the network layer, it deciphers the message and compares the target protocol address with its own. If there is an address match, it then makes an entry in the ARP cache, the sender media address and sender protocol address. It then creates an ARP response message.

Practical 18: Duplicate IP addresses

Overview

This practical illustrates how TCP/IP takes into account duplicate IP addresses and how it resolves these conflicts.

Objectives

This practical introduces to you one of the uses of address resolution protocol over a TCP/IP network.

Hardware/software required

The hardware required for this practical is as follows:
- Two or more personal computers or laptops.
- One network interface card (NIC) or Ethernet card per computer.
- One unshielded twisted pair (UTP) patch cable per computer.
- UTP hub(s) with requisite number of ports.
- Power supplies for the various computers and hubs as required.
- One additional Ethernet card with a UTP patch cable.

The software required is as follows:
- All the computers loaded with Windows 98 second edition operating system (Windows 95, NT, 2000 and ME will also do, this book shall refer to Windows 98 SE).
- Driver software for the NIC as provided by the manufacturer.
- Additional third-party Ethernet packet capturing utility called PacketBoy a part of the NetBoy Suite of applications from NDG software.

Implementation

Disconnect one of the host computers from the network. Set its IP address the same as one of the other hosts on the network, following the procedure as described in Practical 2. Restart the host with the duplicate IP address, leaving it still disconnected from the network.

Startup the PacketBoy utility on one of the host computers still connected to the network and start capture.

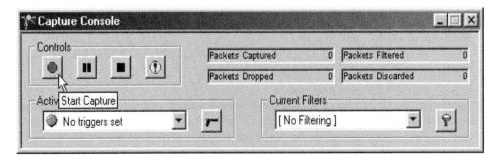

Reconnect the disconnected host computer, with the duplicate IP address, to the network. Within a few seconds, the host whose IP address was duplicated would display a message as below:

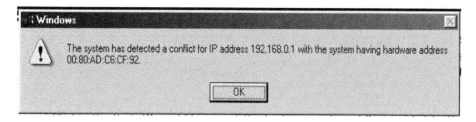

The host with the duplicate IP address would also show a similar message, except that the hardware address would be that of the original host, and that the interface on the host with the duplicate IP address has been disabled in order to avoid future conflicts.

Stop capture and view the packets captured on the host running PacketBoy. Select the first packet to display the various headers within. Resize the sections such that it is convenient to browse through the middle section comfortably, and then expand both the Ethernet Header and the ARP Header.

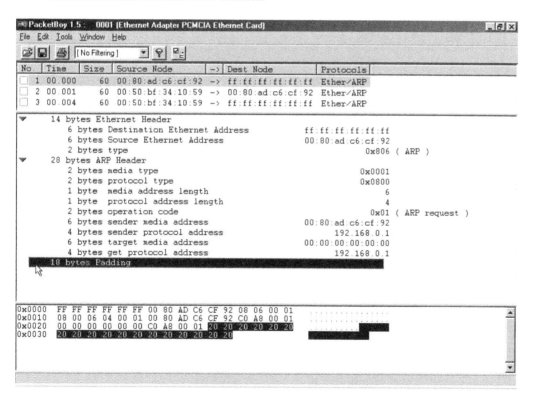

Note the destination Ethernet address and the source Ethernet address within the Ethernet Header, and the sender media address, sender protocol address, target media address and target protocol address, within the ARP Header.

Select the second packet to display the ARP response, expand both headers and note the destination Ethernet address and the source Ethernet address within the Ethernet Header, and the sender media address, sender protocol address, target media address and target protocol address, within the ARP Header.

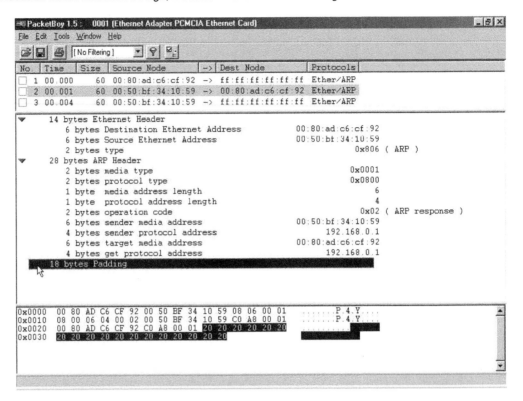

Select the third packet to display the ARP response, expand both headers and note the destination Ethernet address and the source Ethernet address within the Ethernet Header, and the sender media address, sender protocol address, target media address and target protocol address, within the ARP Header.

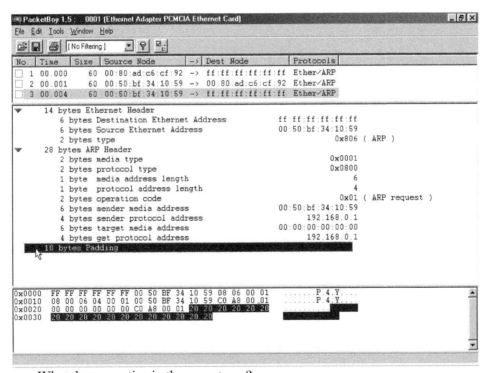

What do you notice in these captures?

Conclusion

As the above screens illustrate, when a host computer is booting up or is being connected to the network, one of the first broadcasts it makes is its identity. That is, the physical media address and its assigned IP address to all other hosts connected to the network, and in specific to any other node that could possibly possess its own IP address. Thus as the user will notice the first ARP message appears to be addressed to itself.

Obviously, as there exists another host with the same IP address, this host immediately sends out an ARP response to the first ARP message. In doing so it also realizes that it seems to be addressing itself, and therefore it understands that there exists a conflict and thus displays the conflict message.

To ensure that the host with the duplicate IP address is aware of the conflict too, the host with the original IP address sends out a third ARP message similar to the first introducing itself to the other. This triggers the operating system's conflict response window seen on the host with the duplicate IP address.

Practical 19: Incorrect subnet mask – misdirected datagrams

Overview

Most of the networks assume that all users reside on the same subnet and therefore usually the default subnet masks, as described in the Practical TCP/IP and Networking workshop, is used. If for some reason, intentional or unintentional, the default subnet mask were to be changed, what would be the outcome?

Objectives

This practical demonstrates to the user the function of the subnet mask.

Hardware/software required

The hardware required for this practical is as follows:
- Three or more personal computers or laptops.
- One network interface card (NIC) or Ethernet card per computer.
- One unshielded twisted pair (UTP) patch cable per computer.
- UTP hub(s) with requisite number of ports.
- Power supplies for the various computers and hubs as required.
- One additional Ethernet card with a UTP patch cable.

The software required is as follows:
- All the computers loaded with Windows 98 second edition operating system (Windows 95, NT, 2000 and ME will also do, this book shall refer to Windows 98 SE).
- Driver software for the NIC as provided by the manufacturer.
- Additional third-party Ethernet packet capturing utility called PacketBoy a part of the NetBoy Suite of applications from NDG software.

Implementation

Three host computers would be required to demonstrate this Practical. For easy identification, they shall be referred to as Tom, Dick and Harry.

Setup the host Tom with the following TCP/IP parameters:

IP Address: 192.168.000.001

Subnet Mask:255.255.255.000
Setup the host Harry with the following TCP/IP parameters:
IP Address:192.168.000.226
Subnet Mask:255.255.255.000
Setup the host Dick with the following TCP/IP parameters:
IP Address:192.168.000.161
Subnet Mask:255.255.255.224
Gateway:192.168.000.001
The procedure to setup these TCP/IP properties has already been covered in Practical 2. To setup the gateway on Dick, after setting up the IP address and subnet mask, click on the Gateway tab as shown.

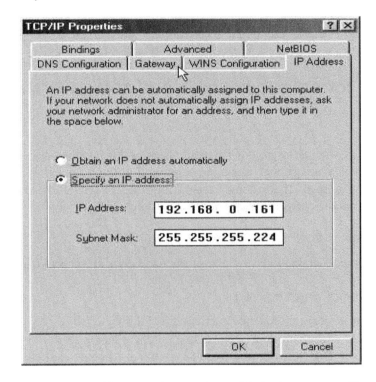

Enter the gateway address as shown, (note this is the same as Tom's IP address), and click on the Add button.

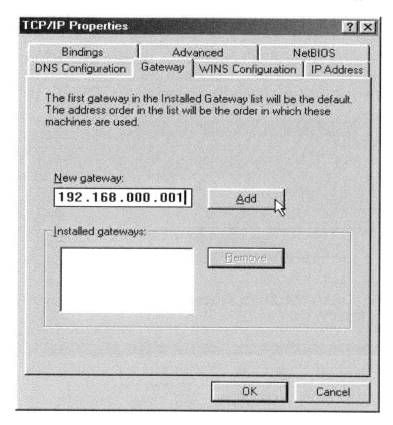

Ensure that the gateway address appears under the installed gateways as shown.

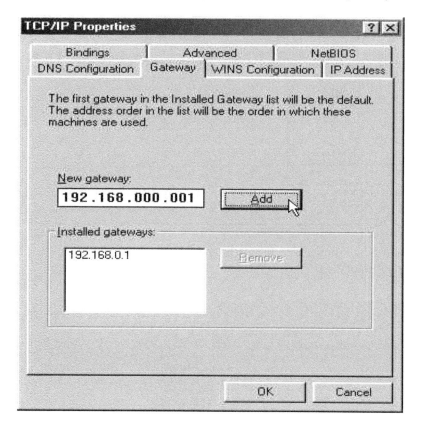

Click on the OK button to exit the TCP/IP properties and again on the OK button to exit the network dialog box. Restart the hosts when prompted to.

In addition to the above TCP/IP properties, an entry into the Window's registry would be required on Tom, as follows. EXTREME CAUTION IS ADVISED!

Click on start and select run. In the run window type REGEDIT and click on OK.

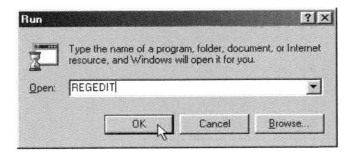

Under the path
HKEY_LOCAL_MACHINE/System/CurrentControlSet/Services/VxD
Locate MSTCP

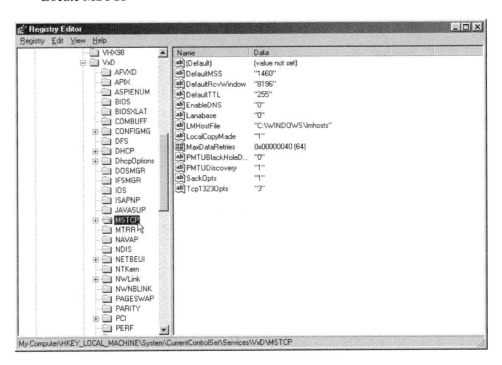

Add a new string value by clicking on Edit in the menu bar then selecting new and picking String Value.

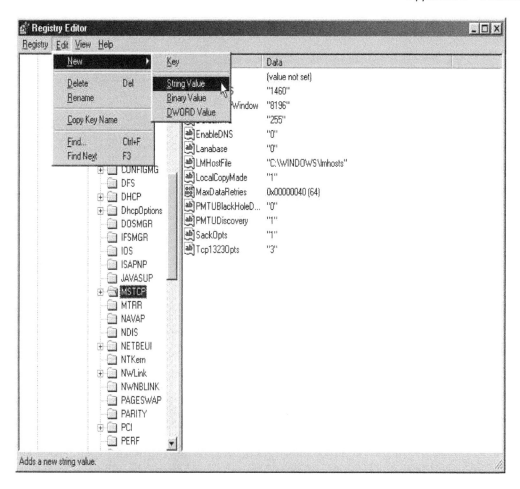

Name it 'EnableRouting.'

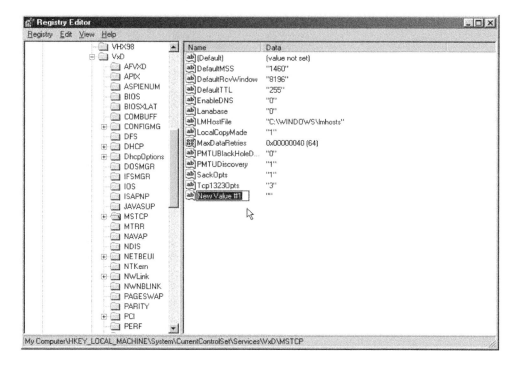

With this string value selected, click on <u>E</u>dit in the menu bar then select <u>M</u>odify.

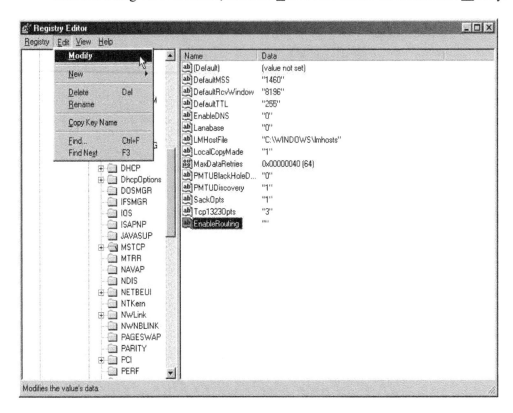

Enter the value '1' and then click on OK to close the Edit String dialog box.

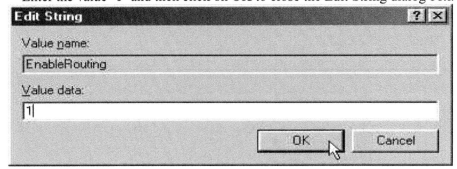

Click on registry and select exit to close the registry editor.

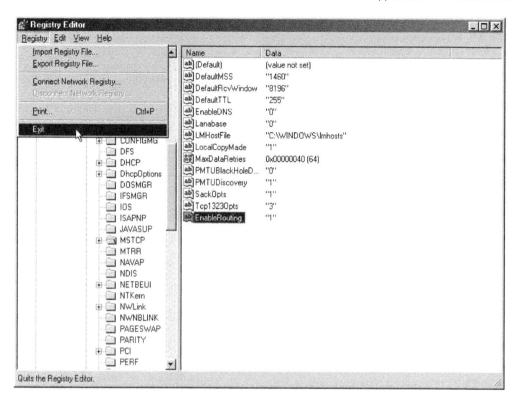

Restart Tom for the registry changes to take effect.

Run WINIPCFG on Tom, Dick and Harry and click on the More Info >> button. Ensure that the correct Ethernet adapter is selected.

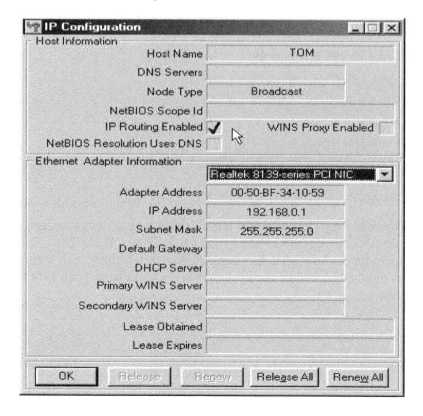

Note the adapter address, IP address, Subnet Mask and default gateway (if present) for all three hosts. Also note that due to the registry entry made earlier, Tom shows a check mark against IP routing enabled.

Startup the PacketBoy utility on Harry and start capture.

At the DOS prompt on Dick type in the following command:

```
C:\>ping 192.168.0.226 -n 1

Pinging 192.168.0.226 with 32 bytes of data:

Reply from 192.168.0.226: bytes=32 time=3ms TTL=128

Ping statistics for 192.168.0.226:
    Packets: Sent = 1, Received = 1, Lost = 0 (0% loss),
Approximate round trip times in milli-seconds:
    Minimum = 3ms, Maximum =  3ms, Average =  3ms

C:\>
```

This would ping 192.168.0.226 (Harry) once with 32 bytes of data. Stop capture and view the packets captured on Harry, which is running PacketBoy. Select the first packet to display the various headers within. Resize the sections such that it is convenient to browse through the middle section comfortably, and then expand the Ethernet Header and IP Header.

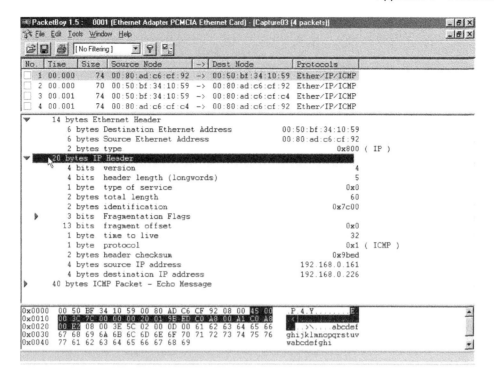

Note the source and destination Ethernet addresses, the source and destination IP addresses and the ICMP packet header information.

Select the second, the third and the fourth messages in turn and again expand the Ethernet Header and IP Header; note the source and destination Ethernet addresses, the source and destination IP addresses and the ICMP packet header information for each.

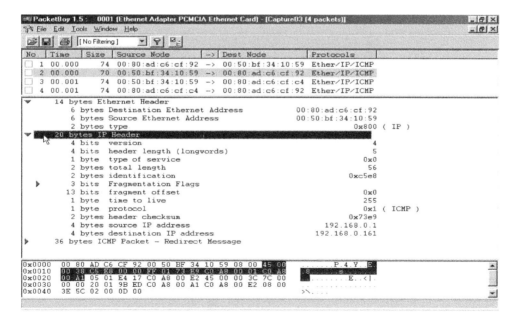

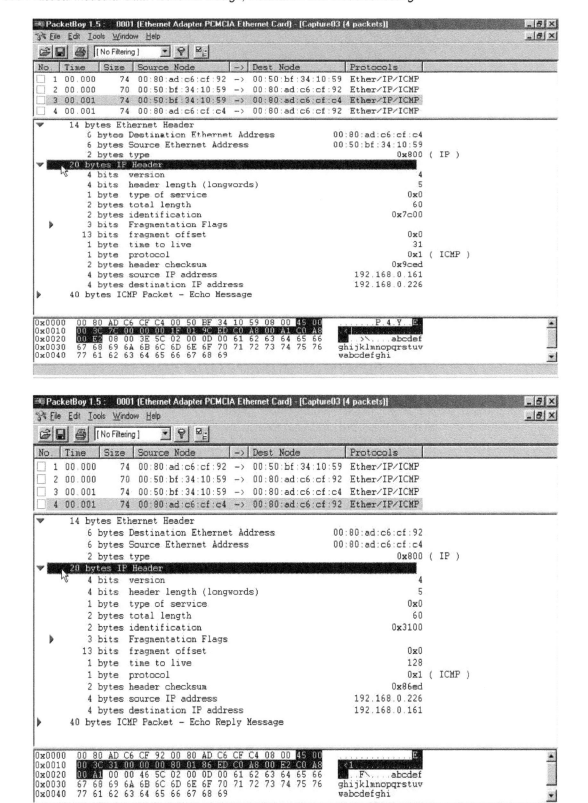

Conclusion

In the first packet, Dick (source IP address 192.168.0.161) is sending the ping packet to Harry (destination IP address 192.168.0.226), as required by the ping command. However, within the Ethernet Header things are different. Here it appears that Dick (Ethernet address 00:80:ad:c6:cf:92) is addressing the packet to Tom (Ethernet address 00:50:bf:34:10:59). The datagram seems to be misdirected.

The fault lies within the subnet mask of Dick. This is the only instrument that any TCP/IP host possesses to verify whether the destination host resides on the same physical network or on another. Each and every host performs the following computations:

The source IP address is ANDed with the local subnet mask and the result is compared to the result of the ANDing between the destination IP address and the local subnet mask. If the results are the same then the host assumes that the destination resides on the same physical network it is connected to. If the results are different, then the host directs the datagram towards its gateway (router) such that the gateway may forward the datagram to the final destination.

The computation performed by Dick is:

192.168.000.161 expressed in binary 11000000.10101000.00000000.10100001
255.255.255.224 expressed in binary 11111111.11111111.11111111.11100000
11000000.10101000.00000000.10100000
11000000.10101000.00000000.10100000 in dotted decimals is 192.168.0.160
Similarly,
192.168.000.226 expressed in binary 11000000.10101000.00000000.11100010
255.255.255.224 expressed in binary 11111111.11111111.11111111.11100000
11000000.10101000.00000000.11100000
11000000.10101000.00000000.11100000 in dotted decimals is 192.168.0.224

The conclusion is that the results are different and therefore the datagram has to be forwarded to the gateway.

Having received the datagram, Tom performs the same computation thus,
192.168.000.161 expressed in binary 11000000.10101000.00000000.10100001
 255.255.255.000 expressed in binary 11111111.11111111.11111111.00000000
11000000.10101000.00000000.00000000
11000000.10101000.00000000.00000000 in dotted decimals is 192.168.0.0
Similarly,
192.168.000.226 expressed in binary 11000000.10101000.00000000.11100010
255.255.255.000 expressed in binary 11111111.11111111.11111111.00000000
11000000.10101000.00000000.00000000
11000000.10101000.00000000.00000000 in dotted decimals is 192.168.0.000
And,
192.168.000.001 expressed in binary 11000000.10101000.00000000.00000001
255.255.255.000 expressed in binary 11111111.11111111.11111111.00000000
1000000.10101000.00000000.00000000
11000000.10101000.00000000.00000000 in dotted decimals is 192.168.0.000
192.168.000.226 expressed in binary 11000000.10101000.00000000.11100010
255.255.255.000 expressed in binary 11111111.11111111.11111111.00000000
1000000.10101000.00000000.00000000
11000000.10101000.00000000.00000000 in dotted decimals is 192.168.0.000
Similarly,
192.168.000.161 expressed in binary 11000000.10101000.00000000.10100001
255.255.255.000 expressed in binary 11111111.11111111.11111111.00000000

11000000.10101000.00000000.00000000

11000000.10101000.00000000.00000000 in dotted decimals is 192.168.0.0

As the results are the same, the fourth packet traverses directly to Dick from Harry, without involving Tom.

Thus we find that for apparent reasons, when in DOS at host Dick, the ping seems to work fine, but where we expect to see two packets, we actually capture four. Needless to say for such a minor error in the subnet mask, we see a doubling of network traffic, that would have a tremendous impact on the network loading.

Practical 20: Radio intermodulation calculation

Overview

The objective of this practical is to give the delegate an understanding of the method of calculating intermodulation frequencies to determine whether any interference could be possibly caused by intermodulation effects from the known transmitters on a site.

Objectives

The delegates will learn how to use the Intermodulation Calculation feature of the TELEDES software package to establish whether any interference is possible.

Hardware/software required

- One computer with TELEDES software.

Implementation

A radio site has the frequencies listed in the following table. Each channel has a frequency tolerance of 12.5 kHz, which means the channel filters will accept any interference within 12.5 kHz of the nominal channel frequency. Use the TELEDES software package to determine whether any intermodulation effects are possible up to the fifth order.

Transmit Frequency (MHz)	Receive Frequency (MHz)
455.250	460.750
457.275	462.325
457.625	462.725
170.325	166.700
171.125	167. 625
173.250	169.425

1. Start the TELEDES program and select Intermodulation Calculations.
2. Use the arrow keys to navigate within the program and select sheet then enter to move the cursor onto the calculation sheet. Use F1 for context sensitive help on the menu items.
3. Enter the above frequencies onto the calculation sheet.
4. Use the ESC key to return to menu bar and save your data using the save feature of the file menu for future reference.
5. Now navigate to the options menu and select config.
6. Here the tolerance frequency (12.5 kHz) is selected as well as the highest order. Initially, set the highest order to 3 using the TAB key to rotate the cursor to the field then the down arrow to select from the pick list. Press

enter to lock in your selection then enter again to return these values to the main program. Note if you press ESC at any selection that data is not changed.
7. Navigate to the calculations menu, then select execute to do the calculations and when finished select results to view the data.
8. A blank result sheet indicates that NO intermodulation components could be generated by the frequencies input.
9. Repeat the calculations up to the required order (5).

Conclusion

The selected frequencies indicate the possibility of interference at the fifth order. However, no serious interference could arise as both interfering frequencies are at the carrier frequencies, which carry no useful data.

Had interference been indicated in the sideband frequencies then this can be confirmed by turning off the offending transmitter(s) in turn and noting whether the interference problem disappears.

Note the program cannot determine the expected level of intermodulation interference, which requires considerable detail (often unknown!) on the coupling effectiveness of the various transmitters and receivers, such as their antennas, power levels, proximity etc.

Appendix D

Miscellaneous industrial protocols overview

D.1   General overview

MODBUS/TCP is a variant of the MODBUS family of simple, vendor-neutral communication protocols intended for supervision and control of automation equipment. Specifically, it covers the use of MODBUS messaging in an 'intranet' or 'Internet' environment using the TCP/IP protocols. The most common use of the protocols at this time is for Ethernet attachment of PLCs, I/O modules, and 'gateways' to other simple field buses or I/O networks.

It was introduced by Schneider Automation in the early 1990s as a variant of the widely used MODBUS protocol, which had been implemented in turn by almost all vendors and users of automation equipment.

Ethernet is a local area network (LAN) protocol that was originally developed to link computers. Invented by Bob Metcalfe at Xerox Palo Alto Research Center (PARC) and later refined by Xerox, DEC, and Intel, the Ethernet technology specification was later adopted by the IEEE as standard 802.3. The original Ethernet specification called for a bus topology over several media types, including coaxial cable. Today's common Ethernet implementations utilize a twisted pair wire commonly referred to as category 5 cabling and provide a raw data transfer rate of 10 or 100 Mbps. A new gigabit Ethernet standard, supporting data rates up to 1000 Mbps, was approved in 1999.

In the Modbus/TCP variant, instead of having a dedicated cable between the client (master) and server (slave), an Internet standard TCP 'connection' is used instead. A single device may have many such connections active at the same instant, some acting in the role of client, some acting in role of server. These connections may be established and broken on a repetitive and continual basis, or they may be left active for long periods. However, they are always broken down and reestablished as a method of investigating and recovering from disruptions such as those caused by power failure or loss of communication with other components.

The key advantages of this protocol can be summarized as follows:

1.   It is scalable in complexity. A device that has only a simple purpose need only implement one or two message types to be compliant.
2.   It is highly scalable in scope. A collection of devices using MODBUS/TCP to communicate can range up to 10 000 or more on a single switched Ethernet network.
3.   It is simple to administer and enhance. There is no need to use complex configuration tools when adding a new station to a Modbus/TCP network.
4.   There is no vendor-proprietary equipment or software needed. Any computer microprocessor with Internet style (TCP/IP) networking can use ODBUS/TCP.
5.   It is very high performance, limited typically by the ability of the computer operating systems to communicate. Transaction rates of 1000 per second or more are easy to achieve on a single station, and networks can be easily constructed to achieve guaranteed response times in the millisecond range.
6.   It can be used to communicate with the large installed base of MODBUS devices, using conversion products which require no configuration.

With reference to the ISO's OSI-RM, and the TCP/IP models, Modbus/TCP appears as illustrated in Figure D.1 below:

OSI model	TCP/IP standard	Modbus/TCP
Application layer	Application layer	Modbus (modified)
Presentation layer		
Session layer		
Transport layer	Transport layer	TCP
Network layer		IP
Data link layer	Physical layer	Ethernet
Physical layer		

Figure D.1
Comparison of Modbus/TCP to OSI

You can break the technology down into three major parts, from top to bottom:

Application layer

This is where the application level protocol resides. Examples of application level protocols include Modbus/TCP and hypertext transfer protocol (HTTP). Several application protocols are supported within the industry standard Modbus/TCP compliant devices and all can be used simultaneously. For example, since both Modbus/TCP and HTTP (web browser protocol) are supported, you could retrieve data from the device using a product like Wonderware's Intouch, while simultaneously viewing a web page using the device's web server.

Transport layer

The transport layer encompasses the TCP/IP protocol suite, and is arguably the most important. All messages transferred between a host and the SNAP Ethernet brain are encapsulated in a TCP or user datagram protocol (UDP) packet. Therefore, any application layer protocol that works with a socket interface (and therefore TCP/IP) can be used with this transport layer without regard to the physical layer. Furthermore, any physical layer component that works with TCP/IP can be used (again, without regard to the top application layer).

Physical layer

The physical layer includes only the actual physical connection to a device, such as Ethernet, fiber optic, or serial. Usually most devices support Ethernet category 5 twisted pair cabling and sometimes also provide an RS-232 serial port for use with modems. Fiber optic support may be available through an external fiber-to-copper transceiver.

As opposed to the Modbus Plus implementations, which is based around a token passing protocol, that assures equal opportunity to all nodes connected on the network, time to transmit, in an ordered deterministic manner, Ethernet, purely defined, is not deterministic. It is important to note, however, that determinism is sacrificed only during a collision event. However, the implementation of Ethernet determines performance. There are two ways to achieve near-deterministic performance with Ethernet:

- In applications that require high speed, such as those used in control environments, you can choose to implement Ethernet I/O devices in a dedicated network architecture. Using traditional Ethernet hubs in this manner will not provide determinism, but traffic in a dedicated network segment would typically be low enough that collisions would be rare. Furthermore, while the current standard speed for most Ethernet networks is 10 Mbps (megabits per second) far exceeding most industrial bus network speeds 100 Mbps networks are becoming common. Most of the Ethernet I/O devices work on both networks, auto negotiating network speed as needed. This increased speed further improves performance, and reduces the likelihood of bandwidth saturation and therefore, collisions.
- For even higher speed requirements, a near-deterministic Ethernet solution can be achieved with network switches. A switch is a network device that quickly routes signals between ports on the hub. It repeats a packet only to the port that connects to the destination for the packet.

The IEEE 1394 specification describes a physical medium and transport mechanism based largely on technology developed by Apple Computer and marketed under the trademarked name 'FireWire'. (Sony is currently marketing the same technology under the trademark 'iLink'.) FireWire was adopted by the Institute of Electrical and Electronic Engineers as IEEE 1394-1995 to be an industry-standard serial data bus. One of the really cool things about FireWire/IEEE 1394 is that it is a platform-independent technology. FireWire/IEEE 1394 works with both Macs and pcs. One of the biggest benefits of FireWire/IEEE 1394 is that it can support very, very fast speeds up to 1600 Mbps.

MODBUS/TCP has been around a while – yet most people have not been introduced to the absolute simplicity of its use for multi-vendor TCP/IP networks. It is used to communicate to Modbus enabled PLCs with Ethernet adapters, but it is not a complex new protocol, it merely is MODBUS/RTU in a TCP/IP packet.

The real advantage of MODBUS/TCP is in how fast it can be implemented. Any organization with the following three resources can implement MODBUS/TCP in a few days: source code for an existing MODBUS/RTU serial driver, a TCP socket library and a competent programmer.

Thus, vendors can now have a MODBUS/TCP driver almost instantly, plus all the application programs that already support the data access paradigm of MODBUS/RTU can also support MODBUS/TCP instantly. In addition, direct MODBUS/TCP access offers good performance.

Some of the various network protocols supported by most Modbus/TCP compliant hardware, that also form a part of the TCP/IP suite of protocols for network communication are: ARP, UDP, TCP, ICMP, Telnet, TFTP, DHCP, and SNMP. For transparent connections, TCP/IP (binary stream) or Telnet protocols are used. Firmware upgrades can be made with the TFTP protocol.

The IP protocol defines addressing, routing and data block handling over the network. The TCP (transmission control protocol) assures that no data is lost or duplicated, and that everything sent into the connection on one side arrives at the target exactly as it was sent.

For typical datagram applications where devices interact with others without maintaining a point-to-point connection, UDP datagram is used.

IP addressing forms an important part of any standard TCP/IP network and therefore must also be considered when using Modbus/TCP. Every device connected to the TCP/IP network including the multi-serial device servers must have a unique IP address. When multiple Modbus devices share a single IP, then Modbus/TCP includes an additional address called the unit ID.

When the multi-serial device server is receiving Modbus/TCP messages from remote masters, the unit ID is converted to use in the Modbus/RTU message as the slave address.

When the multi-serial device server is receiving Modbus/RTU messages from local serial masters, a user-defined lookup table is used to match the 8-bit Modbus slave address to a remote IP address. The Modbus slave address received is used as the unit ID.

While the Modbus/TCP standard specification requires Modbus/TCP masters/clients to only issue 1 poll at a time, the full-duplex flow-controlled nature of TCP/IP allows them to issue more than one at a time and the TCP socket will happily buffer them. The Modbus bridge will fetch them 1 at a time and answer each in turn.

Appendix E

Local services, regulations and standards

E.1 AUSTRALIA

E.1.1 Regulatory licensing requirements for radio frequencies

In Australia, the governing authority for the control and issuing of licenses for the use of frequencies in the radio frequency spectrum is the Australian Communication Agency (ACA).

Telemetry frequency allocations will fall into three categories – the majority of links falling into the first category, which is fixed services below 1 GHz; the second being low powered services and the third is base/mobile.

E.1.2 Fixed services frequency regulation

The fixed services category is basically broken down into two sections. Point-to-point and point-to-multi-point services.

The ACA requires that the licensee pays a fixed annual fee for use of any allocated frequency. A fee is paid for each location where that frequency is used.

Two frequency point-to-point

The ACA has allocated a limited number of frequencies for use with point-to-point services below 1 GHz.

These are as listed below:

Master site	Remote site
150.05–151.3935	154.65625–156.00
404.00–405.00	413.45–414.45
45–.50–451–50	460.00–461.00
852.00–853.00	928.00–929.00
852.00–853.50	929.00–929.50

Table E.1

Lower frequencies can be obtained in the low band VHF and the HF bands, if the requirement can be suitably justified in a special application to the ACA.

Single frequency point-to-point

For links that require one way communications only, the ACA has allocated a small number of frequencies specifically for point-to-point applications. These are as follows:

410.55–411.55 Mhz
857.00–861.00 Mhz

Table E.2

Point-to-multi-point

For point-to-multi-point applications the ACA has specifically allocated frequencies in the UHF 400 and 900 MHz bands. These are as follows:

Master	RTU
461.00–462.00	451.50–452.50
929.50–930.00	853.50–854.00

Table E.3

Again frequencies in the HF and VHF bands can sometimes be obtained if special applications are made to the ACA. Successful allocation would be very dependent on location and requirements. Allocation of VHF frequencies is more likely to be successful in remote areas.

E.1.3 Fixed services physical requirements

Output powers

For all point-to-point links, the maximum power allowed at the antenna input is 1 watt.

For point-to-multi-point systems, the maximum power allowed at the antenna of an RTU is 1 watt and the maximum power allowed at the antenna at the master site is 5 watts.

Antenna requirements

To control the emission of unwanted RF energy in these limited frequency bands the ACA has firstly limited the transmitter output power as was noted above, and has also put restrictions on the radiation characteristics of the antennas used.

The following table lists the antenna requirements for point-to-point services.

Frequency (MHz)	Maximum beamwidth in E plain (degrees)	Minimum front/back radio (dβ)	Mid-band gain (dβ) (typical)	Typical antenna
148–174	55	16	12	6 element YAGI
403 520	36	17	13	9 element YAGI
820–960	30	20	16	15 element YAGI

Table E.4

The polarization of the antennas used in point-to-point links shall be horizontal unless special permission is obtained from the ACA.

For point-to-multi-point systems, omnidirectional antennas may be used at both master and remote stations.

E.1.4   Low powered devices

The ACA has allocated a range of frequencies that can be used for telemetry applications that have restrictions on their transmitter output power but do not require licensing. Because the allowed output powers are so small the possible applications for telemetry purposes are limited to very close vicinity systems. The maximum allowed output power is given as an effective isotropical radiated power (EIRP) from the antenna.

Some manufacturers will provide equipment that operates at these low power outputs. If equipment is not available, attenuators may be placed in line with the RF output of transmitters to lower the EIRP to the restricted level. The following is a list of the frequencies.

Item	Permitted Operating Frequency (MHz) (Lower Limit Exclusive, Upper Limit Inclusive)				Maximum EIRP
1	0.009	to	0.014		200 µW
2	0.014	to	0.01995		50 µW
3	0.02005	to	0.07		7.5 µW
4	0.07	to	0.16		3 µW
5	0.16	to	0.285		500 nW
	0.325	to	0.415		
6	3.025	to	3.155		7.5 nW
7	3.5	to	3.7		30 nW
8	3.7	to	3.95	vf	7.5 nW
	4.439	to	4.65		
9	13.553	to	13.567		100 mW
10	21	to	21.89		10 mW
11	26.957	to	27.283		1 W
12	29.7	to	29.72		100 mW
	30	to	31	or	
	36.6	to	37	vf	
	39	to	40.66		
	40.7	to	41		
13	40.66	to	40.7		2.5 mW
14	54	to	56		2.5 mW
15	70	to	70.24375	or	100 mW
	77.29975	to	77.19375	or	
	150.7875	to	152.49375	or	
	173.29375	to	174		
16	225	to	242	vf	10 µW
	244	to	267	or	
	273	to	303.95		
	304.05	to	326.6		
	335.4	to	399.9	or	
17	915	to	92R		3 mW
18	10,500	to	10,550		100 mW
	24,000	to	24,250		
21	Biomedial transmission in a telemetry transmitters originate in the license area of a operating in the same channel		174 to 204 IF TV channel does not TV broadcasting station		10 µW
22	472.0125	to	472.1125		100 mW
23	2400	to	3450		1 W
	5725	to	5875		

Table E.5

E.1.5   Base/mobile frequency

The greater majority of frequencies in the HF, VHF and UHF bands are allocated for use with mobile radio systems. It is not general policy to allocate these frequencies from telemetry applications (except where it is data communication to fleets of vehicles) particularly in urban or heavy industrial areas where there is a perennial shortage of frequencies. ACA will however consider applications for mobile frequencies in the HF and VHF bands for telemetry applications if the location is remote and the environment and distances involved in the application require lower frequencies.

E1.6   Landline options available in Australia

This section looks at the range of analog and digital services that could be used for the provision of landline services for telemetry application.

Until recently, all landline services in Australia were provided by the state owned telecommunications provider Telecom Australia 'Telecom'. However, the second telecommunications carrier – Optus are able to provide some competitive landline services as detailed below.

Refer to 'Practical SCADA', Chapter 4 for details on landline type services.

E1.7   Switched analog lines

The following modem standards should work over the Telecom switched network.

- V26 ter   (2400 bps)
- V27 ter   (4800 bps)
- V32   (9600 bps)
- V32 bis   (14 400 bps)

The charges associated with PSTN analog services are the annual line rentals and the call charges, which are time and distance dependent.

E.1.8   Leased analog lines

The following analog tie line services are available from Telecom. At the time of writing this manual, Optus do not advertise analog tie line services but the author has been informed that if there is a requirement close to one of their terminal boxes they will simply provide an appropriate multiplex card to provide the service.

Two wire ring/loop tie line – (Telecom).

Voicelink C (conditioned).
Best possible data speed about 14.4 kBit/s (V32 bis).

Four wire direct tie line – (Telecom).

Four wire permitted attachment private line (PAPL).
Two types, standard and premium quality.
Data speeds up to 4800 bps (standard) and 9600 bps (premium).

Two wire direct tie line – (Telecom).

Two wire PAPL – (standard quality and premium quality).

Performance parameters are the same as for four wire PAPL circuits except that there is a lower return loss figure for the two wire services. Achievable data rates around 2400 bps (standard) and 4800 bps (premium).

Out of area services.

Telecom provide a number of two wire and four wire tie line services that have signaling facilities with the performance parameters of the PAPL lines. These are generally used for connecting off-site telephone extensions back to a sites PABX. Other terms used for these lines are:

- Outdoor extensions (ODX).
- Network connected leased lines.
- Diverted telephone service lines.

E.1.9   Analog data services

DATEL is a Telecom provided data transmission service over standard analog bandwidth lines. The service can be provided over switched or leased lines and includes the provision of a modem at each end. End-to-end maintenance is also included as part of the service.

The DATEL service is provided on analog network lines. The service is classified according to three parameters:

- The type of line used, i.e. switched or dedicated.
- The speed of operation.
- As point-to-point or point-to-multi-point operation.

E.1.10   Packet switched services

Packet switched services are available from Telecom (Austpac service) or through other service providers such as AAP renters etc. Normally, a dedicated digital data link or a dedicated analog data link is used to provide access from the sites to the packet network.

Call charges for packet switch networks are normally broken into the following elements:

- Call attempt charge (usually the most significant charge component).
- Packet count charge.
- Call duration charge.

E.1.11   Digital data service

Telecom provide a dedicated digital transmission service for point-to-point and point-to-multi-point links referred to as the digital data service (DDS). The DDS offers data rates of 2400, 4800, 9600 and 48 kbit/s for synchronous, leased line services with similar functionality to the DATEL services.

DDS provide services at speeds of 1.2 k, 2.4 k, 4.8 k, 9.6 k, 19.2 k and 48 kbit/s for both point-to-point and point-to-multi-point applications.

Australia is divided into a number of regions, which dictate the costs of the services.

These regions are defined as primary areas, which is all major inner cities and some major isolated industrial areas, secondary area, which are major regional areas around cities and in the country and finally, tertiary areas, which are in smaller provincial towns.

E.1.12   ISDN services

The telecom basic rate access (BRA) ISDN service is referred to as Microlink providing 2 B channels of 64 kbit/s and a 16 kbit/s D channel.

The telecom primary rate service (PRA) is referred to as Macrolink comprising 30 × 64 kbit/s for voice and data, 1 × 64 kbit/sec for signaling and data and 1 × 64 kbit/s clocking channel.

E.1.13   Other services

Both Telecom and Optus provide a straight point-to-point 2 Mbit/s service. The Telecom service is referred to as Megalink. The Optus service is referred to as Datalink 2M.

Optus also provide a number of 64 kbit/s point-to-point service referred to as Datalink 64 k.

E.1.14   Satellite services available in Australia

The Overseas Telecommunication Commission (OTC) was absorbed by Telecom in the early 1990s and no longer exists as a separate corporation. OTC was the only provider of international telecommunications services in Australia until the industry was deregulated in 1992. The services they provided before are now provided by Telecom.

Besides a large range of international services, which are not of relevance to telemetry applications, there are a number of services available domestically through the two international satellite networks, Intelsat and Inmarsat.

In addition, the regional satellite service that was originally owned and run by the government corporation AUSSAT, is now owned and operated by Australia's second telecommunications carrier – Optus. They presently have four satellites operational, with several more due for launch in the near future. Optus offer two main classes of service – dedicated leased line service for voice and data and the recently released mobile service for switched voice and data.

E.1.15   Optus satellite services

Austlink

The Austlink service is a dedicated leased line service for point-to-point-to-multi-point services. The service operates in the Ku band using single channel per carrier technology (FDMA).

The service can provide dedicated links between a main earth station (referred to by Optus as major city earth station [MCES]) and a remote earth station (referred to by Optus as a minor earth station [MES]) or between two MESs.

Connection is made into the public switch telephone and data networks at the MCES.

The service can be used for voice and facsimile tie lines, data tie lines and for telephone exchange line extension to remote sites.

An MES can provide up to eight voice or data channels.

The standard voice circuits can also carry 2400, 4800 or 9600 bit/s data using analog modems.

Digital channel data services are available from 64 kbit/s to 768 kbit/s in 64 kbit/s increments.

MOBILESAT

MOBILESAT is a domestic satellite service to mobile earth terminals. It provides telephone and data services to small terminals mounted in terrestrial vehicles or ocean going vessels.

The MOBILESAT service was launched mid-1994 using the new Optus 'B' series satellites.

The MOBILESAT service provides fundamentally the same services as the SATCOM-M service described earlier except that voice services are coded onto a 6.6 kbit/s bandwidth data channel to the satellite using P/4 QPSK modulation. The voice signal is coded at a data speed of 4.2 kbit/s on the channel.

MOBILESAT will be ideal for telemetry applications in remote areas. It will provide high quality and relatively noise free digital communications. The very low frequencies used by the satellite will provide high reliability (over 99%). Access will be available to terminals up to 200 km out from the coast line. HF radio in comparison, which is the only alternative to very remote locations, has an availability of 70–80% at best (non diverse system), poor quality reception, high noise levels and high BERs.

E.1.16 List of telemetry suppliers

Note that this list is not all inclusive.

ALMOS Systems

52 Garthrie St
Osborne Park WA
09 244 2346

Control Systems International

32 Delhi St
West Perth   WA 6005
09 322 1812

DATRAN

47–49 Carlotta St
Artarmon NSW 2064
02 439 6733

ELPRO Technologies

8 Cox Rd
Windsor
Brisbane 4030
07 357 5344

HARRIS Controls

PO Box 80
Bentley W.A. 6102
09 470 2088

INTROL

19–21 Olive St
Subiaco, Perth WA 6008
09 388 3144

MIRI Engineering

Unit 6, 519–521 Walter Rd East
Morley 6062 WA
09 378 2388

Motorola Australia

666 Wellington Rd
Mulgrave Vic.
Melbourne
03 566 7766

OMNITRONICS

15 Hector St
Osborne Park 6017
Perth W.A.
09 445 2633

ROMTECK

4 Collingwood St
Osbourne Park
WA 6017
09 244 3011

TEKNIS Electronics

Angas Mews
75A Angas St
Adelaide
South Australia 5000

UNIDATA Australia

3 Whyalla St
Willeton, Perth WA
09 457 1499

VECTOR International

Unit 3, 59 Walters drive
Herdsman Business Park
WA 6017
09 242 3396

Micro Control Engineering

59 Collingwood St
Osborne Park, Perth WA 6017
09 445 254

Index

Printed and bound by CPI Group (UK) Ltd, Croydon, CR0 4YY

03/10/2024

01040331-0013